OPTICS

EXPERIMENTS AND DEMONSTRATIONS

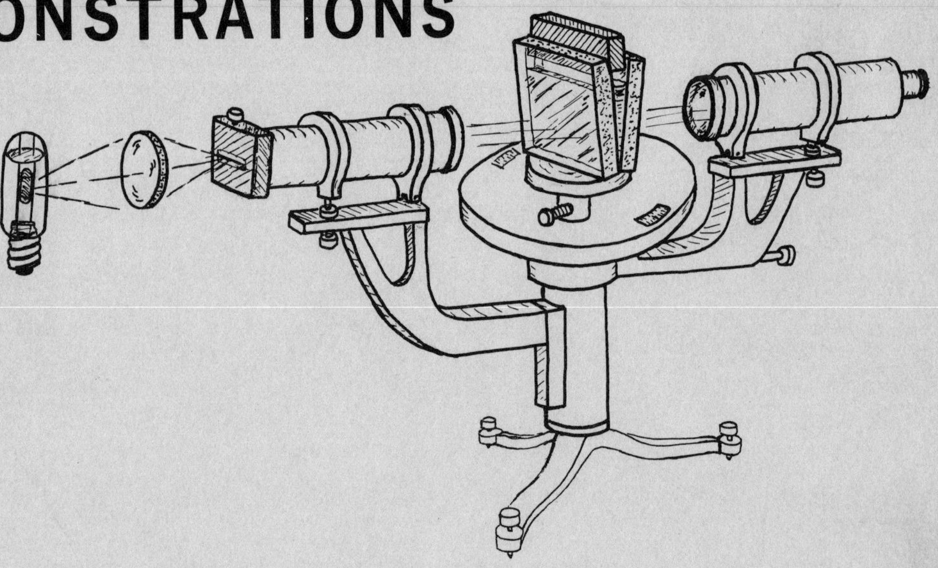

C. Harvey Palmer
THE JOHNS HOPKINS UNIVERSITY

OPTICS

EXPERIMENTS AND DEMONSTRATIONS

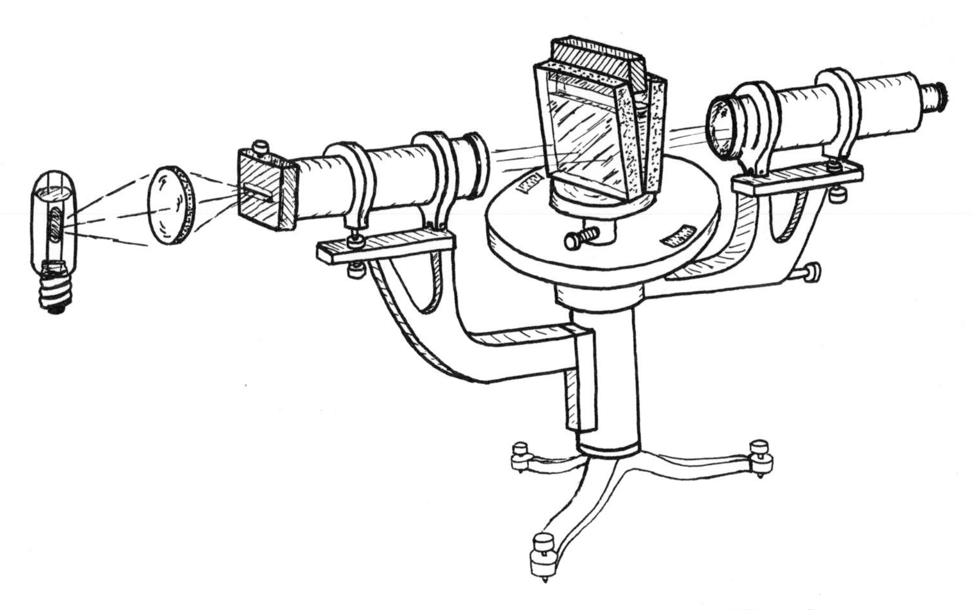

C. Harvey Palmer
THE JOHNS HOPKINS UNIVERSITY

THE JOHNS HOPKINS PRESS
BALTIMORE • 1962

OPTOMETRY

Free copies of this book have been sent to approximately two thousand colleges and universities in the United States through the courtesy of the National Science Foundation. Additional copies may be ordered from The Johns Hopkins Press at a price of $4.95

Illustrated by the author.

Printed in the United States of America
Library of Congress Catalog Card Number 62- 17546
NSF Grant Number G 9298

PREFACE

This book is intended to fill the need for a reasonably comprehensive, up-to-date laboratory manual in optics. It illustrates recent progress in the science as well as the more significant older discoveries and offers a wide choice of experiments to suit the requirements of both small colleges and the larger universities.

It is hoped that this manual may contribute in a small way to the rescuing of optics from its current state of undeserved neglect. The tremendous importance of this branch of science is undeniable. Historically the study of optics has played a dramatic role in the advancement of science -- in its theoretical structure it is rightly called "the mother of physical concepts." Today it is becoming an increasingly powerful tool in both pure and applied research as well as being a notable science in its own right. We may mention in passing the important applications of infrared techniques in practically every branch of science and technology from biology to astronomy; the fact that the standard meter has been replaced by the wavelength of orange-red krypton (Kr^{86}) spectral line; the increasing application of laboratory curiosities such as zone plates to current millimeter wave instrumentation. Despite the importance of optics, however, the student today may find it treated almost as a dead science with all the attention focused on the newer and more glamorous fields of space science, nuclear physics, solid state physics, and electronics. Yet surely the dramatic and colorful nature of optics, its high precision, its broad applications of other fields of science and engineering, can -- and should -- appeal strongly to the student.

The nature of this book may, perhaps, be made clearer by a few examples. The experiment on moire fringes, for instance, which is based on a very recent research in England by Jones and Richards (see Experiment 20) illustrates how mirror rotations of less than 10^{-8} radians may readily be detected by purely ray optic means. Such an angle corresponds to only a ten thousandth of the angular dimensions of the diffraction pattern of the mirror used. The required apparatus includes a few generally available electrical components and only very inexpensive optical parts which may be mounted with beeswax and rosin. The triangle path interferometer (Experiment B6), recently invented for precision measurements of end gauges, can be constructed of plywood and surplus optical parts. With it white light interference bands can be found in a matter of seconds, and they may be projected on a screen across the room if desired. The construction of an infrared spectrometer (Experiment D3) requires optical components costing several hundred dollars, but with this crude instrument one can measure the major near infrared atmospheric absorption bands. (Incidentally, the same instrument with the addition of a fluorescent screen will show some of the ultraviolet spectrum of mercury vapor). The experiments on the human eye (A18, A19, B16, and C2) require but a few dollars worth of apparatus but they demonstrate several extraordinary phenomena, still not properly explained.

The experiments in this book have been designed so as to keep the cost of the required apparatus at a minimum. Many excellent surplus components -- lenses, prisms, and other items -- are currently available at low cost, and they are freely used in the experiments. Much of the necessary apparatus can be readily constructed in the student shop or elsewhere. Some basic apparatus (an inexpensive student spectrometer, an optical bench, and light sources, etc.) is also required, of course. An attempt has been made to allow a wide range in the selection of components where practical; lenses, for example, are specified by giving a range of focal lengths and diameters. Enough experiments are included in this manual to permit selection according to the level of the course, the departmental budget, and the interests of the instructor and the students.

The experiments fall roughly into three groups: precision experiments (starred in the Table of Contents) giving results to 0.1% or better, those giving 1 to 10% results, and a number of purely qualitative demonstrations. Since precision experiments often consume excessive time, it is suggested that only a very few be undertaken, and that the emphasis be placed on both qualitative demonstrations and the experiments of moderate precision. In this way the student will be able to cover more ground in the laboratory and thus learn more optical principles, procedures, and techniques than he could otherwise. It is very desirable for the student to set up the apparatus for at least some of his experiments, even selecting the components if possible; he thereby acquires invaluable experimental skill. Most of the demonstrations in this book, therefore, are designed so that the student can assemble them for himself in the laboratory instead of merely watching someone else manipulate them at the other end of the lecture room. In the selection of experiments it should be borne in mind that the time required to perform them may vary rather widely. Long ones may be shortened if need be by having the apparatus partly set up by the instructor or in some cases by omitting parts of the experiment. The student should not be so rushed that he cannot reflect on the significance of what he does.

The experiments may be performed in almost any reasonable order, beginning with those in either ray optics, wave optics, or crystal optics depending on the nature of the course. (The order in the book obviously reflects the author's prejudice in this respect). It is highly desirable, however, that two basic experiments, A1 and A5, be mastered early in the course.

It is assumed that the student has done the required reading for an experiment before coming to the laboratory, or that he has consulted such references as are available where references are given. The basic theory of the experiment is not given in this manual if it is readily found in the reference texts. With the exception of two books, all references include the author, title, edition, date, and publisher as well as page or section. The two exceptions are J. Strong's Concepts of Classical Optics, (1958, Freeman) and F. A. Jenkins and H. E. White's Fundamentals of Optics, 3rd ed. (1957, McGraw-Hill), which are referred to as "Strong" and "Jenkins and White" respectively. It is assumed that both these are readily available -- with preferably one or the other in the student's possession. Nearly every experiment includes a rather complete list of apparatus needed. The procedure is given in considerable detail only in the earlier experiments of a given section; in later or more advanced experiments the details are left for the student to fill in. Several appendices and an index, which it is hoped will be of use to the reader, are indicated in the Table of Contents.

This book has been made possible by a grant from the National Science Foundation which has also provided for the free distribution of one copy to every college and university in the United States. The author takes pleasure in expressing his indebtedness to the National Science Foundation. In addition, he is immeasurably grateful to Dr. John Strong who inspired this project, and to Mr. Frederick R. Stauffer who patiently labored through the text making invaluable criticisms and corrections. Warm thanks are also due to Dr. John S. Thomsen for his editorial help, to Mr. William Plummer who set up a number of experiments, and to Mrs. Shirley Gundry, typist par excellence. The author is also grateful to the other members of the Johns Hopkins Laboratory of Astrophysics and Physical Meteorology who contributed in various ways to the making of this book. Most of the experiments were assembled at the Johns Hopkins University, the rest at Bucknell University.

C. Harvey Palmer

Baltimore
April, 1962

CONTENTS

SECTION C. POLARIZATION AND CRYSTAL OPTICS

SECTION D. SPECTROSCOPY

APPENDICES

→

Section A

RAY OPTICS

moire fringes

EXPERIMENT A1. THIN LENSES AND THIN LENS COMBINATIONS.

READING: Jenkins and White, Ch. 4; or Strong, Ch. 13.4, 13.5, 13.6.

Although on logical grounds this experiment is out of order, it is nevertheless placed first because most of the experiments in this book make use of lenses in one way or another. Thus, whether the student begins his study of the science of light with ray optics or with wave optics, it is most desirable that he be thoroughly familiar with at least the simplest properties of thin lenses and thin lens combinations. This experiment is designed, then, to give the student some practical experience in the use of lenses, and, therefore, most of the refinements, appropriate to a more advanced stage, are omitted.

THEORY

For a lens in air whose thickness is small compared to its focal length, the following equations hold:

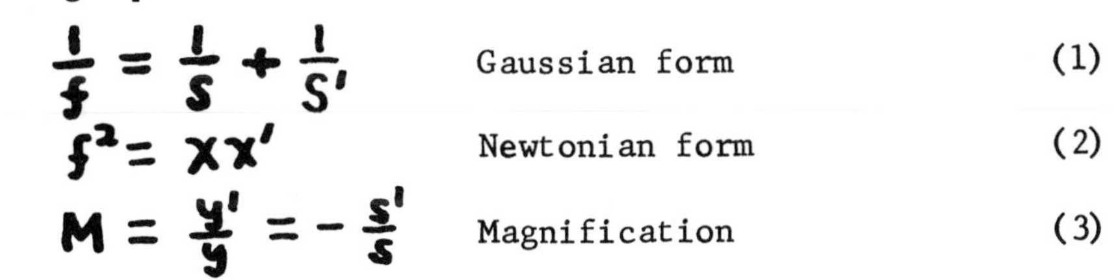

$$\frac{1}{f} = \frac{1}{s} + \frac{1}{s'} \qquad \text{Gaussian form} \qquad (1)$$

$$f^2 = xx' \qquad \text{Newtonian form} \qquad (2)$$

$$M = \frac{y'}{y} = -\frac{s'}{s} \qquad \text{Magnification} \qquad (3)$$

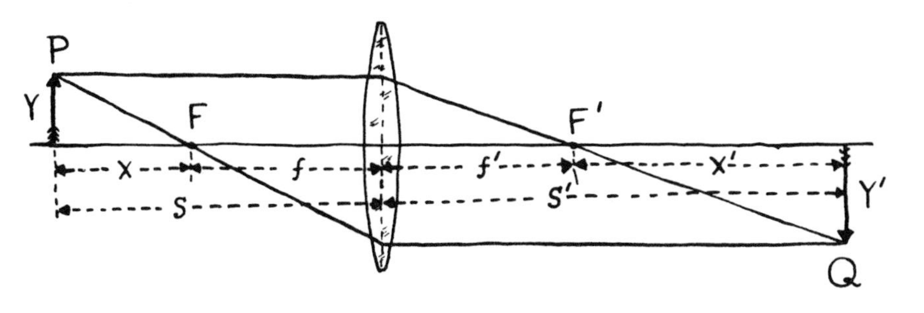

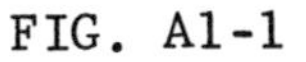

FIG. A1-1

Significance of the symbols in equations 1 to 3: P-object; Q-image; F-primary focal point; f-primary focal length; F'-secondary focal point; f'-secondary focal length; S,S'-object and image distances from lens center; X,X'-object and image distances from focal points.

These equations relate object and image distances, focal length, object and image dimensions, and magnification. The meaning of the symbols, defined in the references, is illustrated in Fig. A1-1. The sign conventions are given in Jenkins and White, pp. 33-4. Alternative forms of the equations and corresponding sign conventions are given in Strong's book. The equations are based on several assumptions: negligible lens thickness; paraxial rays; and monochromatic light. The formulae can be used successively for two or more lenses in series. Both object and image may be either real or virtual as described in the references.

APPARATUS

1) One meter optical bench.
2) Adjustable slit for mounting on the optical bench.
3) Light source (either incandescent or arc).
4) Transparent metric scale for measuring magnification.
5) Meter stick and preferably also a half meter stick.
6) Lenses to test, 2½ to 6 cm dia. Two lenses are needed, a convergent lens of 5 to 15 cm focal length, and a divergent lens of at least (minus) twice this focal length. The lenses may conveniently be the two components of an achromatic doublet if they are separable. The approximate focal length of a convergent lens is easily found by forming with it an image of a distant landscape or light source. The focal length of a divergent lens can be determined by combining it with a convergent lens and finding the overall focal length.
7) Auxiliary convergent lens of 10 to 25 cm focal length.
8) Plane mirror about two inches square which need not be of the highest quality.
9) Ground glass screen to mount on the optical bench.

Provision must be made for mounting the lenses either singly or in combination on the optical bench; tackiwax may be used if necessary.

PROCEDURE

Before making any measurements, label or otherwise identify the lenses to be tested.

I. Focal length of convergent lens:

A. Object-image method.

At one end of the optical bench, mount the adjustable slit (preferably vertical) and illuminate it so that a divergent cone of light passes through it with its axis directed along the optical bench.

Place the ground glass screen at the other end of the optical bench and mount the lens on a slide between the slit and screen. Make sure the heights of the lens and slit above the optical bench are about the same. Find a position for which the lens produces a sharp image of the slit on the screen. It may be desirable to alter the slit width to obtain the best results. Measure both object and image distances to center of lens rim, estimating as well as possible to a fraction of a millimeter. From this data, calculate the focal length of the lens, and record the results. Next slide the lens to the other possible position where an image is formed on the screen and repeat the measurements and calculations. How are the two positions related?

B. Autocollimation method (primary focal point).

Use the calculated focal length to position the lens so that the slit will be at the primary focal point of the lens. The beam emerging from the lens should now be very nearly parallel. Is it? Next mount the plane mirror so as to intercept the parallel beam emerging from the lens and reflect the beam back through the lens to form an image a little to one side of the slit jaws. Make any necessary adjustment in the position of the lens to produce the sharpest possible image -- note how far it must be moved for this purpose. Again measure the focal length and record the results. This method of measuring focal length is referred to as autocollimation, and it is a most important and useful procedure.

C. Secondary focal point.

With the auxiliary convergent lens, produce a parallel light beam along the optical bench. Use the autocollimation method to position this auxiliary lens. If the lens is plano-convex, place the plane side toward the slit to reduce spherical aberration. Now focus this beam with the test lens to obtain an image of the slit on the screen. Measure the focal length again and record the results.

II. Focal length of divergent lens:

Mount the convergent and divergent lenses in contact (avoid scratching them!). Measure the focal length of the combination using two different methods. Compute the focal length of the divergent component.

III. Separated lenses:

Mount the two lenses so that they are separated by a fixed distance -- 2/3 to 3/4 the focal length of the convergent element. The combination is mounted on one carriage and moved as a unit. Set the combination to form an image of the slit on the screen. Measure the distances of the slit to the nearest lens and from the image to the other lens. Also measure to a fraction of a millimeter the distance between the centers of the two lenses. Be sure to note which lens is nearer the slit. Use the thin lens formula twice in succession to calculate the distance between the second lens and the screen and compare the result with the measurement.

Turn the lens combination around so that the relative positions of the two lenses are interchanged and repeat the experiment. The significance of the interchange in terms of principal planes will be clarified in the experiment on thick lenses and lens systems.

IV. Lateral magnification:

Remove the divergent lens. Open the slit wide (or remove it) and as an object, use a transparent illuminated metric scale. Position the screen so that the source and screen are separated by a distance of 4½ to 6 times the focal length of the lens. Measure the magnification produced for both the enlarged and reduced images. It is convenient to compare the image on the screen with another millimeter scale to determine the magnification. Also, measure the separation of the source and screen. From the focal length of the lens and the separation of the object and image, calculate the two possible values of magnification and compare with the measured values.

Reassemble the separated lens combination of part III and measure the magnification for the enlarged image (a) with the convergent lens toward the object, and (b) with the divergent lens toward the object. Do the results indicate that the lens combination may be treated as having an effective focal length?

RESULTS

Tabulate all the results with estimated uncertainties in the measurements. Indicate clearly in each case what the measurement represents (i.e. focal length by autocollimation, etc.). Discuss briefly the significance of the results.

QUESTIONS

1) Is a collimated beam of light necessarily parallel? What is meant by the term "parallel beam?"

2) Is the entire lens area needed to form a complete image?

3) In the section on apparatus, a simple method of selecting suitable lenses is suggested. How far from the lens must the object be if the image distance and focal length must differ by less than 5%?

4) Under what conditions is it possible to form two, just one, or no real images of a source on a screen?

5) Are any of the results of this experiment limited by the approximations made in the formulae of the experiment? Explain.

6) Why are the three assumptions: (1) negligible lens thickness, (2) paraxial rays, and (3) monochromatic light needed to obtain the simple lens formulae given above?

EXPERIMENT A2. REFRACTIVE INDEX BY THE MICROSCOPE METHOD.

READING: Jenkins and White, Ch. 2 through 2.6; or Strong, Ch. 13.4.

An object lying on the bottom of a stream or pond appears to be closer to the observer than it really is. This apparent decrease in distance is directly related to the refractive index of the medium. Quantitative measurement of this change in distance provides a simple method for measuring the refractive index of a transparent liquid or a polished slab of solid.

THEORY

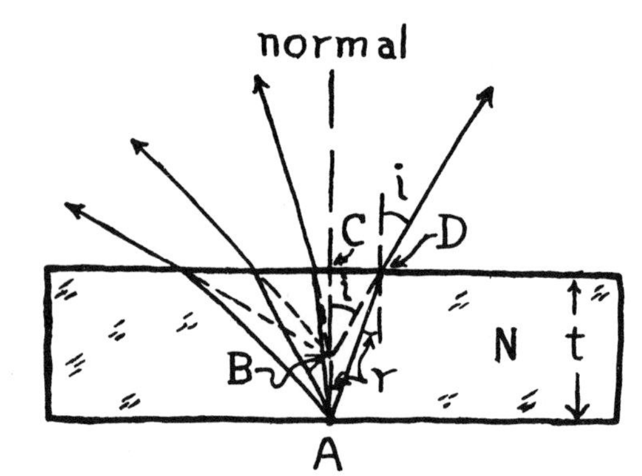

Fig. A2-1. Refractive index by the microscope method.

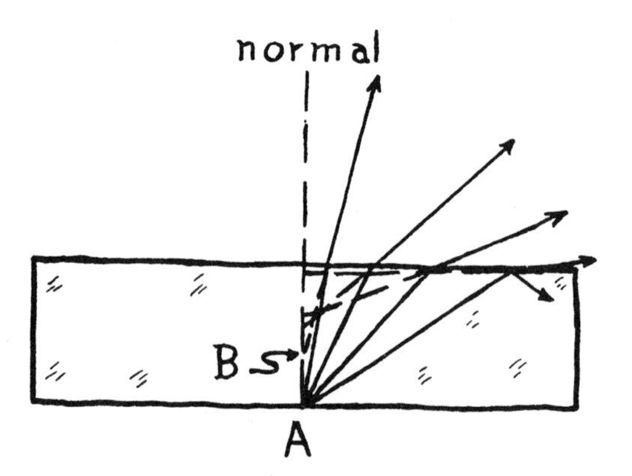

Fig. A2-2. Realistic situation for large angles.

Figure A2-1 shows an object A which, when viewed through the optically dense medium appears to be located at B (the virtual image). Let the refractive index of the medium be N and its thickness t. In order to make a convenient approximation, we assume the object to be viewed near normal incidence. The right side of the figure shows a typical ray leaving point A at angle r with respect to the normal. This ray emerges from the medium at angle i and apparently originates at point B. The following relations hold:

$$N = \sin i/\sin r \approx \tan i/\tan r \quad \text{for small angles}$$

Also $\tan i = CD/CB \qquad \tan r = CD/CA$

Thus $N = CA/CB = t/CB =$ thickness/apparent depth.

So if t and CB are measured, the refractive index of the medium can be readily calculated. A low power microscope equipped with a depth scale allows these measurements to be made.

As an illustration, an inexpensive ($15) microscope was used. No depth scale was provided and the focusing mechanism did not have enough travel for the sample measured (the microscope tube was simply slid by hand beyond the limit of the rack and pinion). A cheap millimeter scale was waxed in the side of the instrument. Using a magnification of 50X crude measurements gave the following result:

$t = 2.39$ cm, $CB = 1.57$ cm, from which $N = 1.52$.

APPARATUS

1) Low power (20 to 50X) microscope, equipped with a depth scale and vernier.
2) Glass sample one or two inches on the sides and about 1/4 inch thick.

PROCEDURE

On the stage of the microscope, put a piece of paper with a black cross or dot in the center of the field of view. Focus the microscope carefully on the mark and record the reading of the depth scale. Next insert a glass plate about 1/4 inch thick over the mark on the paper. With care,focus the microscope on the virtual image of the mark. Be careful not to allow the microscope objective to touch the glass plate and perhaps to become scratched. Record this second position of the microscope. Finally, focus the microscope on the top surface of the plate -- either on a scratch on the glass or on some dust. From the data, calculate the refractive index of the glass plate. Repeat the procedure at least twice more.

Is it necessary to know the actual distance between the microscope objective and the points measured? How about the magnification of the microscope?

RESULTS

Tabulate the results of the several measurements and record the average value with the estimated probable error.

Consider now the errors in the experiment. The chief error lies in the assumption made that the sines of the angles of incidence and refraction are equal to their respective tangents. Figure A2-1 is based on this assumption. The more realistic situation is illustrated in Fig. A2-2. The error implicit in the assumption is minimized by restricting the field of view to small angles. But if this is done, the determination of the exact point of focus is more difficult, for the depth of focus of the microscope is increased, (as in photographic practice, stopping down the lens increases the depth of focus). If reasonable values of magnification and of sample thickness are used, the error due to the approximation is not great, but it does seriously limit the possibility of obtaining high accuracy with this method.

Are there any other important errors in the experiment? Explain.

EXPERIMENT A3. REFRACTIVE INDEX BY PFUND'S METHOD AND BY BREWSTER'S ANGLE.

READING: R. W. Wood, Physical Optics, 3rd ed., (1934, Macmillan), pp. 70-72.

Two inexpensive methods can be used to determine refractive indices to about 1/2%. Pfund's method is based on a determination of critical angle. This method is suitable for the measurement of liquids having a refractive index less than that of the measuring plate. About one cc of sample is required for the measurement. The Brewster's angle method can be used not only for transparent substances but also for opaque (non-metallic) ones. Either liquids or solids with one polished face can be measured.

Although far more accurate methods are available -- a few of them described in this book -- they require far more elaborate apparatus.

I. PFUND'S METHOD

THEORY

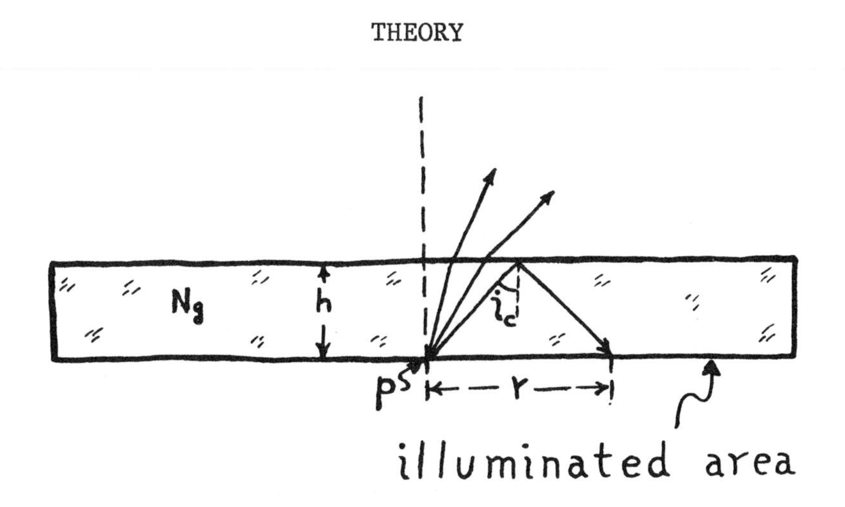

FIG. A3-1

Principle of Pfund's critical angle refractometer.

In the Pfund critical angle experiment, a small bright spot of light falls on the painted lower surface of a glass plate. A cross-section of the plate is shown in Fig. A3-1. Light falling at point P is scattered upward in all directions by the painted surface. Some of this scattered light strikes the upper surface of the plate at angles of incidence less than the critical angle and is partly transmitted and partly reflected. The light striking the upper surface at the critical angle, i_c, or greater, is totally reflected and forms a well defined bright circle of light on the painted surface within which is a darker area as shown in Fig. A3-2.

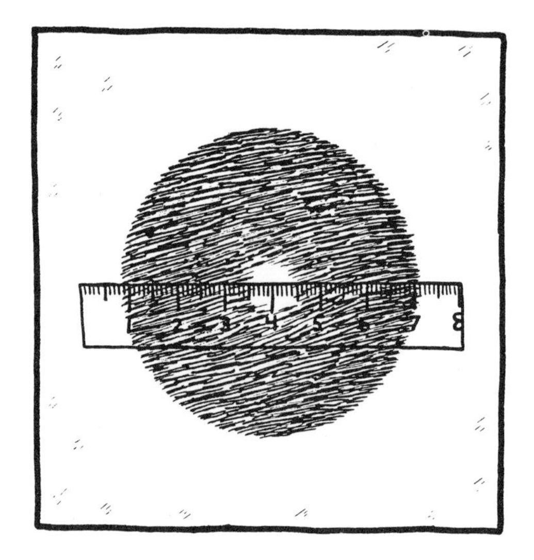

FIG. A3-2

Appearance of illuminated plate.

The diameter of the dark area is dependent on the thickness of the glass plate, its refractive index, and the refractive index of the liquid (if any) spread on its upper surface. With the glass plate alone, the refractive index of the glass, N_g, is given by

$$N_g = \sqrt{D_g^2 + 16h^2} / D_g \quad (1)$$

in which D_g is the diameter of the dark circle and h the thickness of the glass plate. With a liquid layer on the plate, the refractive index of the liquid, is given by

$$N_\ell = N_g D_\ell / \sqrt{D_\ell^2 + 16h^2} \quad (2)$$

in which D_ℓ is the diameter of the new circle with the liquid on the plate.

APPARATUS

A special plate and an illuminator are needed for this experiment. The plate is made of a piece of plate glass about 4 in.square and 1/4 in.thick. (If it is desired to measure liquids of higher index than that of plate glass, dense flint glass could be used). A thin millimeter plastic scale is cemented to the lower surface of the plate. Care must be exercised with some cements as they dissolve the plastic. The scale edge should pass through the middle of the plate. After the scale is attached, the entire lower surface (including scale) except for one corner, is painted with flat white paint. The unpainted corner is used to measure the thickness of the plate.

Figure A3-3 shows a simple illuminator constructed of metal tubing and containing a plano convex lens of about 2 in. focus at one end and a 21 cp auto lamp at the other mounted in a bakelite ring. The inside of the tube and a diaphragm fitted over the lens to reduce the spherical aberration are painted with flat black paint to reduce reflections. All but a small aperture at the end of the lamp should be blackened with paint (which tends to burn brown) or with a mixture of lampblack and water glass solution. A simple stand, if desired, is illustrated in Wood's Physical Optics.

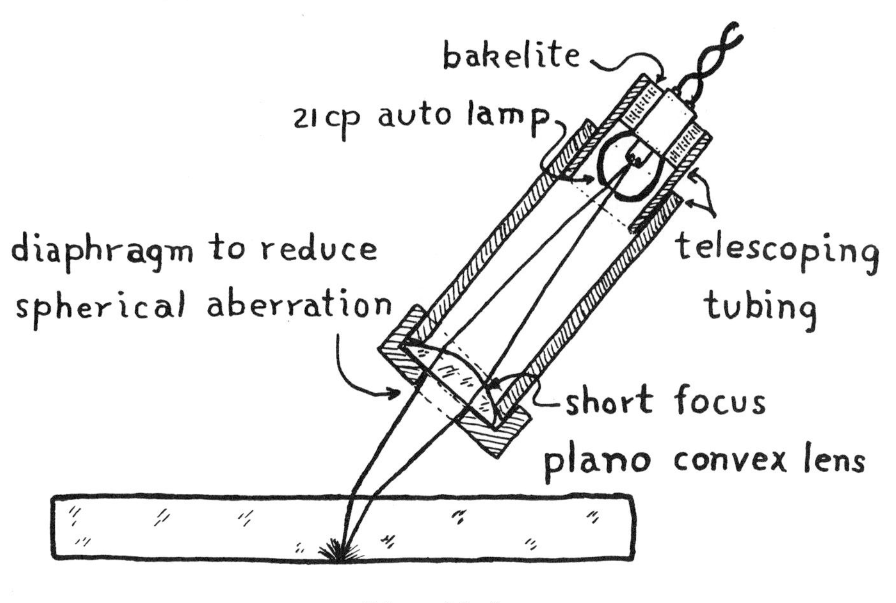

Fig. A3-3

Set-up for Pfund Refractometer.

PROCEDURE

Set up the apparatus as in Fig. A3-3. Care should be used to eliminate as much stray light as possible or the rings may be difficult to see in some cases. Clean the surface of the glass plate with acetone or other solvent to remove any grease film or other contamination. The illuminator should be adjusted to give the smallest possible bright spot at the center of the painted surface of the glass plate -- right next to the scale. A ring of light as in Fig. A3-2 should now be evident. Measure its diameter as accurately as possible, using a magnifier, if desired. Next, pour about a cc of pure water on the plate with a medicine dropper and measure the new diameter of the ring. Repeat with other liquids such as methyl or ethyl alcohol, kerosene, glycerine, or turpentine. Make sure that the surface is carefully cleaned before each new liquid is applied.

Be sure to measure the thickness of the plate with a good micrometer.

If convenient, build a wax wall around the edge of the plate so that a layer of water may be poured over the glass to a depth of a few millimeters. Look for a faint additional ring of light. Note that the diameter of this ring depends on the thickness of the water layer. How can this ring be explained?

RESULTS

Derive the equations used to determine the refractive indices of glass and of the liquids. If possible, explain the origin of the extra ring obtained with a thick layer of liquid.

Calculate the refractive index of the glass and of the liquids. Discuss the sources of error and make an estimate of the accuracy of the results obtained.

II. BREWSTER'S ANGLE METHOD

THEORY

This method depends on the disappearance of a reflected ray when light of the proper polarization is incident on a dielectric at Brewster's angle. Although Fresnel's equations for reflection show that the reflection minimum at Brewster's angle is not very sharp, it is possible to determine refractive indices surprisingly well in this way.

APPARATUS

1) 21 cp auto lamp or equivalent.
2) Polaroid sheet or Nicol prism.
3) Meter stick.
4) Samples: glass plate (preferably with one side blackened to avoid extra reflections), a piece of black glass or a slab of polished coal, and a watch glass about 3 inches diameter for liquids.

PROCEDURE

Clamp the auto lamp to a tripod 6 to 12 in. above the table. Put the sample to be measured on black cloth or paper on the table in such a place that the angle of incidence is not far from Brewster's angle. Look at the reflected light through the Polaroid and rotate the Polaroid for minimum illumination. Now slide the sample toward or away from the light so as to change the angle of incidence to approach Brewster's angle. The minimum should be quite sharp -- light from one part of the lamp filament should be much less bright than from another near the minimum. Using a paper marker for instance, determine the point on the glass surface where the best minimum occurs for the center of the lamp filament. A complete extinction will not be observed, chiefly because the Polaroid is not a perfect analyzer.

Measure the distance -- horizontal and vertical -- from the point of minimum reflection on the glass to the lamp filament. The refractive index, of course, is given by $N = \tan B$. Change the height of the lamp and repeat the measurement.

Measure the refractive index of the glass plate used for Pfund's experiment and the other samples provided.

RESULTS

Calculate the refractive indices of the various samples. Discuss the sources of error and estimate the accuracy of the results.

Compare the indices found by this method with those found in the previous method, if any of the same substances were measured.

Can you suggest any simple improvements in the technique?

EXPERIMENT A4. DEMONSTRATION OF THE REDIRECTIVE PROPERTIES OF PRISMS AND PLANE MIRROR DEVICES.

READING: Jenkins and White, Ch. 2, also D. H. Jacobs, Fundamentals of Optical Engineering, (1943, McGraw-Hill), Ch. 11; or L. C. Martin, Technical Optics, Vol. I, (1950, Pitman), pp. 56-66 and pp. 256-7; see also F. W. Sears, Optics, (1949, Addison-Wesley), 3rd ed., Ch.2.

Prisms or plane mirror combinations may be used as redirective elements to modify either the direction or the orientation of a light beam. In many cases, (though not all) the desired redirection may be achieved by either a prism or a mirror combination. A knowledge of these redirective properties is important in the design of optical systems, and several of the most important arrangements will be considered briefly.

I. PLANE MIRRORS

One of the simplest and most common of all optical devices is a plane mirror. Its imaging properties could scarcely be simpler -- the law of reflection governing its optical behavior has been known since antiquity.

Sometimes it is convenient to use mirrors which reflect from both sides. One example is a simple but ingenious device for signalling. A beam of sunlight reflected by a plane mirror can be seen at great distances. But the problem is how to orient the mirror to reflect the sunlight in the desired direction. Figure A4-1 illustrates the principle of the device. The mirror reflects on both sides and has a small, unsilvered area at the center. To use it, the sender sights at the target through the unsilvered aperture. At the same time, the sunlight shines on the mirror and a small portion passes through the aperture and forms a spot of light on the sender's hand or a card. The sender, looking in the mirror, can see this spot by reflection, and, if he tilts the mirror to put this reflected image in the direction of the target seen through the aperture, he can be sure that the mirror is reflecting sunlight toward the distant target. The student should draw a diagram to convince himself that this is true. Such a device is easily made from an inexpensive pocket mirror from which the backing paint has been removed with paint remover to provide reflecting surfaces on each side. Kleenex should be used to wipe off the paint so as not to scratch the reflecting surface. A pencil eraser will remove the silvering where desired to form the aperture -- a cross with arms about 1/8 inch wide is convenient.

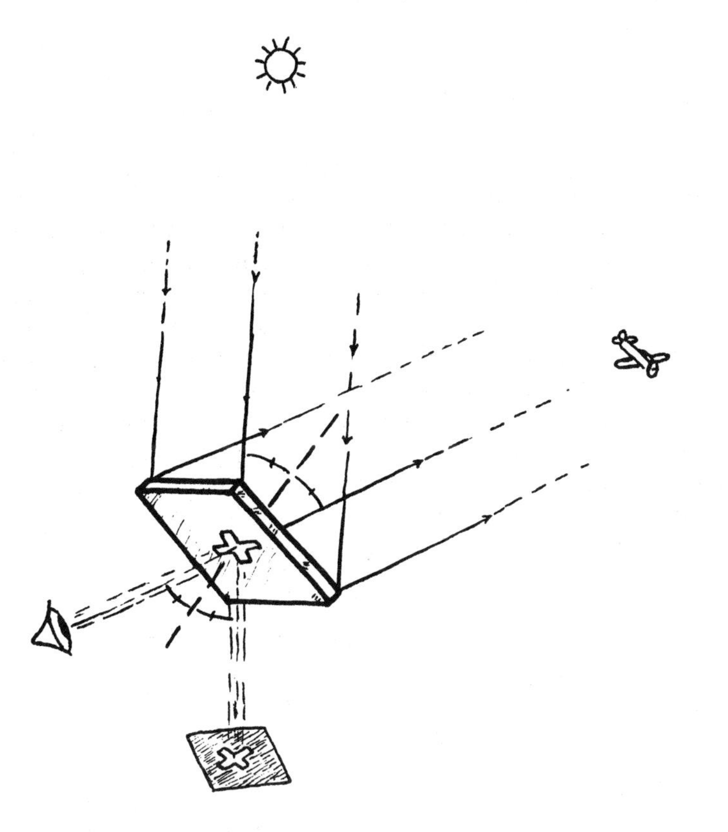

FIG. A4-1

Signalling mirror device.

Unsilvered or partially silvered mirrors provide both transmission and reflection. They may, therefore, be used either to combine two light beams or fields of view or else to split one light beam into two parts. The thickness of the metal coating depends on the desired use of the mirror.

Unsilvered glass plates may be used to superpose scales or cross hairs over the images formed by microscopes, telescopes, periscopes, or gunsights. Brightly illuminated scales are easily seen despite the low reflectivity, and, at the same time, little light is lost (by reflection) from the main field. In the Gauss eyepiece described in the next experiment, an unsilvered glass plate is used to enable the observer to set into coincidence the telescope cross hairs and their reflected image.

Half silvered mirrors are often used to divide light beams into equal parts as in a monocular microscope arranged for binocular vision or, as in interferometers, both to divide light beams and to recombine them.

Mirrors which are silvered to reflect 90% or more of the light are often called "one way mirrors." They are used in nursuries, for instance, so that the children, in a brightly lighted room, may be observed without their knowledge. Are such mirrors really "one way?"

Yet another type of plane mirror is the dichroic mirror (properly treated under the heading of wave optics). This device is produced by coating the glass surface with thin layers of dielectrics of such thicknesses that some wavelengths are reflected and others transmitted. Such mirrors find application in three color processes such as color photography or color television when it is desired to image the red light at one place, and the green and blue at other places, respectively. The first mirror set at 45° reflects one of the primary colors and transmits the other two; a second such mirror reflects one of the remaining primary colors and transmits the third. A recent application of the principle was the use of a dichroic mirror by Professor Strong to reflect the visible light from Venus to the observer in a high altitude balloon flight and to transmit with little loss to the spectrometer the infrared radiation.

II. RIGHT ANGLE PRISM

Perhaps the most common prism type is the simple right angle prism used, for example, to redirect the cone of light in a Newtonian telescope to one side for convenient observation with an eyepiece. Figure A4-2 shows the appearance of such a prism, not as usually represented, but more realistically to show the internal reflections. (How are the various lines within the figure explained)? This prism performs the same function as a 45° plane mirror which frequently is used in its place. On a white card, mark a red arrow and at right angles to it, a blue one. Observe the card through the right angle prism looking through one of the faces forming the right angle.

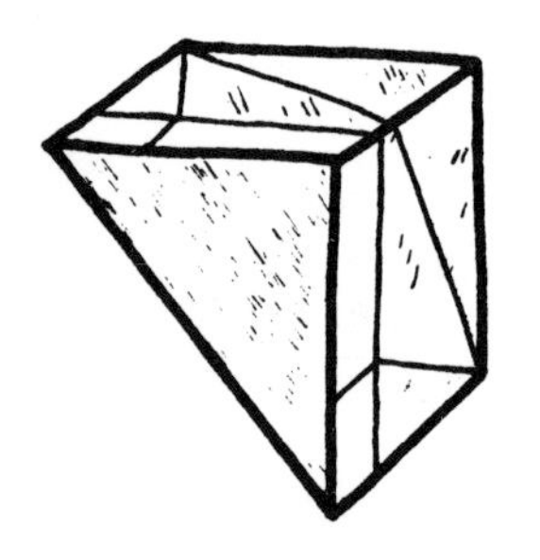

FIG. A4-2

Right angle prism showing internal reflections.

Note which of the arrows is reversed. Is there any difference between the action of the prism and that of the equivalent mirror? When the prism is used in this way, the light beam suffers total internal reflection. If, however, the eye is moved so as to look into the hypotenuse face, the object can still be seen. Explain. The hypotenuse of some right angle prisms is silvered. Under what conditions might this be useful?

A pair of right angle prisms placed as in Fig. A4-3 constitutes a Porro prism pair. These are commonly used in prism binoculars to erect the inverted image produced by the objective lens and eyepiece lens and to reduce the length of the instrument. A mock up of half a prism binocular system is easily made from an objective (preferably achromatic) of about 8 to 12 inch focal length, a pair of right angle prisms and a plano-convex eye lens. The parts may be held in place on a board with tackiwax. The focal length of the objective should not be more than five or six times that of the eye lens unless the system is very rigidly held (high power binoculars are annoyingly difficult to hold in one's hand).

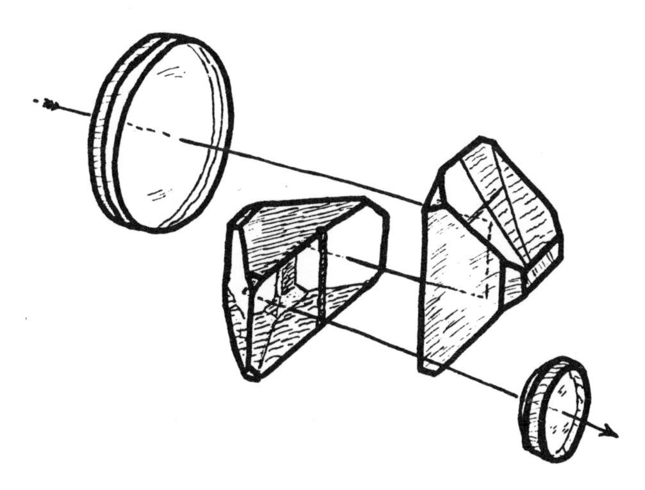

FIG. A4-3

Porro prism pair as used in binoculars.

III. AMICI PRISM

Another useful type of prism is the Amici or "roof" prism. Two views are shown in Fig. A4-4. The prism is basically a right angle prism, but the hypotenuse side has been modified into a "roof" (see upper left of figure). This roof, which is a pair of total reflecting surfaces (or mirrors) placed at right angles, has the property of interchanging right and left or "reverting" the image. Thus the Amici prism both redirects the beam and reverts the image. Combined with an objective and eyepiece as shown in the lower part of the figure, it is used to make an elbow or right angle telescope. A pair of mirrors placed at right angles will illustrate the principle. (To check that the angle is just 90°, look at the reflection of your eye -- if the angle is not correct, the iris will not appear circular).

IV. PENTA PRISM

The penta prism and its mirror equivalent are illustrated in Fig. A4-5 (a) and (b). Two reflections produced by mirrors inclined at 45°

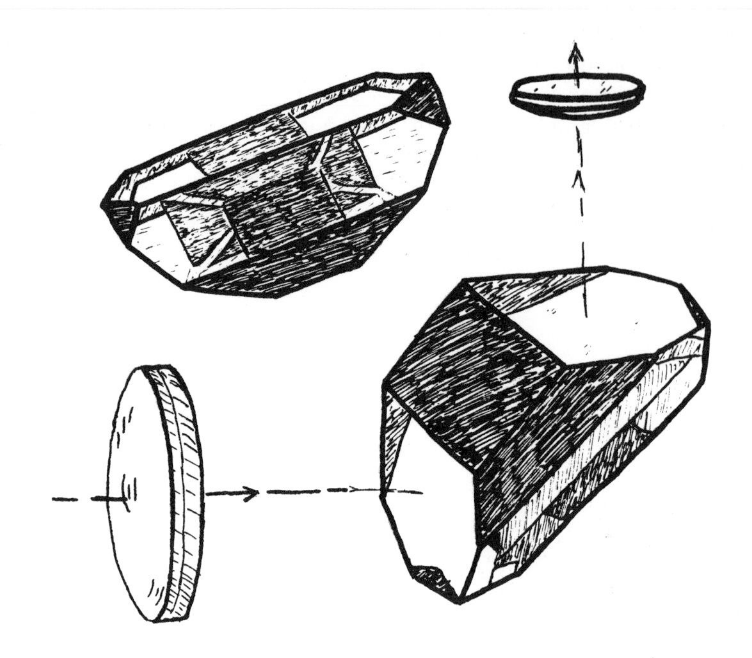

FIG. A4-4

Amici prism in elbow telescope system.

deviate the incident beam by 270° irrespective of the angle of incidence in the plane of the figure. Prove that this is also the case for the prism where refraction is involved. Such prisms are obviously useful in any situation where a light beam must be deviated at exactly 90°. They are also especially useful in the design of rangefinders where very small angles must be measured accurately, for they minimize the error introduced by slight warping and bending of the rangefinder tubes. The properties of these prisms or of the readily constructed mirror equivalent should be observed.

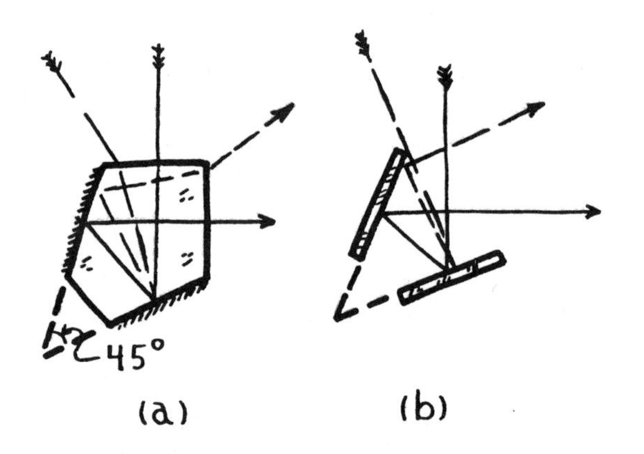

FIG. A4-5

Penta prism (a) and mirror equivalent (b).

V. DOVE PRISM

Sometimes it is necessary to rotate an image by an arbitrary angle. For this purpose, a Dove prism or the mirror equivalent, Fig. A4-6, is used. Such prisms are used in some camera viewfinders, in panoramic telescopes, and as erecting elements in telescopes. The prism rotates an image through twice the angle, through which it is turned. For large aperture rotating devices, the mirror equivalent is clearly preferable. Since the prism is effectively a large right angle prism with the right angle corner cut off, evidently a right angle prism will show the rotation property. In what respect is a Dove prism preferable to a right angle prism for beam rotation purposes?

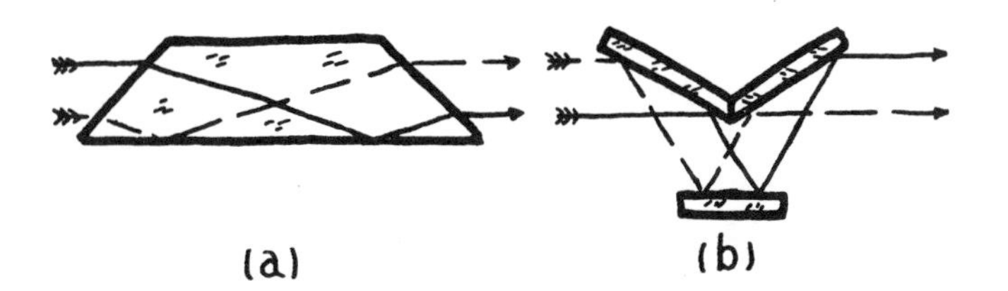

FIG. A4-6

Dove prism (a) and mirror equivalent (b).

VI. TWO AND THREE MIRROR CORNER REFLECTORS

In part III on the Amici prism, it was pointed out that the addition of the two reflector roofs added a reversion of the image to the deviation of the right angle prism. Two mutually perpendicular (orthogonal) mirrors or reflecting surfaces used in the plane orthogonal to both surfaces, constitute a two mirror corner reflector. A two mirror corner (a) reverses the original direction of the rays, (b) reverts the image, and (c) displaces the rays to one side. The action is identical to that in one Porro prism.

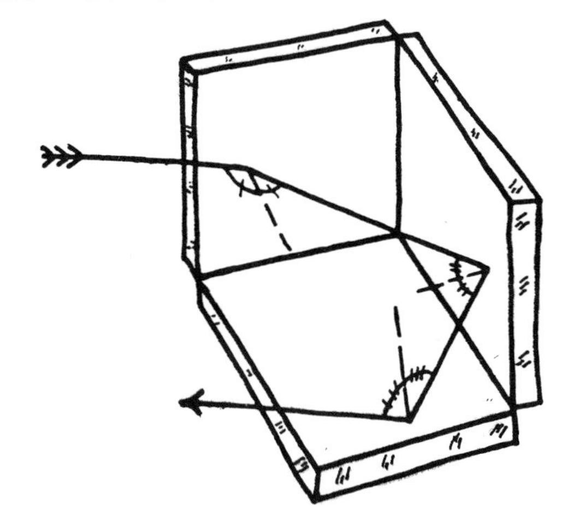

FIG. A4-7

Three mirror retrodirective corner reflector.

If a third reflecting surface orthogonal to the other two is added, one has a true corner reflector or corner cube prism. This device will reverse the direction of incident rays provided that the rays are reflected off all three surfaces. The mirror arrangement is illustrated in Fig. A4-7. Corner reflectors are widely used for automobile tail lights, for bicycles, and for road signs. They find use in aircraft landing aids, in interferometers, and elsewhere. At long wavelengths, they are extensively used for radar navigational aids and for tracking balloons, etc.

VII. OTHER TYPES

In addition to the prism and mirror types considered above, there exist a large number of special purpose varieties. Many are described in surplus optical parts catalogs and include such types as octagon prisms which can be used to substitute for the shutter in a movie camera, Lehman prisms used in binoculars, special deviating prisms or field dividing prisms used in range finders, photometers, and so forth.

EXPERIMENT A5. ADJUSTMENT OF THE STUDENT SPECTROMETER AND MEASUREMENT OF PRISM ANGLE.

READING: G. S. Monk, Light, (1937, McGraw-Hill), Appendix IV, pp. 426-430; or J. Morgan, Introduction to Geometrical and Physical Optics, (1953, McGraw-Hill), Ch. 8.16; also see R. A. Sawyer, Experimental Spectroscopy, (1946, Prentice-Hall).

The student spectrometer, as it is commonly called, is basically a mechanical-optical angle measuring device. It does not have the convenience or precision of many more elaborate or specialized spectroscopic instruments, but it does have greater versatility. It consists of a collimator, a telescope, a prism table, and an angle measuring scale with verniers. In this way, it is quite similar to a goniometer which is used to measure the angles between the various faces of a crystal. The purpose of the collimator is to produce an accurately parallel beam of light. The deviation of this beam by a prism, grating, double slit, etc., can be studied with the telescope (which forms images of the collimator slit). Better spectrometers are provided with various adjustments for focusing and levelling the collimator and telescope. The telescope is generally provided with a Gauss eyepiece. The prism table is also adjustable in height, and tilt. Both the telescope and the prism table are provided with clamps and vernier screws so that they can be rotated through accurately measureable angles.

APPARATUS

1) Student spectrometer with adjustments for levelling the telescope and collimator.
2) Gauss eyepiece.
3) White light source.
4) Flashlight or other light to use with Gauss eyepiece.
5) 60° prism polished either on two or three sides.

PROCEDURE

I. ADJUSTMENT OF THE SPECTROMETER

The immediate purpose of the following set of directions is to enable the student to adjust a common type of spectrometer properly. In addition, it is hoped that they will illustrate or suggest the kind of procedure needed to put into adjustment any kind of spectrometer. Thus, it is essential that the purpose of each step be understood. The procedure given here is not the only possible one, but it is a satisfactory one.

To put the spectrometer into adjustment, it is necessary to achieve several conditions:

(A) The telescope must be focused for parallel light and the crosshairs must be at the secondary focus of the telescope objective.

(B) The collimator must be focused for parallel light.

(C) The optical axes of the telescope and the collimator must be perpendicular to the mechanical instrument axis.

A. Focus the telescope:

If no Gauss eyepiece is provided, the telescope can be focused by sliding the eyepiece to bring the virtual image of the crosshairs in sharp focus and then sliding the unit, including both crosshairs and eyepiece, so that a distant landscape is in focus. Note that the virtual image of the crosshairs should be at infinity, not at normal reading distance. If the focus is proper, slight movement of the observer's eye back and forth behind the telescope will not cause the crosshair image to shift with respect to the distant object.

Incidentally, in using optical instruments, it is unnecessary to wear glasses which correct near or far sightedness, for a mere change in the focus of the eyepiece will accomplish the same purpose as the glasses. But the eyepiece will not correct for the observer's astigmatism. The best way to focus the telescope accurately is to use the Gauss eyepiece illustrated in Fig. A5-1.

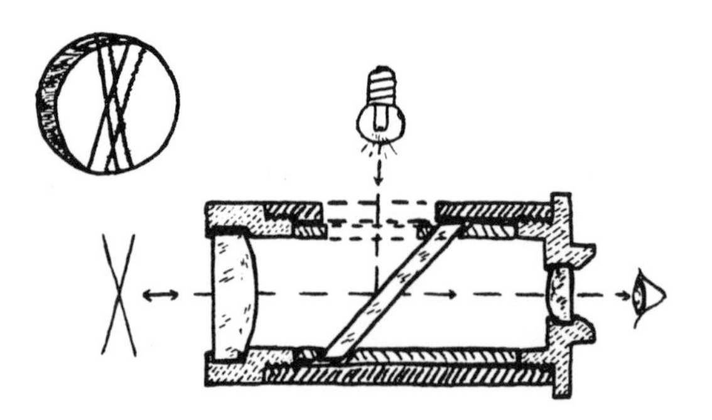

FIG. A5-1

Construction of Gauss eyepiece. Insert shows appearance of field with reflected crosshair image slightly out of focus.

In addition to the lenses, the Gauss eyepiece contains a hole in its side (also there is a corresponding hole in the telescope tube) and an inclined clear glass plate. The purpose of these additions is to allow a light beam to be introduced from the side of the telescope to permit observation of a dark "image" of the crosshairs by reflected light. This light illuminates the whole area near the crosshairs, and emerges through the telescope objective. Some of the light is returned by another reflecting surface and produces a "dark image" of the crosshairs (see insert in figure) which is seen in addition to the crosshair image seen directly. If the reflecting face is normal to the telescope axis, the two sets of crosshairs, direct and reflected, will coincide.

In focusing the telescope with a Gauss eyepiece, the first step is to focus the eyepiece on the crosshairs as before. Next, the eyepiece is illuminated from the side and the telescope made approximately normal to a plane reflecting surface on the prism table. (This could be one face of the prism, for example). The light must be placed so that a beam is reflected through the telescope objective. Tilt the prism face slightly until a reflected circle (or most of a circle) of light is seen. Some trial and error may be required. When the reflected circle of light is reasonably centered, a fuzzy set of crosshairs should be seen unless the telescope is very far out of focus. Note: if a 60° prism with three polished surfaces is used, several sets of crosshairs may be seen. The extra crosshair images arise from internal reflections in the prism. They may be eliminated by blackening one face of the prism. (Draw a diagram to explain the origin of the extra images. Are the images inverted and/or reverted)?

After the reflected crosshair image is put in focus, check for parallax to see that as the eye is moved back and forth, the direct and reflected crosshair images remain together. This adjustment is impossible if the real crosshairs are not in one plane.

B. Focus the collimator:

The collimator is now illuminated with a light source. If the source has appreciable area as, for instance, a frosted tungsten lamp, no auxiliary lens is needed, but if the source has small area, then an additional lens is needed. The lens must be able to deliver to the collimator a cone of light fat enough to illuminate both the full area of the slit and the full area of the collimator lens. (An appendix to the experiment is concerned with the proper illumination of the collimator).

Nearly close the collimator slit and look through the telescope directly into the collimator. Obtain the sharpest possible slit image by sliding the slit holder toward or away from the collimator lens. If the telescope and collimator are now properly focused, the slit image should remain coincident with the crosshairs when the eye is moved slightly back and forth from the axis of the telescope.

C. Make telescope and collimator axes perpendicular to mechanical axis:

1. Make the telescope and collimator parallel

The collimator slit should now be turned horizontal (if not already so). Adjust either the telescope or the collimator leveling screws to bring into coincidence the horizontal slit image and the intersection of the crosshairs. The telescope and collimator axes are now parallel, but not necessarily perpendicular to the mechanical instrument axis -- see Fig. A5-2.

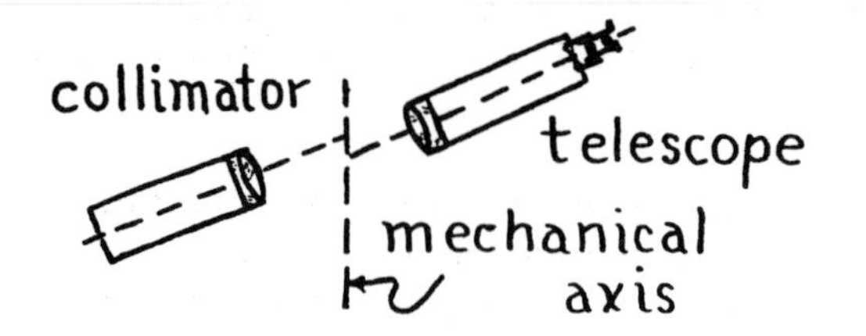

FIG. A5-2

Exaggerated view: telescope and collimator axes parallel but not perpendicular to the mechanical axis.

2. Make one prism face normal to the mechanical instrument axis*

Move the telescope so that it makes approximately a right angle with the collimator axis. Set one face of the prism to reflect light from the collimator into the telescope. Make the slit image coincide with the crosshairs by tilting the prism face. This face of the prism is now normal to the mechanical axis. If the collimator is tipped up from the normal, by, say, 2 degrees, the telescope which was parallel to the collimator must be down 2 degrees. The prism face bisects this vertical angle as well as the horizontal angle between collimator and telescope, and hence is normal to the instrument axis. Draw a ray diagram to demonstrate this fact.

3. Make the telescope axis normal to the adjusted prism face

Now rotate the prism table about the instrument axis so that the telescope can be adjusted normal to the reference prism face. Use the Gauss eyepiece to adjust the telescope tilt. The telescope is now in complete adjustment.

4. Again make the collimator and telescope axes parallel

Again look directly into the collimator through the telescope (over the prism, if desired). Adjust the collimator until coincidence between the slit and crosshairs is obtained. Now the collimator and telescope are both adjusted; they are parallel to each other, and the telescope is normal to the instrument axis.

5. Make the collimator slit vertical

The slit may be rotated to a vertical position with reasonable accuracy by eye or else by a reflection method after the prism is adjusted, provided the prism is a 60° one with 3 polished sides.

---

*For the purpose of this experiment, it is possible to combine two steps by positioning the prism as indicated in Part II below, and then adjusting two faces (b and c) normal to the instrument axis instead of only one.

The spectrometer is now in proper adjustment except possibly for a final adjustment of the slit to make it more nearly vertical. The foregoing procedure or parts of it will be frequently needed in the following experiments in this book.

II. MEASUREMENT OF PRISM ANGLE

It is suggested that the student use a particular identified 60° prism for this experiment and all other experiments which require such a prism (eg., refractive index and dispersion of a prism, dispersion of water by critical angle, etc.). Ground surfaces of the prism may be readily marked with pencil. Label one angle so that this angle is measured and used throughout.

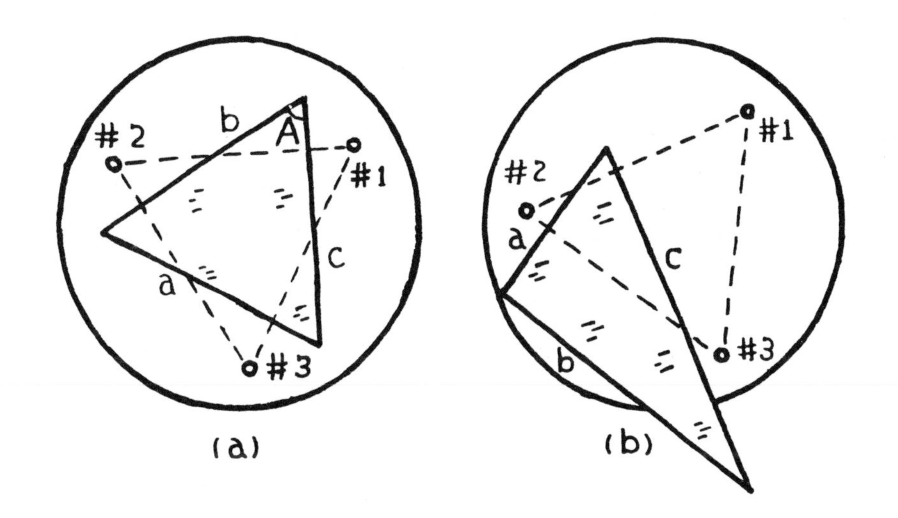

FIG. A5-3

(a) Correct placement of 60° prism on table;
(b) Placement of constant deviation prism.

The prism should be placed on the prism table as in Fig. A5-3(a), with the 60° prism placed as shown. Face c may be tilted with screw #1 while face b is unchanged save for rotation in its own plane. Conversely, face b may be tilted with screw #3 without disturbing face c. A constant deviation prism is placed as shown in Fig. A5-3(b). Face a is adjusted using any screw. Then face b is adjusted using screw #1 which leaves face a undisturbed.

The two prism faces b and c are made normal to the adjusted telescope by the Gauss eyepiece method. Since the prism is rarely in the ideal position on the table, it is imperative to check both faces after adjusting either one. The third face a must remain unadjusted for obvious reasons.

The prism angle A is now to be measured. First, clamp the telescope at a convenient position and rotate the prism table so face b is normal to the telescope (Gauss method). Record the angular reading using both verniers. Now rotate the prism table to bring face c normal to the telescope. Again record both angular readings. Calculate the prism angle from the readings for each vernier and average the two results. Explain why the two measurements do or might disagree. Consider errors in scale calibration and errors in eccentricity of the telescope and prism table axes. Why are two readings 180° apart, desirable? Do such readings tend to cancel some errors?

As an alternative method, but a less accurate one, turn the prism table so that the beam from the collimator, which must be fully illuminated now, is reflected partly off face b and partly off face c. Leave the prism fixed and set the telescope to measure the two reflected images from faces b and c. In each case, record both vernier readings. Calculate the prism angle. This method is subject to much greater error, for errors in focus of the telescope and collimator, and especially any aberration, will affect the measurement. Also, if the prism faces are not truly plane, more errors result.

Record the prism angle and the probable accuracy.

Appendix: Collimator Illumination

Correct illumination of the collimator is essential if the best spectral resolution and brightest spectra are to be obtained. The illumination is correct if (1) the collimator slit is fully illuminated and (2) the collimator lens is fully illuminated. If the source has large enough area and is placed close to the slit, an auxiliary lens is both unnecessary and undesirable. But for a small area source a lens is needed. To select a suitable lens, note the relative size of the source and slit and the lowest magnification needed to form an image which at least covers the slit. This magnification determines the minimum value of image distance to object distance. The lens selected must have a diameter large enough so that the cone of illumination which supplies the collimator will fill the collimator lens. Thus, the auxiliary lens which need not be achromatic (except, perhaps, in rare instances), must have a lower focal length/diameter (f number) ratio then that of the collimator lens. Some of the relations are illustrated in Fig. A5-4. Evidently the lens should satisfy the conditions:

$$S'/S \geqslant \frac{\text{slit length}}{\text{source length}}$$

$$S+S' \geqslant 4f$$

$$S'/d \leqslant F/D \quad \text{(f/no collimator)}$$

and for the special case of unit magnification,

$$S = S' = 2f \quad \text{so} \quad 2f/d = F/D \, .$$

That is, the auxiliary lens must be twice as fast as the collimator lens in this case.

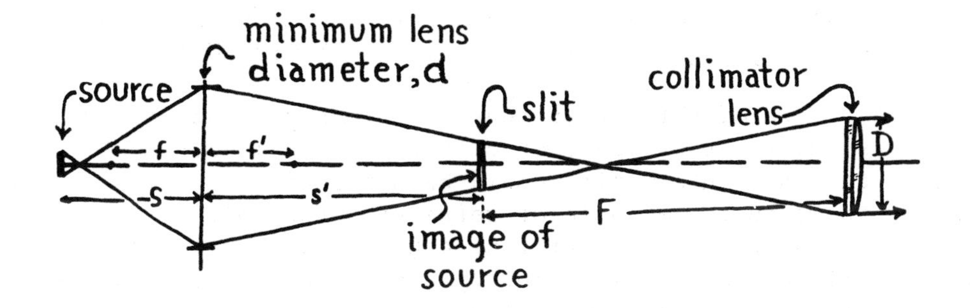

FIG. A5-4

Illumination of the collimator

To put the lens into position, first, omit the lens and center the light beam emerging from the collimator with respect to the rim of its lens. Make sure the distance between source and slit is greater than 4 times the focal length of the lens or equal to the sum of the calculated object plus image distance. Next, insert the lens in the beam and focus the source on the slit using the magnified or reduced image as appropriate. The lens should preferably be oriented for minimum spherical aberration. A white card held to receive the illumination emerging from the collimator will show whether or not the collimator lens is filled with light.

EXPERIMENT A6. DISPERSION OF GLASS AND WATER BY CRITICAL ANGLE

READING: Jenkins and White, pp. 10-11, p. 17, Ch. 23-23.3; for further discussion of refractometers see A. C. Hardy and F. H. Perrin, Principles of Optics, 1st ed., (1932, McGraw-Hill), pp. 359-364.

The object of this experiment is to measure the dispersion curves for both glass and water by means of critical angle measurements. It is evident that the same information could be obtained directly by using a spectrometer to measure the angle of refraction for solid glass or hollow, liquid-filled prisms. But the critical angle method makes for a more interesting experiment and is also important because it forms the basis of various direct-reading commercial refractometers such as the Abbe and Pulfrich types.

Although the Pfund refractometer described in Experiment A3 is based on the critical angle principle, it is insufficiently accurate for the measurement of dispersion. The present method in which angles are measured with a spectrometer is capable of far higher accuracy.

THEORY

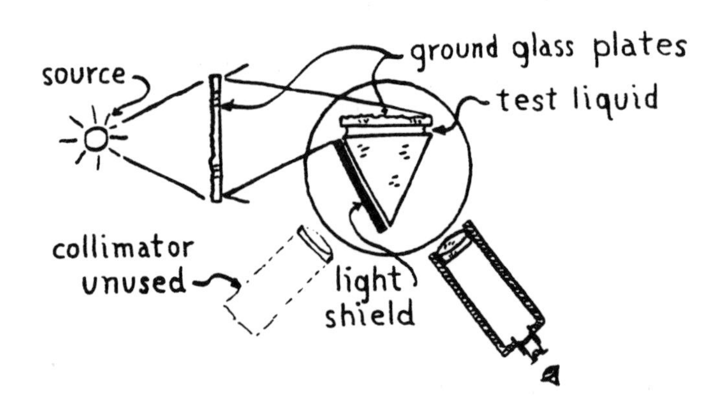

FIG. A6-1

Arrangement of apparatus to measure dispersion by critical angle.

Figure A6-1 shows how the apparatus is arranged. The collimator is not used, and it is turned out of the way. Assume that the source is monochromatic, and that either the prism alone, or the prism with a liquid film, is diffusely illuminated as indicated. Light rays will be refracted by the prism (with or without liquid film) at all angles up to, but not exceeding, the critical angle. Figure A6-2 shows a more detailed diagram of the rays. Ray #1 represents any of the innumerable rays which strike the upper prism face, b, at an angle of incidence less than 90°. Such rays contribute to a broad band of light seen in the telescope. The limiting rays, those which follow a path like that of ray #2, strike the upper prism face at glancing angle and are refracted at the critical angle, r_c. Such rays strike the second prism face, c, at an angle r' and emerge at angle i' as indicated. These rays form a parallel beam and are imaged by the telescope as a line, the line of demarcation between the broad band of light and a dark portion. Thus in the field of view of the telescope, one sees a broad band of light sharply bounded on one side. A knowledge of the angles involved allows calculation of the refractive index.

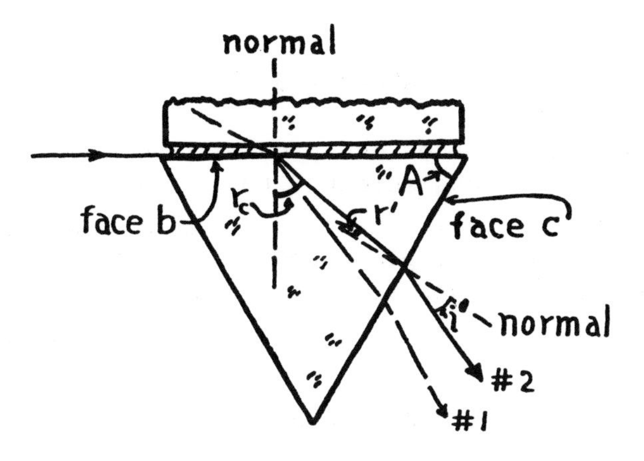

FIG. A6-2

Diagram showing several rays passing through the prism.

From Fig. A6-2 by geometry one has

$$A = r' + r_c$$

From Snell's law it follows that

$$N_\ell = N_g \sin r_c$$

$$\sin i' = N_g \sin r'$$

where N_ℓ is the refractive index of the liquid and Ng the refractive index of the glass.

If we eliminate r_c and r' it follows that

$$N_\ell = \sin A \sqrt{N_g^2 - \sin^2 i'} - \cos A \sin i' \qquad (1)$$

It is important to notice that angle i' may be negative (on the opposite side of the normal to prism face c. In this case, sin i' is negative and the second term in equation (1) must be added instead of subtracted. If there is no liquid layer on the prism face b, then N_l becomes the refractive index of air, and we substitute $N_l = 1$. Solving equation (1) with $N_l = 1$ one obtains

$$N_g^2 = 1 + \left(\frac{\sin i' + \cos A}{\sin A} \right)^2 \qquad (2)$$

Equation (2) is used to calculate the refractive index of the prism, and Equation (1) then gives the index of the liquid.

APPARATUS

1) Student spectrometer equipped with a Gauss eyepiece.
2) Large ground glass screen 3 to 6 inches square.
3) 60° prism.
4) Small ground glass plate about the size of a prism face.
5) Sodium arc and another bright line source such as mercury.
6) Distilled water.

PROCEDURE

Unless the spectrometer has already been partially adjusted by the instructor, the first step is to follow the procedure of Experiment A5 to adjust it. For this experiment, the telescope should be focused for infinity and its optical axis set truly perpendicular to the mechanical instrument axis.

The prism should be clamped on the prism table as in Fig. A5-3a, and faces b and c adjusted so that their normals are perpendicular to the instrument axis. Prism angle A must be known or measured.

After the spectrometer is adjusted, turn it so the collimator is out of the way as shown in Fig. A6-1. Put the large ground glass screen between the source and spectrometer to diffuse the light from the arc source. (For sources of large area, this glass screen may be omitted).

Begin with a sodium arc source placed as in the figure. Put a light shield over the unused face of the prism -- a piece of black paper held to the prism with a drop of water may be used. This shield should block all rays through the unused prism face, but, at the same time, it must not block the desired rays from falling on the prism at grazing incidence.

Looking through the telescope, one should be able to locate the sharp boundary corresponding to the cutoff angle i'. The cutoff is distinguished by the fact that its position is independent of the position (within wide limits) of the light source with respect to the prism. The sharpness of the cutoff boundary, however, does depend on the illumination, and care should be taken to make it as sharp as possible by adjustment of the relative position of the illumination and the prism face. Do not rotate the prism on the prism table -- rotate the table if necessary. (Why)? Now clamp the prism table and leave it clamped. Make sure both scales will be convenient to read.

After the critical angle cutoff has been made as sharp as possible, set the telescope normal to the prism face using the Gauss method and read the angular setting on both verniers. Next, swing the telescope to bring the cutoff line at the crosshairs and read this position also. It is important to measure all angles as accurately as possible to obtain good results.

Substitute a mercury arc (or other line source) for the sodium lamp. (The sodium lamp will be needed again, so it may be preferable not to turn it off). The large ground glass screen will doubtless be needed for this source (even if it was omitted for the sodium arc). The new source, unlike the sodium, has several bright emission lines, and the critical angle cutoffs will be multiple corresponding to the several wavelengths. In the telescope will now be seen a series of colored bands which, however, are not pure spectral colors. Explain. In order to identify the various cutoff positions with the proper wavelengths, it is desirable to use a set of spectral isolation filters to find one cutoff at a time. Determine in this way the cutoffs for the various spectral lines of the source beginning with the more easily visible yellow or green and working out both ways to the red and violet. Since the red and violet are relatively hard to see, great care should be used to obtain optimum lighting conditions.

RESULTS

1) Derive equations (1) and (2) for the refractive index.

2) Compute the refractive indices of the glass and of water for the several wavelengths using six place log tables. In order to plot this data meaningfully, the ordinate must be suitably laid out. It might be well to plot the two curves on separate graphs or else, if one sheet of graph paper is used, to use two scales -- one for glass and one for water. The scales should be sufficiently expanded so that the dispersion is clearly seen.

3) From the two dispersion curves, compute the dispersive powers of both water and the glass prism. That is, evaluate the quantity

$$\frac{1}{\nu} = \frac{N_F - N_c}{N_D - 1}$$

where N_F, N_C and N_D

are the refractive indices corresponding to the Fraunhofer F, C, and D lines. Look up the wavelengths corresponding to these lines to evaluate the dispersive powers.

4) Discuss the sources of error in the experiment and indicate the probable accuracy of the results obtained.

It may be of interest to look at the dispersion curve of water of Rubens and Ladenburg given in R. W. Wood, Physical Optics, 3rd. ed., (1934, Macmillan), p. 509. This curve shows the refractive index from the ultraviolet out to 7 microns in the infrared.

EXPERIMENT A7. DEMONSTRATION OF THE ANOMALOUS DISPERSION OF FUCHSINE.

READING: Strong, pp. 97-98; or Jenkins and White, Ch. 23.4; in addition, read R. W. Wood, Physical Optics, 3rd. ed., (1934, Macmillan Co.), pp. 118-121; and if possible, T. Preston, The Theory of Light, 5th ed., (1928, Macmillan Co.), pp. 513-516.

Dispersion is called normal if the refractive index decreases smoothly and continuously as the wavelength increases. Colorless, transparent substances such as water or glass exhibit normal dispersion (in the visible spectrum). On the other hand, it is possible for the refractive index to decrease at first, then to rise rather abruptly over a short wavelength span, and then to decrease smoothly again. A substance for which this is true is said to exhibit anomalous dispersion. The rapid rise in index, which makes the dispersion anomalous, occurs in a band of wavelengths which the substance absorbs strongly. One such substance is an alcoholic solution of the dye fuchsine, and it absorbs strongly in the yellow green part of the spectrum.

The phenomenon of anomalous dispersion was apparently discovered by Fox Talbot about 1840, but little more work was done for several decades. In 1860, Le Roux found anomalous dispersion in iodine vapor. Beginning with the observation of the anomalous dispersion of an alcoholic solution of the aniline dye fuchsine by Christiansen in 1870, considerable study was made of the phenomenon.

The names "normal dispersion" and "anomalous dispersion" are retained only for historical reasons, for if the infrared and ultraviolet regions are included, then all substances show absorption bands and thus "anomalous dispersion." "Anomalous dispersion" of quartz, for instance, occurs at 8 microns and may be used to separate long infrared radiation from the shorter wavelengths by Wood's method of focal isolation.

PROCEDURE

Perhaps the easiest demonstration of anomalous dispersion is based on the experiments of Christiansen. A solution of 1/2 gram of powdered fuchsine dye is dissolved in about 5 grams (or 6 1/4 cc) of ethyl alcohol. If the solution is much stronger, it is too opaque.

The prism is readily made by clamping together two 1 1/2" square pieces of plate glass about 1/8" thick, separated along one edge by a strip of brass 0.020 to 0.030 inches thick. The edges opposite the wedge should make contact. The glass plates are held together either with a large battery clip or a rubber band and the prism is filled, when needed, with a few drops of solution from a medicine dropper. Surface tension keeps the liquid in the prism. Keep paper towels handy, for the dye stains everything within reach!

The chief difficulty in the demonstration arises from chromatic aberration. Thus a Spencer student spectrometer, although equipped with achromatic lenses, still has too much residual chromatic aberration (secondary spectrum). There are several ways to set up the experiment satisfactorily. One way is to use a high quality spectrometer (such as a Gaertner precision research instrument). Another less expensive way is to omit the collimator altogether.

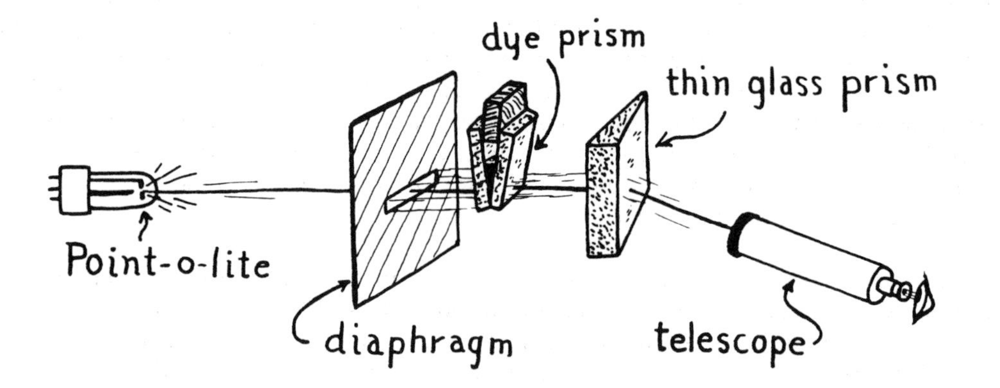

FIG. A7-1

Apparatus to demonstrate anomalous dispersion.

Figure A7-1 shows how the demonstration may be set up without the use of a collimator. A Point-o-lite source is placed at the right height 3 or 4 feet away from a small student spectrometer of moderate quality. On the prism table, mount a prism having a fairly small refracting angle -- (20 to 30°), so that a short horizontal spectrum is obtained. It is important to be able to see the whole visible spectrum at the same time so that the red and blue can be compared. The fuchsine prism should now be mounted just in front of the glass prism with its refracting edge horizontal. Just ahead of this prism, put a horizontal slit, say 1/8" wide x 1" long, to restrict the light rays to those passing through the dye prism as near the edge as possible, but blocking rays not passing through the dye prism. The spectrum now seen should consist of two sections -- two lines which slant in such a way as to indicate that yellow is deviated more than red and red more than blue. The green is strongly absorbed and absent unless the dye film is ultra thin and the source particularly intense.

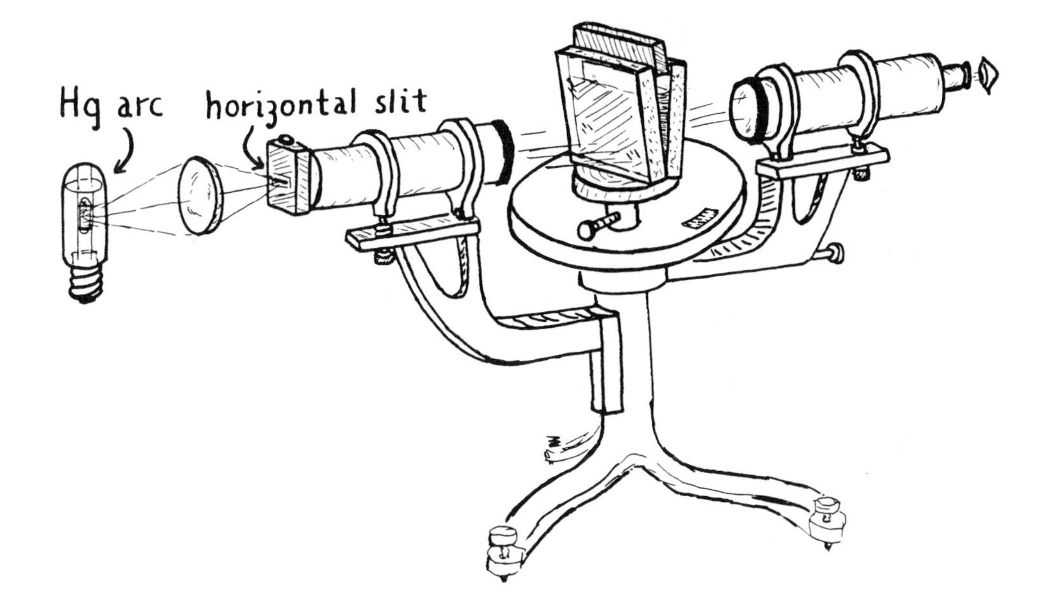

FIG. A7-2

Alternative arrangement to show anomalous dispersion.

Figure A7-2 shows how the demonstration may be set up using a good spectrometer and a mercury arc source. If the arc source is used, it is possible to dispense with the small angle glass prism needed in the first arrangement. In this case, one sees, of course, only a few spectral lines, and the demonstration is perhaps somewhat less dramatic.

EXPERIMENT A8. DEMONSTRATION OF THE ANOMALOUS DISPERSION OF SODIUM VAPOR.

READING: R. W. Wood, Physical Optics, 3rd ed., (1934, Macmillan), pp. 492-7.

The fuchsine dye solution used in the previous experiment has, characteristically, a broad absorption band. Therefore, the anomalous dispersion curve is such that the whole visible spectrum needs to be observed to see the effect. Sodium vapor, on the other hand, being a low pressure gas has very narrow absorption lines in the yellow (D lines). Thus, in this case, the refractive index differs but little from unity save in the yellow very near the absorption lines. The basic phenomenon, the anomalous rise in refractive index, is however, much the same in the two experiments.

Wood's demonstration of the anomalous dispersion of sodium vapor can be very spectacular, but it requires considerable care in the construction of the apparatus. Several modifications of Wood's procedure are suggested.

As in Wood's demonstration, a long tube filled with sodium vapor is used to act as a thin prism (like the fuchsine dye prism). The spectrum is dispersed horizontally with a transmission diffraction grating rather than a prism to give adequate dispersion. Previously the need was for a short visible spectrum -- here we need to be able to distinguish clearly two wavelengths, which differ by only one part in a thousand (5890 and 5896 Angstrom units).

APPARATUS

1) Absorption cell:

The absorption cell described here differs from that of Wood in the method of attaching the windows. Wood sealed the windows in place with wax, but the use of O-rings and flanges is much safer. With Wood's arrangement, the wax tends to melt or soften when the cell is heated, and leaks or possible explosions may result.

The absorption cell is constructed of thin walled steel or brass tubing 1/16 inch thick, 1 1/2 to 2 inches diameter, and about 20 inches long. Figure A8-1 shows two views of the ends of the cell. Two holes are drilled in the side of the tube about 2 inches from either end to accommodate the two side arms for flushing and evacuating the cell. The holes should provide a fairly snug fit for two six-inch lengths of 1/4" O.D. copper tubing to be soldered in place. The ends of the cell tube should be

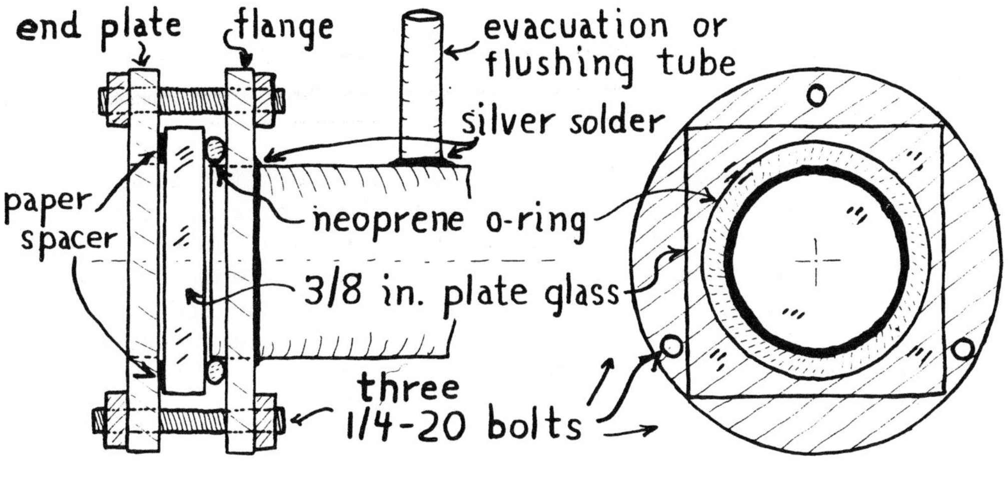

FIG. A8-1

Construction details of the absorption tube.

squared on a lathe or milling machine and the edges smoothly rounded with emery paper. The end plates and flanges, four inches in diameter, are made from four disks 3/16 to 1/4 inch thick. The flanges are to be slid over the end of the tube and, accordingly, have holes machined to give a smooth snug fit over the cell tube. The end plates, to hold the windows in place, require holes which may have the same diameter as the flange holes. The flanges and end plates are then clamped together and three 17/64th inch holes drilled around the circumference as in the figure. One side of the flanges and end plates should be machined or sanded flat and smoothed with emery paper.

It is assumed that O-rings of a size to just slide over the cell tube are available. If they are not, grooves are machined in the flanges to accommodate such rings as are available. In this case, the tubing should not, of course, project through the flange plates as shown in Fig. A8-1.

The cell should now be assembled for silver soldering. The two 6-inch side tubes are fitted into the main tube -- they may project about 1/8th inch inside. The flanges are clamped in place so that the tube projects through them about two thirds of the thickness of the O-rings to be used. The flanges should be carefully squared with the tube. Make sure that the silver solder runs freely over the various joints to ensure that they will be vacuum tight. When the cell is cool, clean it thoroughly and remove the soldering flux.

Windows for the cell are cut from plate glass 3/8 inch or more thick. One might use 3" dia. disks of plate glass (mirror making tools) from Edmund Scientific. Four windows (two sets) are recommended so that while one set is being cleaned the other set covers the ends of the cell. Either round or square windows are satisfactory.

2) Absorption tube heater:

As explained in Wood's book, the absorption tube is heated by a row of gas jets, as shown in Fig. A8-2. The heater is made of a piece of pipe (such as 1/2 in. iron pipe) into which a row of 1/16th in. holes are drilled, about one inch apart. The overall length of the series of gas jets should be about four inches shorter than the body of the absorption tube so that the ends of the cell may be cooled rather than heated. A fitting is provided to allow the heater pipe to be connected by rubber tubing to the gas line. The ends of the heater tube should obviously be capped.

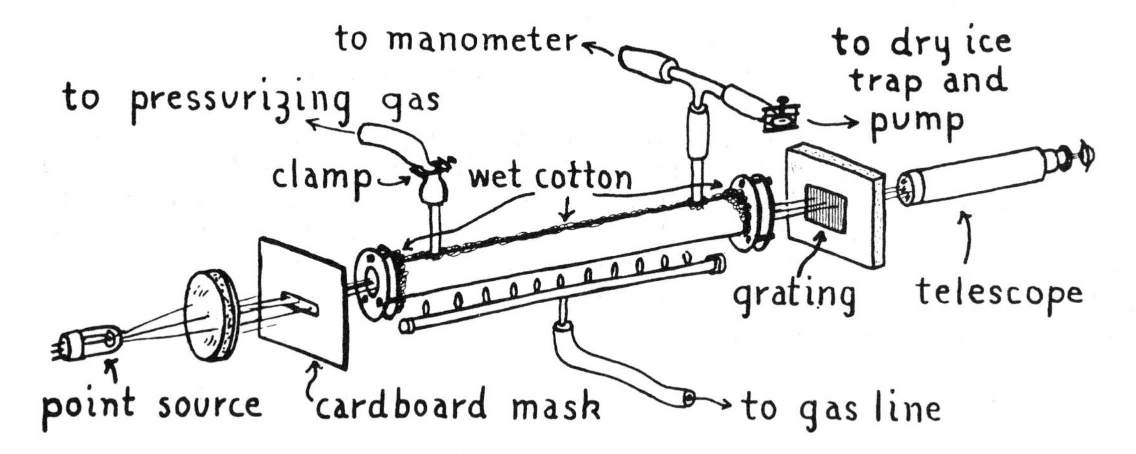

FIG. A8-2

Arrangement for observation of the anomalous dispersion of sodium vapor.

3) Small vacuum fore pump. Any pump which will rapidly reduce the pressure in the cell to less than a millimeter is satisfactory.

4) A dry ice vapor trap between the absorption tube and the vacuum pump is highly desirable to prevent sodium vapor from passing through the pump.

5) Mercury U-tube manometer reading from 1 mm to 1 atmosphere.

6) Several glass stopcocks or good tubing clamps.

7) Eight to twelve feet of good rubber or plastic vacuum tubing.

8) Point source and short focus achromat.

9) Student spectrometer.

10) Diffraction grating of fairly good quality.

11) Chemicals: a few grams of sodium metal, about a liter of kerosene and also of gasoline.

12) A substantial dish, deep enough to submerge the sodium lumps under kerosene while they are cut up.

13) Safety goggles.

PROCEDURE

The absorption tube should first be tested to see that is is vacuum tight. Assemble the cell using neoprene O-rings which may be lightly greased with silicone high vacuum grease. A heavy paper or thin rubber spacer should be used between the end plates and the windows so the latter will not crack when clamped. Use 2 inch x 1/4-20 bolts and tighten the flanges until the O-rings are about 1/3 compressed. The windows should almost, but not quite, touch the ends of the tube.

After the cell is assembled, it should be evacuated and the pressure checked with the manometer. When the pressure is reduced below one millimeter, close the connection to the pump and observe how quickly the pressure builds up. The vacuum system is very small and the pressure will very likely rise to a few cms within an hour or so. If it is no worse, the system is probably sufficiently tight. Any more serious leaks must, of course, be located and sealed. (Glyptal varnish, a General Electric Co. product, is useful for sealing very small leaks).

If the absorption tube is satisfactory, the apparatus may be set up as in Fig. A8-2. The tube is securely clamped in position with one window removed. About a dozen pea size pieces of sodium metal are cut from a larger chunk placed in the dish of kerosene. USE PROTECTIVE GOGGLES. The pieces are removed with tweezers and the excess kerosene blotted off with DRY paper towels (DRY HANDS). These lumps are then rinsed in a beaker of gasoline (which vaporises at a lower temperature than kerosene) and again blotted as dry as possible and put in the absorption cell. They may be distributed along the length of the cell using a rod to push them. The lumps should be kept several inches from either window. The missing window is now replaced and the cell evacuated. After pumping for about 15 minutes, turn on the gas flame and gently warm the cell to increase the vapor pressure of the residual gasoline and kerosene so that as much as possible may be pumped off.

If (unfortunately) the cell windows become fogged with kerosene or deposited sodium vapor, then the cell must be cooled, air pressure restored, and the extra set of windows installed. Make sure there are no wet cotton wads nearby and that the cell is thoroughly dry when changing the windows.

After the kerosene and gasoline have been removed in the above manner, the demonstration may be set up. Light from a point source is collimated with a short focus achromat and directed through the cell. A cardboard mask should be used to restrict the beam to a horizontal sheet -- otherwise, the non-uniformity of the sodium "prism" may spoil the observations.

The light emerging from the cell should be horizontally dispersed by the grating and viewed through a telescope. A student spectrometer with the collimator turned out of the way is convenient.

After evacuating the cell, add hydrogen, methane (illuminating gas) or, as a last resort, even air, to give a pressure of about 3 to 4 cm Hg. Put dripping wet cotton wads around the ends of the cell and along the top of the cell as indicated in the figure. Warm the tube slowly with the gas flame keeping the cotton wet at all times.

With a grating, several orders of spectra should be observable. The spectrum, at first, should be simply a long thin line. As the sodium melts and vaporizes, the sodium prism forms and the spectrum develops a break in the yellow corresponding to a sharp change in refractive index. The blue side of the spectrum bends in the direction corresponding to an index less than unity and the red side the other way corresponding to an index greater than unity.

To observe the deviation between the two D lines as described in the reference, high quality optical parts will be needed and considerable care exercised in setting up the demonstration.

The progressive changes in the density of the sodium "prism" resulting from the heating is evidenced by the gradually increasing deviations of the yellow region of the spectrum.

It is also of interest to turn the grating so that its rulings are horizontal rather than vertical. In this orientation, the grating adds to or subtracts from the deviation produced by the sodium vapor prism alone. For grating orders on one side of the direct beam, one sees exaggerated black absorption lines. For grating orders on the other side of the direct beam, the absorption lines may appear to be absent.

. . . .

After the experiment is finished, and the tube cooled, it may be sealed off at low pressure or it may be filled to atmospheric pressure with nitrogen or even air. If the tube is to remain unused for a considerable period, it is perhaps best to remove the sodium. Use great caution if the sodium is to be disposed of; it reacts extremely violently with water!

EXPERIMENT A9. LONGITUDINAL CHROMATIC ABERRATION AND SECONDARY SPECTRUM.

READING: Strong, Ch. 14.7-14.8; or Jenkins and White, Ch. 9.13-9.14.

In this experiment, we shall measure both the chromatic abberation of a single lens and the residual chromatic aberration or secondary spectrum of a color corrected doublet. Although the dispersive property characteristic of all transparent substances is useful for spectroscopic purposes, it is undesirable in lenses, for it results in a change of focal length with color. This change in focal length means that the image distance for red and blue light is different, and this difference is called longitudinal chromatic aberration. At the same time, because of the different focal lengths, the image dimensions in red and blue will be different in any image plane. The second effect is called lateral chromatic aberration.

The chromatic aberration or chromatism of a simple lens may be intolerable for many uses. As explained in the reading, it is possible to combine two simple lenses of different kinds of glass to form achromatic doublets nearly free of chromatic aberration. In addition, because of the extra surfaces, it is also possible, at the same time, to considerably reduce other aberrations such as spherical aberration and coma. The residual color error of the achromatic lens is termed secondary spectrum. Even secondary spectrum may be very troublesome in some cases, as for instance in astronomical observations. With a well corrected achromatic telescope lens, it is found that the brilliant, white image of the planet Venus is usually surrounded by a colored halo, the color depending on the adjustment of the focus.

Achromatic lenses are designed to have the same focal length for two colors. They correct chiefly for longitudinal chromatism. The two colors selected for correction depend upon the proposed use of the lens. If the lens is to be used for visual observations, the colors are chosen so that the lens performance is optimum in the yellow-green part of the spectrum. On the other hand, if the lens is to be used for photographic purposes, it should, of course, be corrected for the part of the spectrum generally photographed. Lenses are also achromatized for either the infrared or the ultraviolet. The significant point is that outside the spectral region for which the lens was designed, the performance may be quite bad. If the lens must be designed for optimum performance over a rather broad spectral region, it is possible to eliminate most of the secondary spectrum also by a carefully chosen combination of three or more elements. A lens of this kind is called an apochromat.

Another method used to achromatize a lens system is to separate two positive lenses (which may be of the same kind of glass) by half the sum of their focal lengths. The correction, in this case, is for lateral chromatism. This type of achromatization is used in both the Huygens and Ramsden eyepieces, though the separation is generally a little less than that required for complete compensation. (If the theoretical separation is used, the surfaces of the field lens are in focus for the observer and all the dust particles and any scratches on the field lens are also in focus).

In performing this experiment, there is little difficulty in measuring the chromatic aberration of a simple lens, but it requires a bit more finesse to measure the secondary spectrum of a well corrected doublet, for the change in focus amounts only to about 0.1% or less of the focal length.

In setting up this experiment, perhaps the most obvious procedure is to illuminate an object such as a fine reticle scale with various spectral lines in turn and to determine the exact image position with an auxiliary magnifier. The method is poor, however, because the magnifier itself has residual chromatism. Although it is possible to correct for this error, there is a better scheme which avoids it altogether. In this procedure, a knife edge is used to determine with good accuracy the point of crossing of two pencils of light.

APPARATUS

1) Mercury-cadmium arc (or ordinary mercury arc), and if available, a sodium arc for reference.

2) Monochromator

It is rather difficult to select a series of filters which will give adequate spectral purity for all the useful lines of the mercury arc, though filters may be used if need be. The better way to isolate the desired spectral lines is to use a monochromator -- either a commercial one or one constructed from a student spectrometer. Figure A9-1 shows how the monochromator is made with the aid of a constant deviation prism. It is necessary to modify the spectrometer to the extent of removing the eyepiece mechanism and substituting a crude exit slit. Alternatively, one may replace the whole telescope with the collimator of another instrument. The correct positioning of the prism is described in the experiment on adjustment of the spectrometer. The collimator and output tube must be perpendicular to each other when the constant deviation prism is used.

3) Knife edge assembly

This is the crucial piece of apparatus. A razor blade may conveniently be used. The problem is the mount for it. A jeweler's lathe cross slide which has two sets of ways at right angles, is adaptable to the purpose. The ways which provide longitudinal translation must be calibrated;

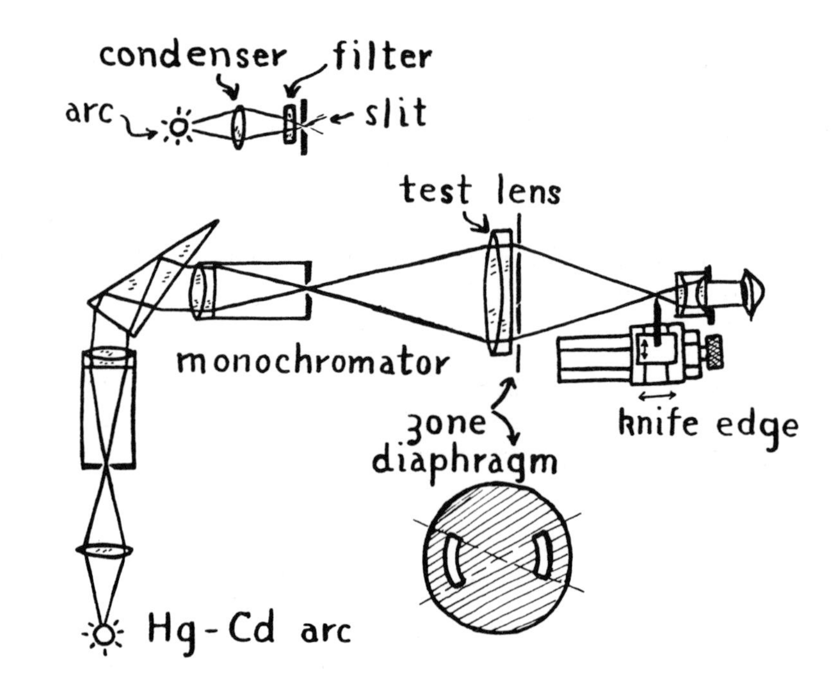

FIG. A9-1

Apparatus to measure longitudinal chromatic aberration and secondary spectrum. Above is an alternative scheme to isolate the various spectral lines with filters. Below is shown the form of a useful cardboard diaphragm to isolate rays for the measurements.

if they are not, a scale may be engraved on the side, or a good machinist's scale cemented on. It is best to replace the crank (if the cross slide has one) with a simple wheel, otherwise, the experimenter may become slightly biased and tend to make all settings accord with the first one. It may also be desirable to remove the screw which provides the cross travel so that the knife edge may be slid into the beam directly rather than slowly cranked into it.

4) Mask

The required mask may be cut from black paper or cardboard, as shown in the lower insert in the figure. Its purpose is to isolate two pencils of light for which the exact position of crossing may be accurately measured.

5) Observing eyepiece

A low power eyepiece placed a little behind the crossover point renders more or less parallel the two divergent pencils passing the knife edge. In this way, both pencils may be seen at the same time for comparison without the necessity of placing the observer's eye too close to the knife edge.

6) Lenses to test: one simple lens and one achromatic doublet, both about 2 inches diameter and 10 inches focal length.

PROCEDURE

First set up the monochromator (or the arc arrangement shown in Fig. A9-1). Line up the arc and collimator slit without any intervening lens and center the beam using a piece of paper to make the beam emerging from the collimator visible. Then insert a simple lens to focus the arc on the slit. Set the monochromator to deliver a beam of green light. Put the test lens (the simple lens) in position with the mask close to it. The test lens may be put at a distance of about twice its focal length from the monochromator exit slit. Make sure that is is uniformly illuminated by the monochromator. Mount the eyepiece (it may be mounted separately from the knife edge unit) so that the two pencils may be clearly seen. The two patches of light should be symmetrical and uniform in brightness. If they are of different shape, it may be that the test lens is tipped or the eyepiece in the wrong place. Finally, the knife edge unit is placed so that the longitudinal ways are parallel to the optical axis of the system.

As the knife edge is slid into the light path, one or the other patch of light will grow dark first, and it will be evident whether the knife edge is inside or outside the crossover point. As the exact point of crossover is approached, the two patches will darken simultaneously. A still better setting can be made by studying carefully the gradual cutting off of the light in the vicinity of the focus. Very close to, but not precisely at focus, the patches both cut off from the same side, and just before extinction, one or the other will appear brighter. A very slight adjustment of the longitudinal position of the knife edge will make both the patches equally bright. With care, it should be possible to set the knife edge consistently to better than 0.1% of the image distance.

Several preliminary settings should be made on the green line to give the observer some feel for the apparatus and some judgment for accurate setting. Several factors are of importance in making the judgment: The observer may select different parts of the two illuminated patches on which he bases his judgment. As his experience increases, he is less and less likely to overshoot the mark, for he begins to remember how the appearance changes as the knife edge is moved. He also becomes dark adapted to the level of the illumination. As a result of these factors and perhaps others, it is not uncommon for the first few readings to progressively increase or decrease. Thus it is well to record five or ten preliminary readings to make sure that the observer's standard in determining the cut off has become constant. Needless to say, the knife edge should always be moved in the same direction to avoid backlash -- if the mark is overshot, return farther than necessary, and approach the cut off again from the same direction.

If the observer has reasonable confidence that all is well, he should begin a series of measurements using first a red line -- that of cadmium in the mercury-cadmium arc is convenient. Take five readings for the crossover point. After each reading, move the knife edge away (longitudinally) and approach the crossover anew. Repeat the measurements using in turn, yellow, green, blue green (cadmium), blue and violet lines.

Average the several sets of readings and determine the change in the image position relative to that for yellow -- sodium if used, otherwise, mercury yellow. Be sure to note whether the changes represent an increase or decrease in distance and also what spectral line is being used.

Measure the object and image distances with a meter stick and record for future use.

Repeat the experiment using the achromatic doublet.

RESULTS

Calculate the percentage change in focal length (not image distance) relative to the yellow focal length (sodium or mercury), for the two lenses. Plot on one full sheet of graph paper $\Delta f/f$ as a function of wavelength. If desired, the scale for the achromat may be different from that of the simple lens.

Discuss the significance of the results and the various sources of error. How might the experiment be improved?

EXPERIMENT A10. SPHERICAL ABERRATION.

READING: Strong, Ch. 14.5-14.6; or Jenkins and White, Ch. 9 to p. 142.

The purpose of this experiment is to make a quantitative study of the longitudinal spherical aberration of two simple lenses and of one achromatic doublet. Understanding the nature of spherical aberration and how to minimize its effects is most important. It is a matter of concern not only in the ray optics of light, but also, of course, in the ray optics of millimeter waves and of microwaves. It is, therefore, a factor in any use of lenses -- whether in wave optics, spectroscopy, electron microscopy, radio astronomy, or microwave communications. With the exception of a few special aplanatic systems, one of which is considered in the next experiment, spherical aberration (S.A.) is invariably produced by a spherical lens or mirror, but the proper choice and arrangement of components will reduce the aberration to a minimum.

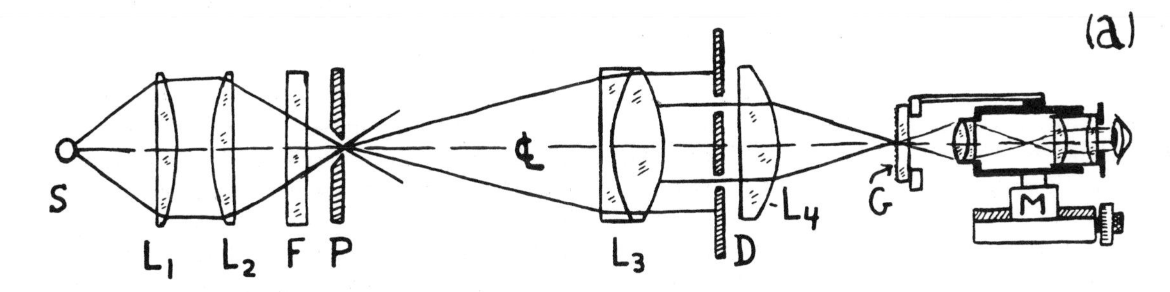

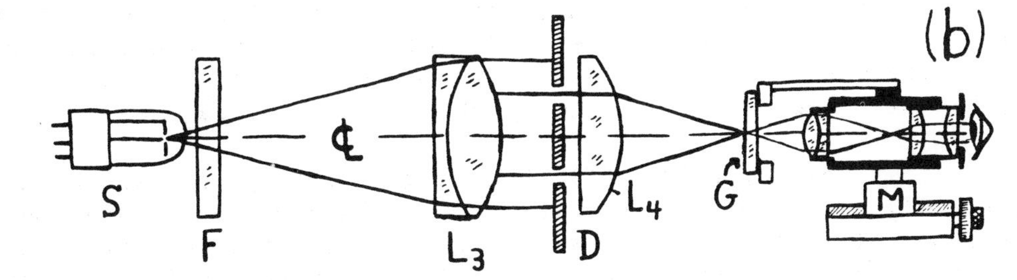

FIG. A10-1

Two arrangements of apparatus to study the spherical aberration of a lens.

The method used to measure the spherical aberration in this experiment is illustrated in Fig. A10-1. The principle is rather simple. A collimated beam of parallel, monochromatic light falls on a specially constructed diaphragm which restricts the illumination to various selected parts of the lens to be tested. The different zones of the lens, each represented by three holes in the diaphragm, are studied one at a time. The lens refracts the three pencils of light for each zone to a common point of intersection which can be accurately measured. In this way, the longitudinal S.A. of the lens is measured.

APPARATUS

1) One meter good quality optical bench.

2) Bright monochromatic point source. Either of the arrangements shown in Fig. A10-1 is satisfactory. If scheme (a) is used, a mercury arc, two plano-convex condenser lenses, a filter to transmit only the green line, and a screen with a small pinhole in it. If scheme (b) is used, only a concentrated arc source and green filter are needed.

3) A good quality achromatic doublet to produce the collimated test beam. (Although freedom from chromatic aberration is not required in scheme (a), the added correction for spherical aberration is needed). This lens should have a diameter appreciably greater than the required diameter of the collimated beam, and a focal length of 15 in. or more -- several times that of the lenses to be tested.

4) A special diaphragm. A convenient diaphragm for the experiment is shown in Fig. A10-2. It may be constructed of sheet metal 1/16th inch thick on which a pattern of concentric circles and three radial lines making 120° angles are scribed. The scribed circles are useful for measuring the radii of the zones. The size of the diaphragm is, of course, adapted to the lenses to be studied. Holes 1/16th inch diameter with the central hole 1/8th inch diameter are suitable. The fixed diaphragm is to be mounted on the optical bench and arranged with a slide as shown at the left of Fig. A10-2. The holes in the slide are about twice the diameter of those in the fixed part so that they will not interfere with the desired light rays, but will block the undesired parts of the diaphragm.

5) Low power (20-40X) observing microscope and ground glass screen. Figure A10-3 shows how the ground glass may be mounted on the microscope. The glass itself is readily prepared from either a microscope slide or a piece of 1/8th inch plate glass by grinding it with the finest available abrasive -- preferably finer than #600 carborundum. (For instructions on grinding, see appendix at the end of this manual). The ground glass is waxed to a frame mounted on the low power microscope with the ground side toward the light. The whole unit -- microscope and ground glass, on which it is to be focused -- is mounted on a traveling microscope carriage or jeweler's lathe cross slide calibrated in thousandths of an

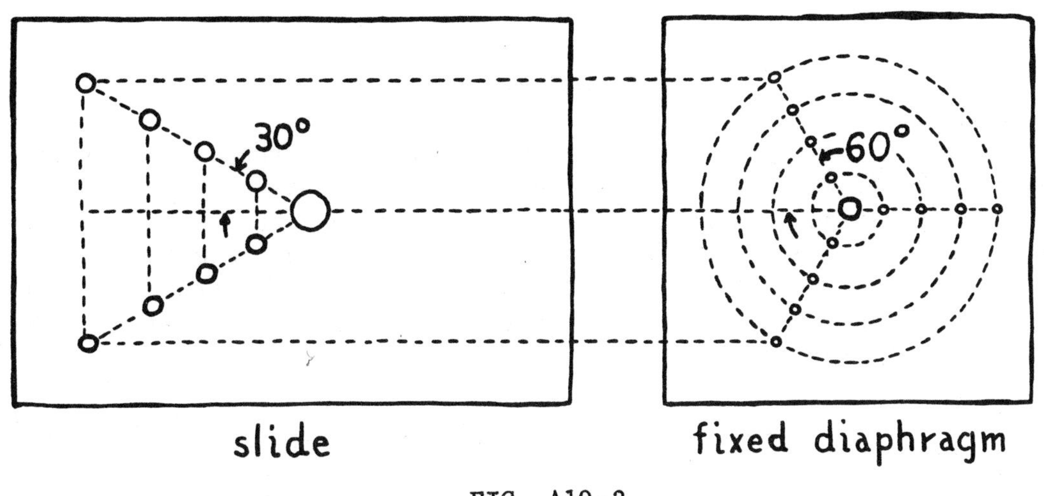

FIG. A10-2

Diaphragm for spherical aberration measurements.

inch or in fractions of a millimeter. It is convenient, but not essential, to mount the whole assembly on the optical bench. It is, however, essential that the support for the observing microscope be rigid. It should be possible to measure the focal point to 1/100th inch.

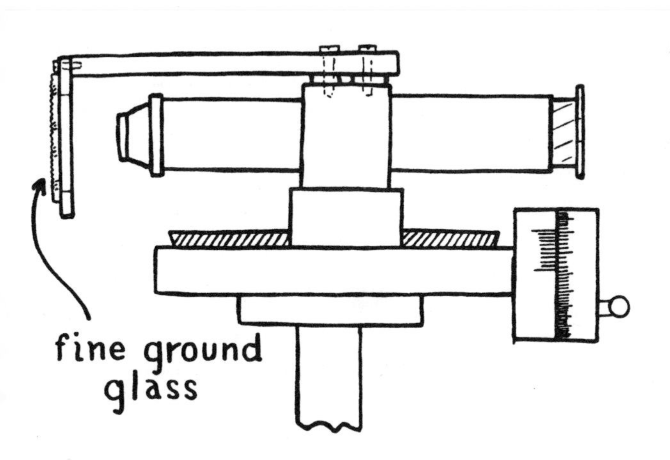

FIG. A10-3

Low power observing microscope arrangement.

6) Lenses to test: The lenses to be tested include (1) a plano-convex lens, (2) a double-convex lens, and (3) an achromatic doublet. All three lenses should have about the same diameter and the same focal length -- say 2 or 3 in. diameter and 6 in. focal length. The spacing of the holes in the diaphragm should, of course, be designed according to the diameter of the test lenses.

If the plano-convex lens is modified by cutting off two segments of the circular lens to form a prism of about 70 or 80 degrees, the refractive index can be conveniently measured with a spectrometer and the S.A. calculated from the measured refractive index and the radius of curvature. Alternatively, the refractive index can be computed from the paraxial focal length and the radius of curvature.

PROCEDURE

Arrange the apparatus as in Fig. A10-1. If arrangement (a) is used, line up S, P, and the center of lens L_3 making sure that P and the center of L_3 are at equal heights from the optical bench so that the collimated beam will be parallel to the bench ways. Now insert the two condenser lenses L_1 and L_2 so as to focus the arc on the pinhole. Their curved sides should face each other (Why?). Insert the green filter and focus the achromat L_3 by autocollimation.

If the simpler scheme, (b), is used, set the point source at the same height as the center of the achromat, insert the filter and focus by autocollimation.

It is important in either scheme to be careful about the exact orientation of the achromat. First, make sure that the flint divergent element of the lens faces the light source (for minimum spherical aberration). Then, make sure that the tilt of the lens is such that no coma or astigmatism is introduced by the lens being off axis.

The diaphragm should be placed symmetrically in the collimated beam with its surface accurately perpendicular to it. The lens under test is, of course, mounted behind the diaphragm and should be rather close to it. Begin with the plano-convex lens oriented with its plane side toward the light. Unless the lens is accurately centered with respect to the diaphragm and carefully oriented, the results will be unsatisfactory. If the lens adjustment is correct, the bundle of rays passing through the center of the lens will be undeviated and undisplaced by the lens (note the position of the light spot from the central hole in the diaphragm before the lens is inserted). In addition, the short radial lines of light originating from the outer zones will cross symmetrically in the field of view of the microscope.

Set the microscope on the focal position of the central zone. Make sure that the microscope is set near the end of its travel (on its slide) so that there will be enough travel to allow setting on the focus for the outside zone. (If the microscope travel is insufficient to measure all zones for the lens, measure the ray intersection points for three zones. Move the microscope unit, and by re-measurement on the last zone previously measured, determine how far the microscope was moved). Make five settings on the paraxial focal point and determine the focal length of the lens by measurement of the distance between the curved surface and the average focal point on the ground glass screen. Is this distance the focal length of the thick lens? Measure the longitudinal S.A. as the change in focus for each zone. Three readings on each of the outer zones should suffice. If the illumination for the outermost zones is insufficient, remove the green filter and determine the green focus by noting the point at which the three intersecting spectral lines have a common green point at intersection -- the crossing of the green rays.

Measure the S.A. in the same manner for the other orientation of the plano-convex lens, and for the other lenses to be tested. Determine in any suitable manner, the focal length of the other lenses. (See experiment on thick lenses).

Finally, measure the radii, h, of the various zones on the diaphragm.

RESULTS

1) Plot on one graph the S.A. data obtained for the single lenses. For comparison, the curves should be plotted as h/f vs $\Delta f/f$. Identify each curve according to shape factor (discussed in the reading).

2) On a new graph, plot only the data for the plano-convex lens, plane side toward the light source. Plot also a theoretical curve $\Delta f/f = ah^2$ which coincides with the experimental curve at the origin and some intermediate zone. Plot also a second theoretical curve $\Delta f/f = ah^2 + bh^4$ which coincides with the experimental curve at the origin and two intermediate zones.

3) Plot the S.A. for the doublet for both orientations. Which component of the achromat should face the light source when used as a telescope objective? Why is the S.A. less for one orientation than the other?

4) If the plano-convex lens has been cut to form a prism as suggested above, the refractive index may be measured with a spectrometer, and the radius of curvature with a spherometer. With this data, one can use either the ray tracing methods of Ch. 8 in Jenkins and White, or the simpler restricted method of the appendix to calculate the S.A. of the lens. The method given in the appendix applies only to the one case where the light is incident normally on a flat surface. The exact ray tracing results should be plotted for comparison with the graph for section B.

APPENDIX: Simplified calculation of long. S.A. for light normally incident on the plane face of a plano-convex lens.

Figure A10-4 shows a ray of light normally incident on the plane face of a lens. It is desired to calculate the distance S' at which the refracted ray crosses the optical axis. From the figure, it can be seen that:

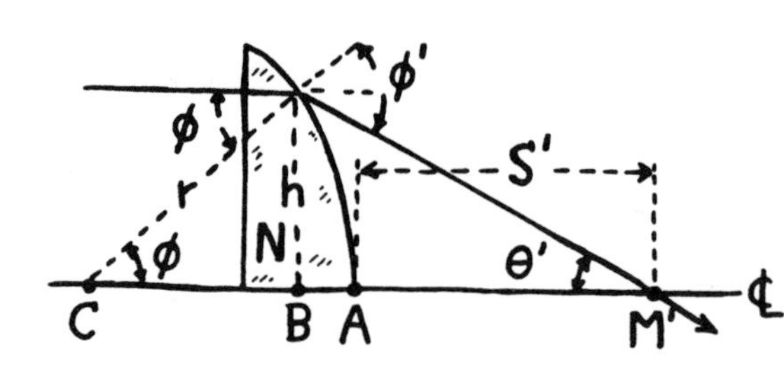

FIG. A10-4

Geometry for calculation of S.A. for special case.

$$S' = BM' - BA$$

$$BM' = h \cot \theta'$$

$$BA = r - CB = r(1 - \cos \phi)$$

$$\therefore S' = h \cot \theta' - r(1 - \cos \phi) \qquad (1)$$

$$\theta' = \phi' - \phi \qquad (2)$$

$$\phi = \sin^{-1} h/r, \quad \phi' = \sin N h/r \qquad (3),(4)$$

Thus, given N, r, and the chosen h, one can use (3) and (4) to calculate θ' and (2) and (3) to find S'.

For the paraxial case, these equations should reduce to the familiar form. From the series expansions for the trigonometric functions, we have for small angles $\sin x = x$, $\cos x = 1 - x^2/2$, $\tan x = x$. Use these expansions to prove:

$$S' = \frac{h}{(N-1)h/r} - r[1 - (1 - h^2/2r^2)]$$

$$= \frac{r}{N-1} - \frac{h^2}{2r} \doteq r/(N-1).$$

Evaluate S' for several values of h using log tables for accuracy.

EXPERIMENT A11. DEMONSTRATION OF APLANATIC SYSTEMS.

READING: Strong, Ch. 14.5, 15.4; or Jenkins and White, pp. 146-147.

This demonstration is intended to show the aplanatic properties of a sphere. Aplanatic systems are lens or mirror combinations which have the property of introducing no spherical aberration or coma for some special position of the object and image. Such systems may use aspherical surfaces or, in a few cases, spherical surfaces. These few spherical systems are the exceptions to the usual rule that spherical surfaces produce spherical aberration. Probably the most important example is that of the oil immersion microscope objective of short focus. Since one of the aplanatic points of a sphere lies within its surface, the oil immersion objective is made of a hemisphere of glass and the object is placed in a drop of oil of the same refractive index as the glass at the aplanatic point, as illustrated in Jenkins and White.

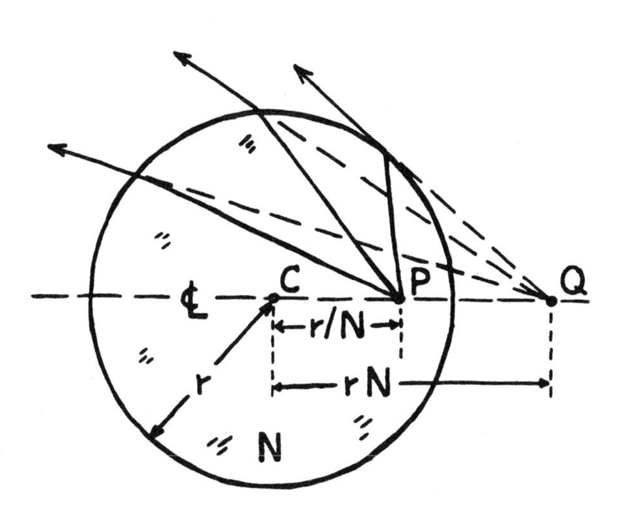

FIG. A11-1

Crossection of a sphere showing the aplanatic points. P is real object point, Q is virtual image point.

Figure A11-1 illustrates the aplanatic points, P and Q, of a sphere. If an object is placed at point P, a virtual image of it is formed at point Q. The distance between the center of the sphere C and the object P is r/N where r is the radius of the sphere and N is its refractive index. The image distance, CQ, is equal to rN. This relation between object and image points is rigorously true even for very large angles, and spherical aberration and coma are absent.

There are two alternative procedures for the demonstration. The straightforward one is often used in elementary courses to trace rays through a prism or lens, and it is convenient for locating a virtual image. The alternative method reverses the object and image positions and allows the convergence of rays to be observed by means of fluorescence.

I. STRAIGHTFORWARD DEMONSTRATION

The apparatus required is simply a container with a cylindrical wall; this represents a crossection of the sphere. It may be a goldfish bowl with a cylindrical side, a large diameter beaker, or a flat bottom petri crystallizing dish.

The container is filled with water, and the object, a straight heavy wire supported vertically, is held at the calculated aplanatic object point. Pins are used to mark on a large piece of paper, the direction of the virtual image as seen from various directions. After the position of both the container and object are marked on the paper, the container and object are removed and lines drawn to locate the position of the virtual image. If the experiment is carefully done, all the lines will cross very nearly at one point. Then put the object just outside the container and repeat the experiment. The spherical aberration will now be quite apparent.

II. FLUORESCEIN DEMONSTRATION

APPARATUS

For this demonstration, the following materials are needed (in addition to the container mentioned above):

1) Five small plane mirrors cut from an aluminized microscope slide or from an inexpensive compact mirror. The mirrors should have reflecting surfaces about 1/8 x 1 in. area.

2) Auto lamp source.

3) Small collimating lens.

4) Fluorescein powder.

PROCEDURE

Since the image formed by the spherical lens (or its two dimensional cylindrical equivalent) is virtual, the aplanatic properties are best demonstrated in reverse. That is, convergent light rays are directed toward the previous virtual image point Q (Fig. A11-1), and upon striking the curved refracting surface, are converged toward the real former object point P. In this way, the ray directions may be made visible.

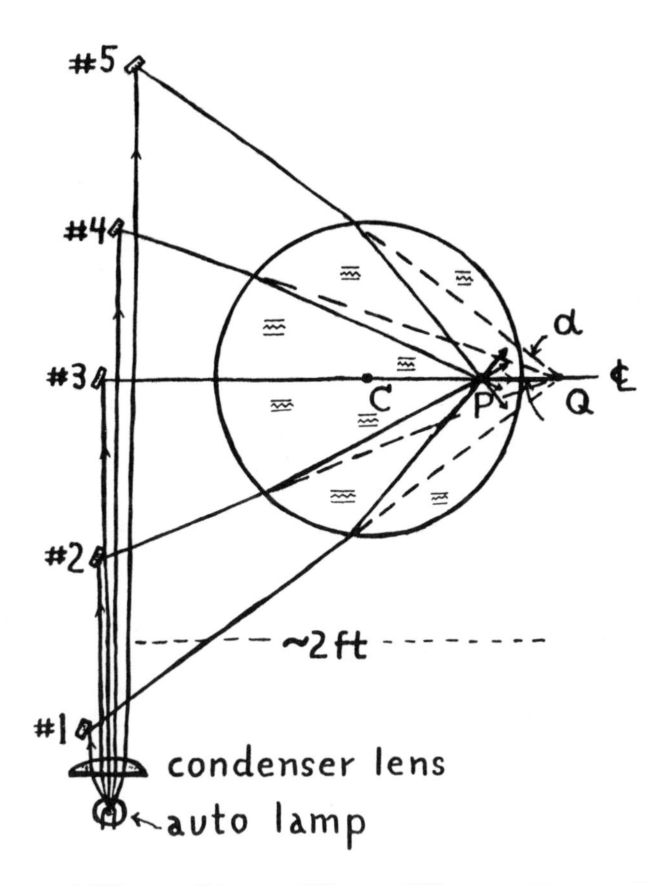

FIG. A11-2

Arrangement to demonstrate the aplanatic points of a sphere. Here P is the real image point, and Q the virtual object.

Figure A11-2 illustrates how the demonstration may be set up directly on a bench top. The auto lamp and its collimator lens are clamped as close to the table top as practical. The lens may be held in place with wax if desired; it is set to give a parallel beam at right angles to the desired optical axis ℄ . The narrow mirrors are arranged to tap off portions of the collimated beam and to direct narrow pencils toward a point Q. (The dish with fluorescein is not used yet). The point Q should be marked on paper or on the table top and mirror #3 placed so that it throws rays along the chosen optical axis. It should be tipped very slightly downward so that the path is clearly visible on the table top near Q. The other four mirrors are then arranged to tap off other portions of the collimated beam to throw light through the point Q at angles of about 20 and 40 degrees with the axis.

Calculate the aplanatic points for the system and the distance between the surface of the container and the point Q. Prepare the fluorescein solution by dissolving in the water enough fluorescein to render light beams traversing it readily visible. Place the container at the calculated distance from the original image point Q, making sure that the center of the dish lies on the path of ray #3. This ray should not be deviated by the dish. If the calculations are correct and the surface of the container is smooth and regular, five rays should be seen in the water, and they should intersect very nearly at one point. A piece of black paper below the container will improve the visibility of the light beams through it. Extraneous room light should be kept at a minimum to improve the demonstration.

If the results are satisfactory, the water "lens" should be moved to other positions to show the appearance of spherical aberration when the aplanatic points are not used. If the center of the water "lens" is placed over point Q, then, of course, there is no spherical aberration either, but, on the other hand, there is no magnification.

QUESTIONS

1) What is the maximum angle which an incident ray could make with the optical axis ℄ in Fig. A11-2 and still be refracted to point P?

2) In Fig. A11-2, consider ray #5 which has reached the "lens" surface and been refracted toward point P. Where does the ray go after it strikes the surface of the "lens" the second time after passing through P?

EXPERIMENT A12. THE FIGURE OF A MIRROR.

READING: Strong, Ch. 13.10, (13.11); see also J. Strong, Procedures in Experimental Physics, (1939, Prentice-Hall), pp. 69-77; additional information on testing is given in the three volumes of Amateur Telescope Making, A. G. Ingalls, Editor, (1933, 1937, 1953, Scientific American).

This experiment is intended to introduce the student to some of the simpler ray-optic methods of testing optical surfaces. Three tests: the Foucault test, the zone test, and the Ronchi test are to be applied to a study of the figure of a concave mirror. Since the differences between a parabolic mirror and a spherical one may be only a few micro-inches, it is evident that the tests must be extremely delicate. The three tests of this experiment are representative of the various ingenious sensitive ray-optic tests available.

The Foucault and Ronchi tests are qualitative tests which are useful in surveying the general smoothness of the surface and in discovering the presence or absence of local raised or depressed areas. The zone test, on the other hand, is a quantitative test which is useful in deciding whether the mirror is spherical, parabolic, hyperbolic, or elliptical.

THEORY

I. Some relations for a parabolic mirror

Since the most frequently desired aspherical mirror shape is a parabolic one, some important relations for such a mirror are given here.

Let the curve $y^2 = 2Rx$ represent a crossection of the parabolic mirror surface as in Fig. A12-1. The center of the mirror surface is at the origin (0,0), and the optical axis ℄ lies along the x axis. If light, parallel to the optical axis, were incident from the right on the mirror, it would all be brought to a focus at the point F (R/2, 0).

In the zone test, however, measurements are made of the position of the center of curvature C of the various zones of the mirror. Thus we need to know the relation between the distance d from the origin to the point C, and the radius y of the zone.

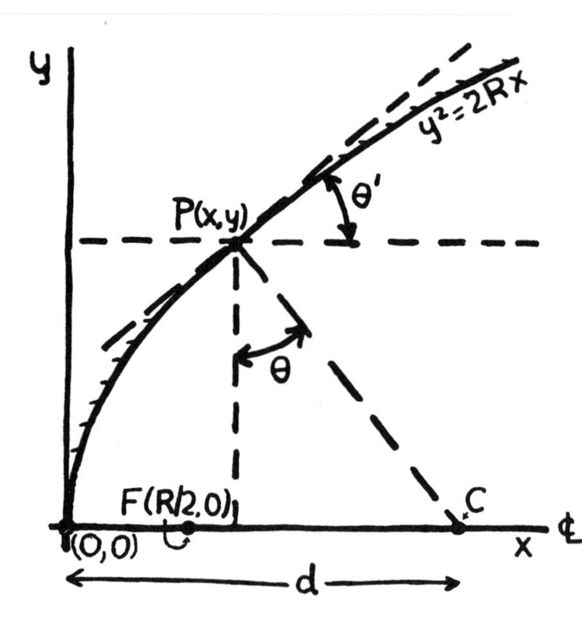

FIG. A12-1

Geometrical relations for a parabolic curve.

From the geometry it follows that $\theta = \theta'$, for the line CP is perpendicular to the tangent to the parabola at P (x,y). Therefore:

$$d = x + y \tan\theta = x + y\frac{dy}{dx} = x + R$$

$$= y^2/2R + R$$

The difference in radii of curvature between any zone and the central zone (which has a radius of curvature R) is therefore: $\Delta = y^2/2R$. This expression as applied to the zone test holds when the knife edge and point source are moved together. If, however, the source is fixed, and only the knife edge is moved, then $\Delta = y^2/R$. (Why?)

II. Difference between a sphere and a paraboloid

The derivation of the formula representing the difference in depth, $\varepsilon \equiv x_p - x_s$, between a parabola and a circle (or paraboloid and sphere) is left to the student. An outline of the steps follows:

Equation of parabola: $y^2 = 2Rx$

Equation of sphere: $y^2 + (x-R)^2 = R^2$

Solve both equations for x. The difference, ε, involves a square root which can be expressed in terms of $(1-y^2/R^2)^{1/2}$. For small values of y/R, the square root is expanded as a series including three or four terms (powers of y up to 4 or 6). The resulting expression for ε is simple.

APPARATUS

1) Point source

The illuminator described in Experiment A3 for the Pfund critical angle apparatus may be used as the basis of an adequate pinhole device for this experiment. Figure A12-2 shows the additional part to be added to the previous illuminator. If the mirror to be tested has very long focus, it may be possible to obtain satisfactory results by simply enclosing an auto lamp source in a small tin can with a pinhole in the side.

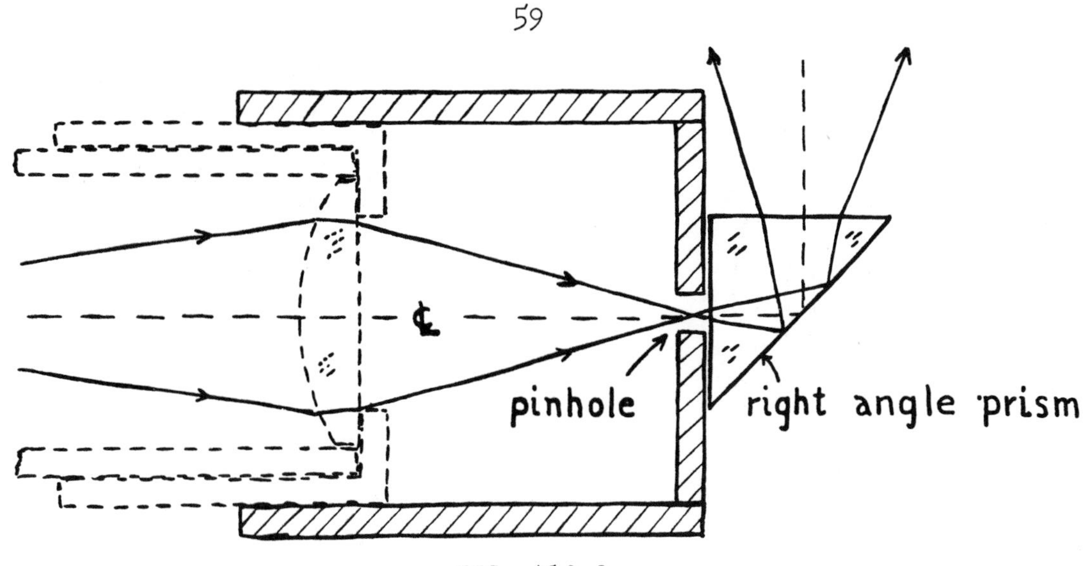

FIG. A12-2

Attachment to illuminator for the Pfund refractometer to provide a point source for the Foucault test.

2) Knife edge cutoff apparatus

This apparatus may be the same device used for knife edge tests in Experiment A9 on chromatic aberration. A more elaborate unit incorporating both the point source and the knife edge is described in the reading.

3) Low power eyepiece

4) Concave test mirror

The mirror to be tested may have a diameter of 4 to 10 inches (possibly made by the student), and it should have an f number, the ratio of focal length to diameter, between 6 and 10 for this experiment. If the f number is lower than this range, the more elaborate test apparatus described in the reading will be needed; if it is higher, the difference between a spherical mirror and a parabolic one becomes inconveniently small. The mirror need not be silvered (nor even completely polished) and it need not have a good figure. It should be rigidly supported on a wood or metal mount.

5) Ronchi rulings

Two pieces of Ronchi ruling (a coarse grating of opaque and clear spaces produced photographically) are needed. They should have about a hundred lines per inch.

6) Zone test diaphragm

A suitable diaphragm for the zone test is shown in Fig. A12-3.

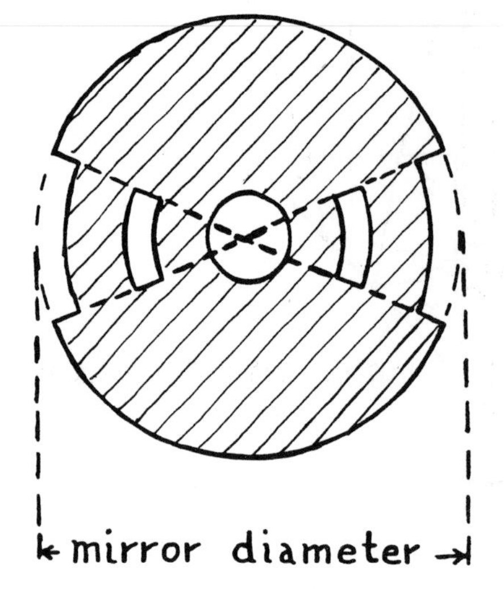

FIG. A12-3

Diaphragm for the zone test.

7) Piece of ground glass two to four inches square.

8) Soldering gun (part Ib). If necessary, a match will do.

9) Straight filament lamp.

PROCEDURE

I. Foucault test

A. Set up the mirror to make the Foucault test as in Fig. A12-4. Determine approximately the radius of curvature of the mirror by reflecting the light from the point source onto a piece of white paper or cardboard. The distance between the source and knife edge should be very nearly the same, and the knife edge should be as close as possible to the point source -- not more than an inch away. (For short focus mirrors of large diameter, it is desirable to use the more elaborate knife edge testing equipment described by Strong in order to avoid serious error arising from separation of the source and knife edge).

Slide the knife edge inside the focal point and observe the appearance of the mirror as the knife edge is moved into the beam. It is usually preferable to use an eyepiece for it is difficult to get the eye close enough to the knife edge to see the whole mirror otherwise. Now slide the knife edge away from the mirror in steps, observing each time, the appearance of the mirror as the knife edge cuts into the beam. Continue until the knife edge is outside the focus.

With the knife edge near the focus for the central part of the mirror, study the shadows as it cuts into the beam and determine whether or not the surface is smooth and free of any zones. Is the mirror surface spherical or not? If the mirror is silvered, hold a piece of ground glass plate behind the knife edge and watch the cutoff shadows on the glass. (If the mirror is unsilvered, the light may be insufficient to observe the cutoff this way. Obviously, it is possible to photograph the cutoff shadows by simply substituting a camera bellows and plate holder. No lens is needed). Sketch the appearance of the cutoff shadows for five positions of the knife edge -- about an inch inside focus; near focus, but just inside; at focus; near focus, but just outside; and about one inch outside focus. Diagram the test arrangement used.

B. While the Foucault pattern is being observed at focus, have a heated soldering gun or iron held in the optical path or just below it.

Record your observations, and explain the effect.

C. Hold a piece of Kleenex or lens tissue over one finger and hold carefully against the surface of the mirror near the top in order to warm the surface locally. Ten seconds or so should be enough to warm the mirror locally. Make sure there is no resulting fingerprint on the mirror. Now observe the Foucault pattern at focus. The local distortion of the mirror surface should be apparent at once. Explain.

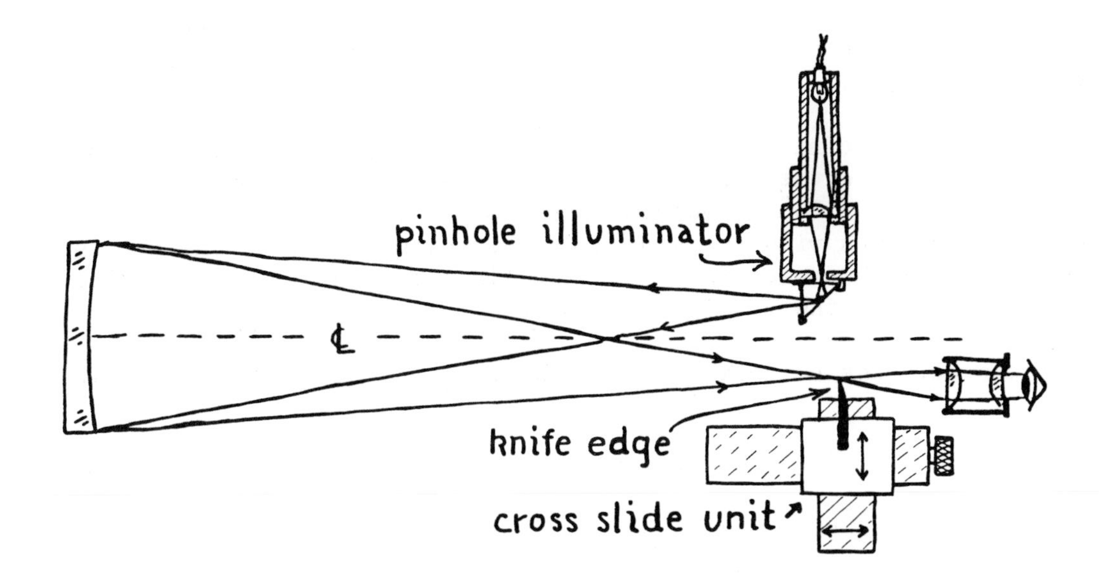

FIG. A12-4

Foucault knife edge test set up.

II. Zone test

Place the zone diaphragm over the mirror and make observations to determine the center of curvature of the different zones. For convenience, it may be desirable to block off the areas for the zones not being measured. Observations should be made both with the eyepiece directly behind the knife edge and with the ground glass farther behind it. Record the positions of the knife edge cutoff for each zone. Begin with the outside zone and make five readings for each zone and ten for the central one. The readings should be made to the nearest tenth of a millimeter. Also measure and record the average radius y for each zone which is measured. A convenient arrangement for the data uses the following headings:

Zone #	Average Radius	Scale Reading	Average Reading

Measure the approximate radius of curvature R to the nearest millimeter.

III. Ronchi test

A. Move the knife edge out of the way, leaving the point source where it is. Cross the two Ronchi ruling plates to form a grid or screen. The sides of the plates which have the photographed lines should be placed in contact. Hold this screen combination in the beam just inside focus with one set of lines vertical. Observe the bands on the mirror. As before, one can either put his eye close to the screen to observe or look at the projected image on the ground glass screen. Sketch the appearance of the bands inside focus, at focus, and outside focus. Also use just one ruling alone with the lines vertical and repeat the observations. Note specially the shape of the bands.

B. Replace the point source with a straight filament lamp. (It may be desirable to operate the lamp at reduced voltage to cut down the excess light). The straight filament should be supported vertically. Observe the Ronchi rulings both with the single Ronchi ruling and with the crossed pair. Describe the observations made. The lines of the ruling should be accurately parallel to the straight filament.

C. Again hold the soldering gun in the beam and describe the results.

RESULTS

1) Describe the results of the Foucault tests and the observations made. Include all the apparatus used and the detailed procedure followed.

2) Compute for the zones measured in part II the theoretical values of Δ_{th} for a parabolic mirror having the measured radius of curvature. From the measurements also, determine Δ_{exp} and make a table to show the comparison of the two sets of Δ values. It may be convenient to plot the results.

3) Derive the expression for ε . Compute for the several zones the difference in depth between a true parabolic mirror and a spherical one having the same diameter and focal length as the one measured.

4) From the observations made, discuss the figure of the mirror tested.

EXPERIMENT A13. THE OFF-AXIS ABERRATIONS OF LENSES.

READING: Strong, Ch. 16; Jenkins and White, Ch. 9.7-9.12; see also the advanced discussions in M. Born and E. Wolf, Principles of Optics, (1959, Pergamon), and L. C. Martin, Technical Optics, Vol. 1, (1948, Pitman), Ch. 4.

This experiment is a qualitative demonstration of the four monochromatic off-axis aberrations of third order theory: coma, astigmatism, curvature of field, and distortion. Like spherical aberration, they all represent deviations from the first order Gaussian theory. The distinguishing features of these aberrations are discussed in the reading; some of the interrelations between these aberrations and the added effects of diffraction are discussed in the last two references.

APPARATUS

1) Monochromatic point source: Either a concentrated arc with a green filter or a mercury arc with a pinhole aperture and filter for the green line.

2) Collimator lens: A good lens of 2 1/2 to 4 inch diameter and 30 cm. or more focal length is needed. This lens is preferably an achromatic telescope objective corrected for spherical aberration.

3) Calibrated optical bench.

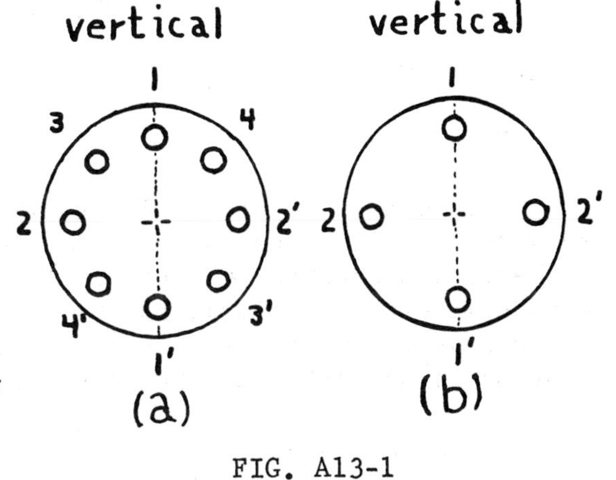

FIG. A13-1

Diaphragm for observation of the comatic circle (a) and for astigmatism (b).

4) Observing screen: The screen should be about 8 inch square, and made either of fine ground glass or of heavy, white-faced cardboard.

5) Test grid: A convenient test grid may be drawn on heavy white-faced cardboard such as is used to mount photographs. It should be about 12 x 18 in., and may be ruled with heavy india inklines, 1/8 in. wide and spaced 1 in. apart, horizontally and vertically.

6) Diaphragms: A diaphragm for demonstrating a comatic circle is illustrated in Fig. A13-1a. Another diaphragm for studying astigmatism

is shown in part b of the figure.

7) Stops: Either a large iris diaphragm which can be mounted at various places with respect to the lens under test or a set of cardboard apertures of diameters 1/2, 1, 2 and 3 cms can be used.

8) A lens mount which can be rotated through known angles or a homemade equivalent.

9) Lenses to test:
a) double-convex, 2-3 in. dia., 8-12 in. focus.
b) achromatic doublet, 2-3 in. dia., 8-12 in. focus.
c) strongly convergent plano-convex or double-convex, any dia. or focus.
d) strongly divergent plano-concave, any dia. or focus.
e) meniscus convergent, 2-3 in. dia., 6-12 in. focus.
f) (if possible) good camera lens 6-10 in. focus.

PROCEDURE

I. ABERRATIONS

A. Coma

Pure coma is difficult to demonstrate because it is almost always mixed with spherical aberration. A simple lens with spherical surfaces, has spherical aberration except in the case of aplanatic lenses, but these have no coma either. (Coma, on the other hand, may be entirely absent for a simple lens if the shape factor is correct). Coma mixed with spherical aberration is not difficult to demonstrate, however.

Use the point source and collimator lens oriented for minimum spherical aberration to obtain a parallel beam of monochromatic light along the optical bench. Mount lens (a) on the rotatable lens mount so that it is fully illuminated by the monochromatic beam (no diaphragm). Orient the lens so that it is as nearly as possible on axis, and focus the beam on the observing screen. Now, without moving the lens or screen position, rotate the lens slowly to an angle of perhaps 20°. Observe the image carefully to see the onset of coma. If the lens is not equi-convex, rotate it 180° and repeat the experiment. The coma observed will be mixed with spherical aberration as mentioned above, and at large angles, it will also be affected by astigmatism.

Place the coma diaphragm (Fig. A13-1a) over the lens with holes 1 and 1' in a vertical line and observe the out of focus image with the lens on axis again. Once more rotate the lens slowly about 10° and watch the changing dot pattern. Return the lens to the axial position and focus the pattern to the smallest possible spot. Without changing the image distance, rotate the lens to the 10° position and cover all the holes of the diaphragm except the top and bottom ones, 1 and 1'. Do the two transmitted pencils fall on one sharp spot on the screen, or, if not, how far must the screen be moved (note postion readings of the screen) to bring the spots together? Next, cover all but holes 2 and 2', and repeat. Try holes 3 and 3'. Are any differences observed from the previous cases? Record also observations made with holes 4 and 4'.

Substitute the achromatic test lens (b) and repeat the observations with and without the diaphragm using both orientations of the doublet. What do the results indicate about the correct orientation of the doublet when used either as a collimator or telescope objective?

B. Astigmatism

Replace lens (a) with the more curved side facing the parallel incident light. With no diaphragm rotate the lens to an angle of 20 or 30 degrees. Adjust the position of the screen to find the tangential and sagittal focal lines. What is the effect of increasing the angle still more?

Substitute lens (b) oriented for minimum spherical aberration and coma. This time, make measurements of the positions of the tangential and saggital focal points. Begin on axis and record the image position in terms of the optical bench reading. Rotate the lens in steps of 10 degrees, each time recording the image positions. Continue until readings cannot be made. (How does the length of the two line images vary with the angle of rotation of the lens)? Measure also for the on axis position, the image distance so that the optical bench readings can be converted into lens-to-image distances. Note that these distances are not the proper image distances S'.

Convert the measurements to the proper image distances (S'), and plot on rectangular or polar paper the tangential and saggital focal curves. Are these results in accord with expectation? Explain.

Turn the lens 180° and make qualitative observations only. Is there any difference in the results? Explain.

Turn the achromat to the original (minimum S.A.) orientation and rotate it to get clear astigmatic images. Mount the diaphragm of part b of the figure as close to the lens as possible. By covering up different holes and adjusting the screen, ascertain which holes in the diaphragm contribute to the centers and ends of the tangential and saggital image lines. Describe the results.

C. Curvature of field

Remove the collimator (or slide it out of the way), and place the large grid on the optical bench with the long dimension horizontal.

Illuminate the grid with a bright light -- a photoflood is possible. Mount the test achromat (b) oriented for minimum spherical aberration at a distance of about three times its focal length from the test grid. Slide the screen to obtain an image of the grid on the axis. The unwanted light should be masked out by pieces of cardboard, black cloth, etc. Now adjust the focus for the central part of the image and make it as sharp as possible. Note that the margins of the image of the grid are not sharp. How must the screen be moved to obtain the best focus for the periphery of the image? Discuss the results.

D. Distortion

Both barrel and pin cushion distortion are observed in extreme form if a short focus positive or negative lens is used to view a sheet of graph paper. Hold the convergent lens (c) within the focal distance and observe the virtual image for both orientations. Which kind of distortion is observed? For which orientation is it worse? Hold the lens several times its focal length away from the paper and the eye far enough away from the lens to see the image. What kind of distortion is observed?

Hold the divergent lens (d) close to the paper in both orientations and describe the results.

II. PHOTOGRAPHIC LENSES

A. Landscape lens

Mount the meniscus lens (e) on the optical bench with the concave side facing the illuminated test grid. Make sure that most of the unwanted illumination is marked out. Note that the image is hoplessly fuzzy. Mount the iris diaphragm (or use various cardboard holes) in front of the lens and find the maximum size aperture which gives reasonably good definition of the image. Determine the approximate f number of the beam. By adjusting the distance of the diaphragm in front of the lens, find the optimum placement and record the position. Why does the diaphragm improve the image?

Consider the image near the center of the lens. Use the largest stop which will give a reasonably sharp image of the grid. How much can the screen be moved without the focus of the image becoming badly deteriorated? Substitute a very small diaphragm 1/2 cm diameter (if there is enough light, otherwise use 1 cm stop), and repeat to find the new depth of focus. Discuss the relation between f number and depth of focus.

B. Mounted photographic objective (optional)

If a good camera lens (f) of reasonably long focal length is available, mount it on the optical bench and make tests for spherical aberration, coma, astigmatism, curvature of field, and distortion.

EXPERIMENT A14. THICK LENSES AND LENS SYSTEMS.

READING: Jenkins and White, Ch. 5; and also Strong, Ch. 14.2, 14.3. Further information on thick lenses and lens systems and on the measurement of focal lengths is found in A. Hardy and F. Perrin, The Principles of Optics, (1932, McGraw-Hill), esp. Ch. 4.

An optical system of thin or thick lenses can often be replaced for purposes of calculation by a set of principal planes and focal points. In this experiment, two independent methods are used to locate these cardinal points, and the results are compared to the theory. Other experiments concerned with particular optical systems will introduce additional methods of measuring focal length.

THEORY

I. MEASUREMENT OF FOCAL LENGTH BY NEWTONIAN DISTANCES

The focal length of a lens system can be determined in a manner reminiscent of the methods of Experiment 1. In that case, the lenses were treated as "thin" which is to say that their thickness could be neglected. For the present purpose, of course, the thickness is not negligible, and therefore a more complicated scheme is required. The method is outlined in the second reference.

By following a three step procedure, one can determine the two distances X and X' whose product is the square of the focal length of the system. The procedure is illustrated in Fig. A14-1. The lens system, mounted on an optical bench is set by autocollimation to produce parallel light (step 1) so that the primary focal point is in the plane of the slit. The focal length is, however, not determined, for the location of the primary principal plane is as yet unknown. In step 2, the lens system is moved to a new position where it forms an image of the slit P in the plane of the other slit Q. It follows that the system must have been moved through the distance X of the Newtonian formula. Step 3 involves moving the system a further distance X' which is required to place the secondary focal point of the system in the plane of the slit Q. The latter condition is determined also by autocollimation, but this time, with the slit Q illuminated and the mirror on the other side of the system. By these steps, then, the two distances X and X' are determined, and clearly the focal length is the square root of their product.

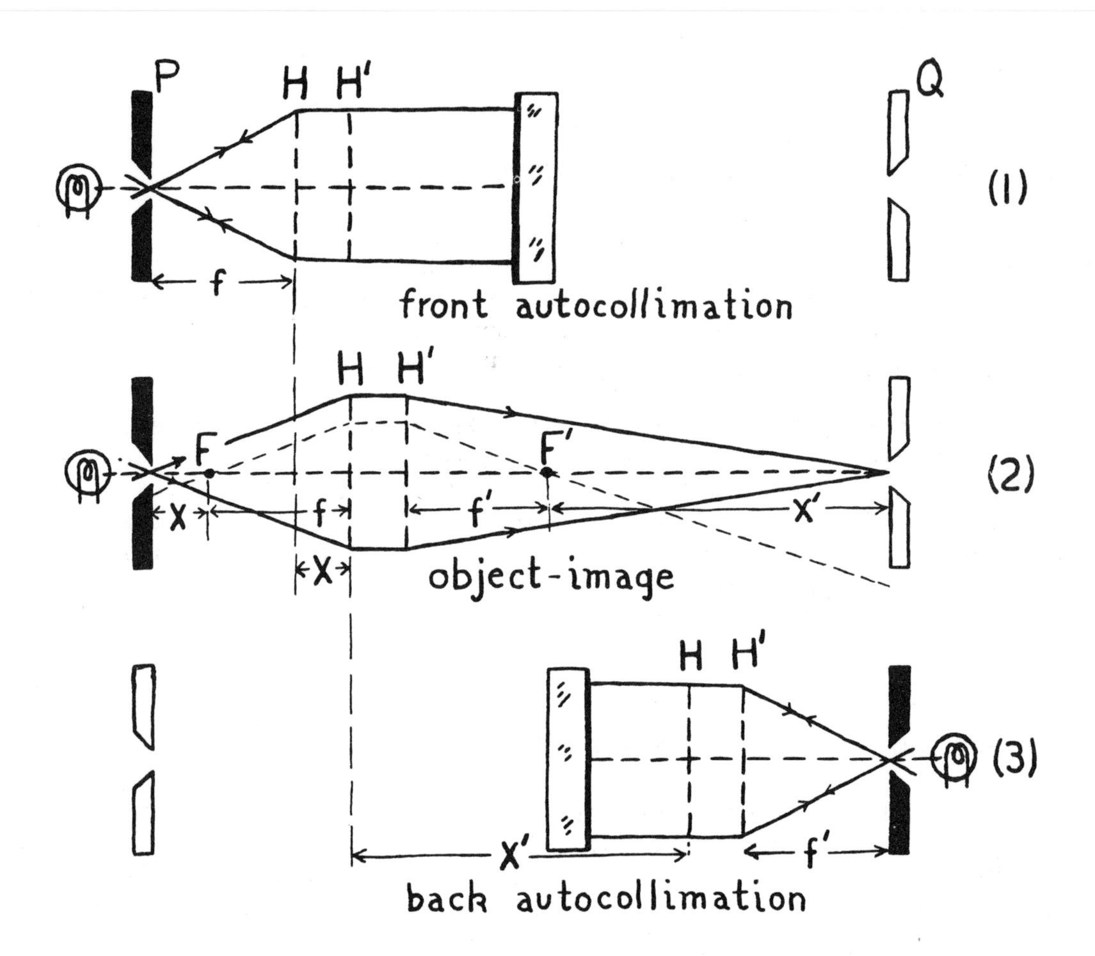

FIG. A14-1

Three step method for finding the Newtonian distances, X and X'.

It is further evident that the principal plane may now be located. If the lens system is set as in step 1, the principal plane H is located by measuring one focal distance from slit P in the direction of the system. In a similar manner, the other principal plane is determined.

II. CARDINAL POINTS BY THE NODAL SLIDE

The nodal slide (see Jenkins and White) is a device which makes it possible to locate the nodal points of a lens system. If the medium in which the lens system is immersed is the same on both sides, the nodal points will coincide with the principal points, and then, if the lens system is rotated about a nodal point, the distance between the axis of rotation and the image point is the focal length.

The nodal slide can be set up in either of two ways, one based on autocollimation as in Jenkins and White or the other based on parallel light. The two alternatives are illustrated in Fig. A14-2. The upper arrangement has the advantage of simplicity in that no auxiliary lens is needed. The lower arrangement, although it requires a collimator lens, may be more useful if either or both focal points are only slightly outside the surfaces of the lenses. In this second scheme, the image may be viewed through a very small ground glass screen which may be placed as close to the lens surface as desired.

APPARATUS

1) Calibrated optical bench.

2) Two slits:

The slits for this experiment may be made from a 2 x 3 in. piece of sheet metal or cardboard. A quarter inch hole is drilled in the sheet metal or a quarter inch square cut in the cardboard with a razor blade about 1/2 in. from one end the long way, and centered along the 2 in. dimension. A double edged razor blade broken in four pieces, form two pairs of slit jaws which are placed over the holes and secured with tape or wax to form a narrow fixed slit. The surface of the blades should be painted white either with flat white paint or with white poster paint. The slits are to be mounted on the ends or rods as indicated in Fig. A14-3. They should preferably be arranged as in Fig. A14-4, so that they are above the slide which carries the optical system, which is to be tested. It is then possible to use the nodal slide for the lens carrier for both parts of the experiment.

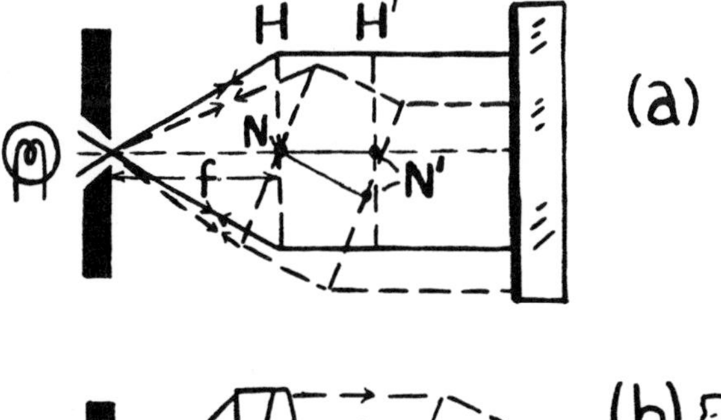

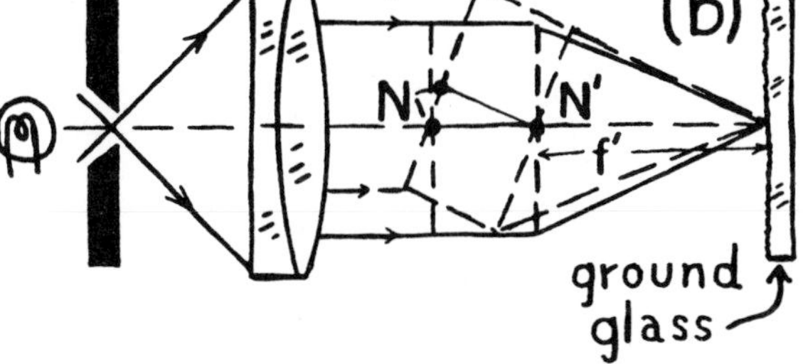

FIG. A14-2

Two arrangements for the nodal slide: (a) without collimator and (b) with collimator.

3) Auto lamp source or flashlight to illuminate slits.

4) Green filter.

5) Mirror for autocollimation.

6) Nodal slide:

The nodal slide may be purchased or may be rather easily constructed. A 4 inch length of aluminum channel 1 1/2 inch wide and 1/2 inch deep may be bolted to the end of a rod (which fits the optical bench carriers), as in Fig. A14-3. A 3/8 to 1/2 inch thick block of aluminum or wood, 2 in. long, which slides smoothly and freely in the aluminum channel is used as the slider. It is desirable that the block have a rather deep V cut in it to position the lens or eyepiece tube (see below). A cross slot may be useful too, as shown in the figure. The position of the axis of rotation should be marked with a punch and also lines scribed on the sides to indicate the position of the axis too (see figure).

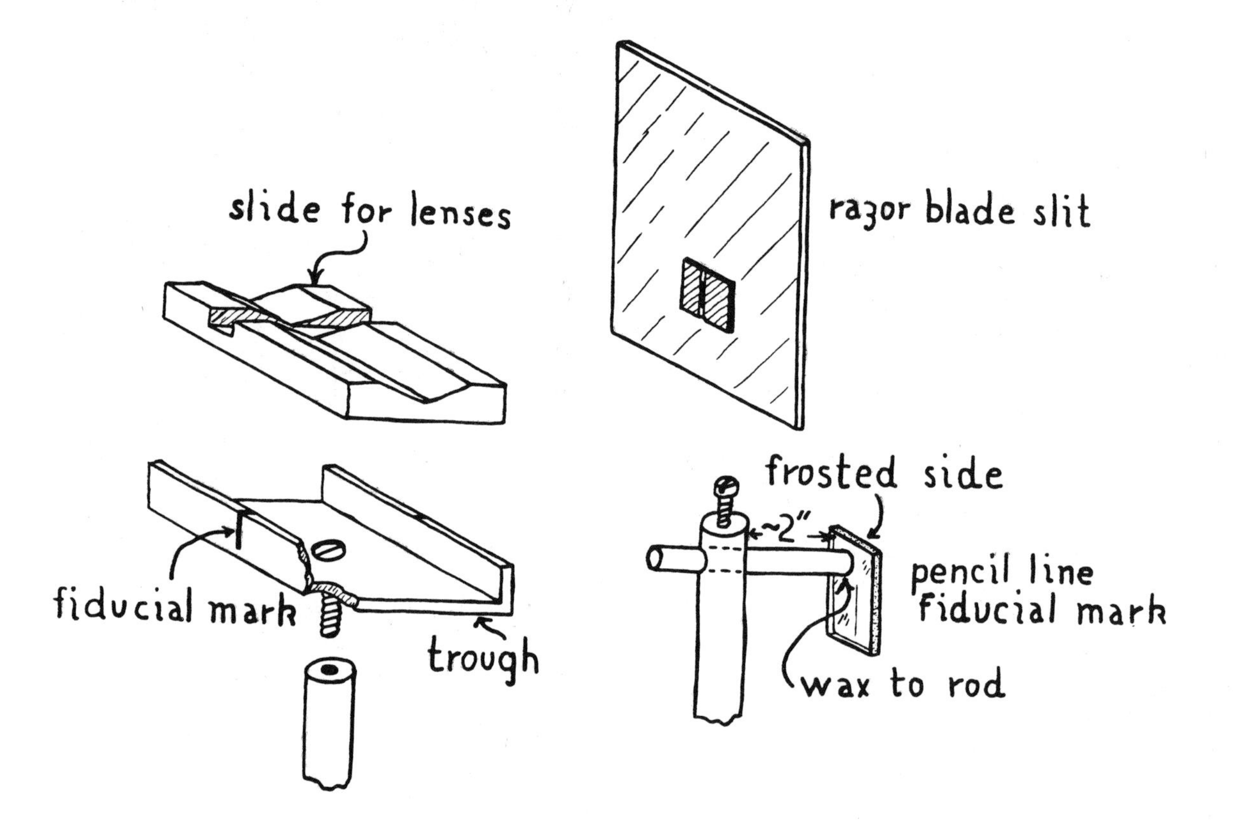

FIG. A14-3

Construction details for Newtonian distance and nodal slide apparatus.

7) Point source.

8) Collimator lens 2 1/2 to 3 inches diameter (achromat preferred).

9) Observing screen:

The screen should be very fine ground glass and need be no bigger than 1/2 x 1 inch. It should be waxed to the end of a rod with hard wax (bees wax and rosin mixture) as in Fig. A14-3. Here again, it should be mounted above the nodal slide. A fine vertical pencil line is used as a reference to tell if the image moves side to side in the nodal test.

10) A pocket magnifier to observe the images in the various tests.

11) Lenses to test:

A. Thick plano-convex lens of short focus (1 in.) about 1 inch diameter. The lens should have a wide rim for ease in mounting and marking the principal planes.

B. Ramsden eyepiece of about 1 inch focus. The eyepiece should be selected with an eye to mounting on the slide. It may be desirable to cut a cross slot in the slider to accommodate the larger rim of the eyepiece at the end (see illustration).

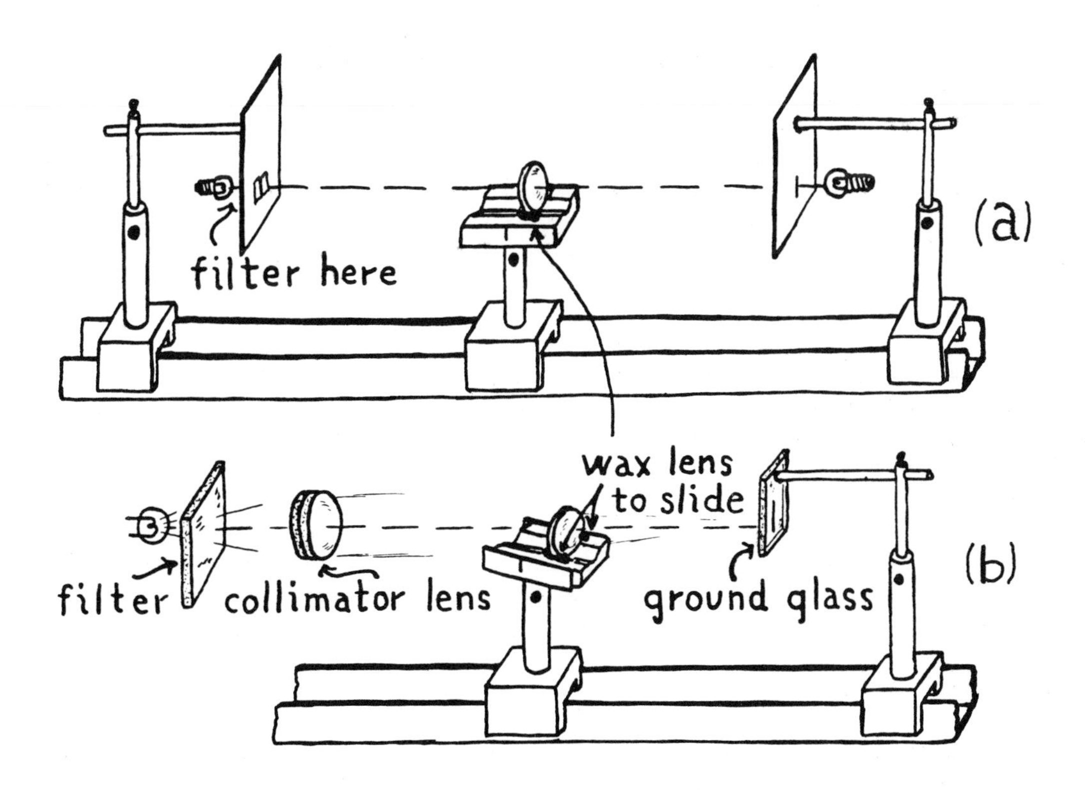

FIG. A14-4

Arrangements for Newtonian distance and nodal slide measurements.

PROCEDURE

I. NEWTONIAN DISTANCE METHOD

On the optical bench, mount the two slits, as shown in Fig. A14-4a. The lower edges of the slit mount should be higher than the lens system mount so that the lens system may be moved as close to the slit as may be necessary. The separation of the two slits should be about six times the focal length of the systems to be tested. Provision must, of course, be made to illuminate either slit as shown.

A. Thick lens

Mount the thick lens on the holder using tackiwax if necessary. Make sure that the axis of the lens is parallel to the axis of the bench. Follow steps 1, 2, and 3 described above to determine the focal length. After a quick preliminary check to make sure everything is properly placed, make accurate readings of the three positions, and calculate the focal length. Next locate the principal planes, and record their position. (A piece of tape may be fastened to the rim of the lens for this purpose). Measure and record the thickness of the lens.

B. Ramsden eyepiece

Remove the thick lens and place the Ramsden eyepiece on the system mount. Determine the focal length and principal planes for the system. (Be careful to distinguish which plane belongs with which focal point). Now remove the field lens of the eyepiece and measure the focal length of the eye lens. Similarly, measure the focal length of the field lens. Determine the separation of the two lenses as mounted in the eye-piece.

II. NODAL SLIDE

Set up the nodal slide arrangement as in Fig. A14-4b. Use a point source and green filter together with a properly oriented collimator lens. For autocollimation, it is necessary to allow for the presence of the filter, and the light should be reflected back through the filter to a card held in the same plane as that of the point source. The observing ground glass screen should be clamped at a height where the image can be seen but high enough so that it will clear the nodal slide. Unless this is done, it may be impossible to get the screen close enough to the lens system when the latter is located at the nodal point.

A. Thick lens

Mount the thick lens on the nodal slide and locate the nodal point. It is suggested that the observing screen remain fixed and the nodal slide moved and rotated. First, with the lens system at any arbitrary position on the slide, move the slide carriage to produce a good focused spot on the observing screen. Adjust the screen so that this image falls on the vertical pencil mark used for reference. Rotate the slide and observe which way the image moves. A little thought will indicate which way to move the lens system on the slide. (If the lens system were simply translated at right angles to the beam, it is obvious which way the image would move. Imagine the lens to be much too far from the axis of rotation either way and consider which way the image would move).

By careful adjustment, the exact nodal point should be located. Note the principal plane of the lens. Remove the slide with the mounted lens system and measure the distance from the axis of rotation to the screen, using a millimeter scale and a celluloid triangle if desired.

Replace the slide and determine in like manner the other principal plane and focal length. Are the two focal lengths the same?

B. Ramsden eyepiece

Substitute the eyepiece for the thick lens and determine the focal length and principal planes for the eyepiece.

As a check, it may be desirable to measure again the focal lengths of the two component lenses of the eyepiece.

RESULTS

1) Show that the thick lens formulae for locating the principal planes and finding the focal length, are greatly simplified for a plano-convex lens. In particular, show that one principal plane lies at the vertex of the curved surface and one lies at distance t/N inside the flat surface (t being the thickness). Assume an approximate value for the refractive index of the glass and calculate the position of the second principal plane.

2) For the thick lens, compare the location of the principal planes and focal points determined by the two methods and the estimated position of the second principal plane calculated in (1).

3) Use the thick lens formulae as applied to a system of thin lenses and calculate the focal length and principal planes for the Ramsden eyepiece using the measured focal lengths of the component lenses and their separation. Compare these results with those measured by the two methods.

QUESTIONS

1) In the Newtonian method of determining the focal length, there are two possible positions of the lens system in step 2.

How are they related?

2) Part 1 of the Results indicates that a determination of the principal points and focal length of a plano-convex lens permits the refractive index of the glass to be calculated. What drawbacks are there to this method?

3) Are there any limitations for either of the two methods? What are they?

4) What is the minimum possible separation of the slits in the Newtonian distance method?

EXPERIMENT A15. THE TELEPHOTO LENS, THE GALILEAN TELESCOPE, AND THE BARLOW LENS.

READING: Strong, Ch. 15.6; or Jenkins and White, Ch. 10.6; also see A. G. Ingalls, Ed., Amateur Telescope Making, Book III, (1953, Scientific American), pp. 215-217, (A. E. Gee); pp. 277-286 (C. R. Hartshorn).

This experiment is intended to illustrate the basic principles of the telephoto combination used in the telephoto lens, the Galilean telescope, and the Barlow lens system. The slight differences between the three are illustrated in Fig. A15-1. Part (a) shows the optical system of the telephoto lens; the effective focal length of the convergent lens is greatly increased by the use of a divergent lens, and thus the image on the film is enlarged accordingly. In part (b) are shown the elements of a Galilean telescope. The primary difference between this and the telephoto lens is that here the image is virtual, and located a minus infinity for convenient observation by the eye, whereas, the image for photographic purposes must be real and located at the film surface. The Galilean optics are used for opera glasses where a useful magnification of two or three can be easily achieved with a short length instrument. The Barlow system, part (c), may be regarded as a telephoto system used as a telescope objective. In this case, the real image formed by the telephoto combination is examined by a magnifying eyepiece. The divergent Barlow lens may also be regarded as decreasing the effective focal length of the eyepiece system, and therefore, as a modification of the eyepiece rather than of the objective.

It should be evident that the single element divergent lenses shown in Fig. A15-1 would cause appreciable chromatic aberration, and therefore, in high quality optical systems of this sort, negative achromatic doublets are used. In the past, some of these systems were made to have adjustable spacing between the front lens and the divergent combination so that the magnification could be varied, but now they are generally made with fixed spacing because better correction for the monochromatic aberrations can be obtained in this way.

APPARATUS

The apparatus of the previous experiment used for measuring the focal length and principal points by the Newtonian distance method is required. In this experiment, however, any conveniently available slits may be used. The nodal slide is not used.

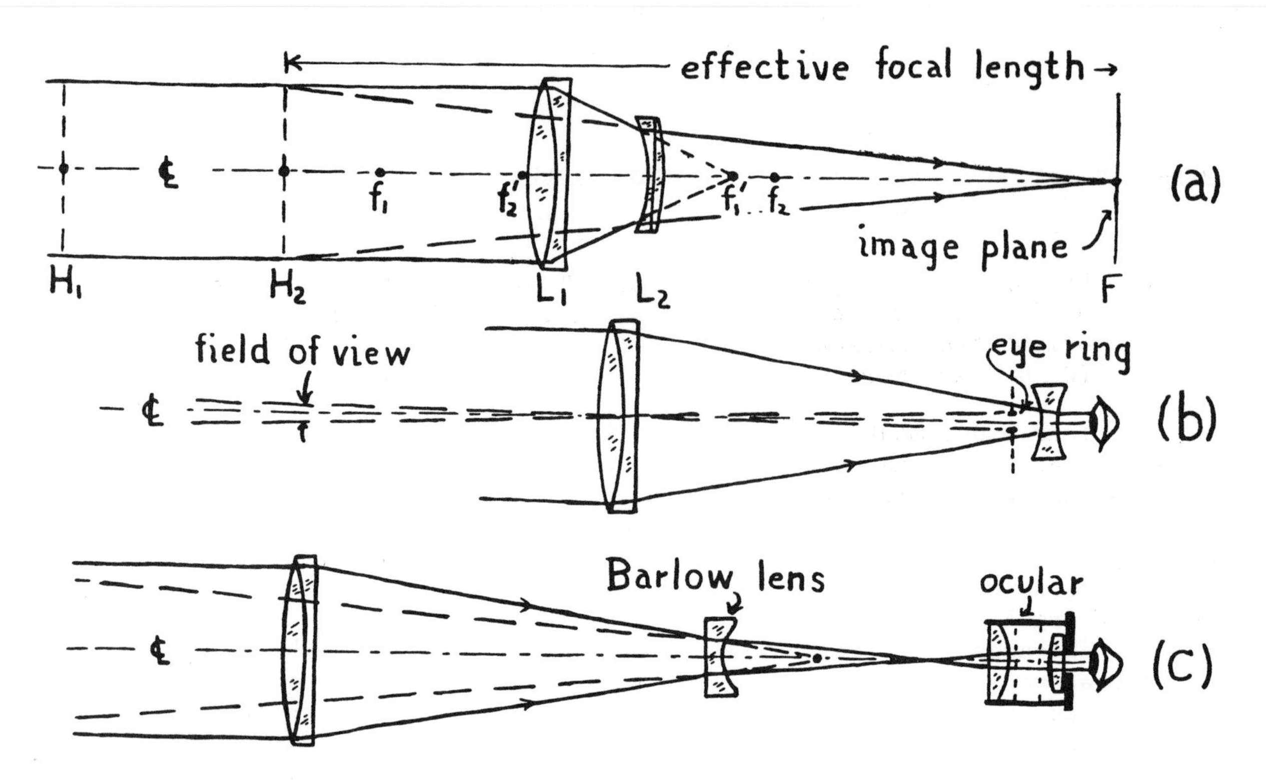

FIG. A15-1

(a) The telephoto lens system showing the principal planes and focal points, (b) the Galilean telescope, (c) the Barlow lens system.

A coarse target with heavy black lines on white paper or cardboard is needed. The target described in Experiment A13 is convenient.

The additional equipment needed for this experiment includes:

(1) Convergent achromat of 5 to 7 in. focus and 1/2 to 2 1/2 in. diameter.

(2) Divergent (single element) lens of about half (negative) the focal length of the achromat and about 1 inch diameter.

(3) Plano-convex lens of about 2 in. focus.

PROCEDURE

1) Fix the brightly illuminated target, a meter or more beyond the end of the optical bench and in line with it. Mount the convergent achromat on a slider on the bench (properly oriented), and form an image on

the screen. Measure the object and image distance and also the magnification. Calculate the focal length, both from the object-image distance and from the magnification.

(2) On a second slider, mount the divergent lens and slide it somewhat inside the focal position of the convergent lens alone, to a point where an image, enlarged about twice, is formed. Check to see which orientation of the divergent lens is better. Clamp a horizontal rod to both sliders to maintain this separation of the lenses. Measure the spacing of the lenses, the object distance from the convergent lens (record if unchanged) and the image distance from the divergent lens. Also measure the magnification.

3) The location of the image for the convergent lens alone is known or easily calculated. Regard this image as a virtual object for the divergent lens, and, taking account of the separation of the lenses and the final image location, calculate the focal length of the divergent element.

4) Use the thick lens formulae (see references in the previous experiment) to calculate the focal length of the combination, and, also, the principal planes.

5) Set up the optical bench for measurements by the Newtonian method of the previous experiment. Here the two slits (which may be of any variety) are placed at the two ends of the optical bench. (If the bench is not long enough, one slit may be mounted beyond the end of the bench). Measure the overall focal length and the location of the principal planes.

6) Remove the slit arrangement for the measurement of focal length and principal planes. Illuminate the target again, and use lens #3 to view the target. Is the target inverted or erect?

7) Remove the auxiliary eyepiece lens of part 7 and slide the divergent lens closer to the leading convergent one so as to form a Galilean telescope. Again view the target and note whether the image is erect or inverted. Note particularly the small field of view, a consequence of the location of the field stop considered in a later experiment.

RESULTS

1) Record all observations and calculations including the formulae used. Indicate the probable accuracy of the various results.

2) Make a table showing the comparison of the theoretical and experimental results.

3) Make a careful diagram of the telephoto system indicating the

position of the lenses, their focal length, the location of the principal planes, and the overall focal length.

4) Determine from the data of part 2 of the procedure, the object and image distances from the principal planes of the system. From these distances, calculate the magnification. Compare this result with the measured value.

QUESTIONS

Which orientation of the negative lens was found to give the best image? Which way should this lens be oriented for minimum spherical aberration?

EXPERIMENT A16. THE REFRACTING TELESCOPE.

READING: Jenkins and White, Ch. 10.11-10.16, Ch. 7-7.11; or Strong, Ch. 15.6-15.11; see also B. K. Johnson, Optics and Optical Instruments, (1960, Dover), Ch. 4; D. H. Jacobs, Fundamentals of Optical Engineering, (1943, McGraw-Hill), Ch. 12; L. C. Martin, Technical Optics, (1950, Pitman), Vol. 2, Ch. 2; A. C. Hardy and F. H. Perrin, The Principles of Optics, (1932, McGraw-Hill), Ch. 5.

This experiment is concerned with measuring the magnification, field, and aperture of astronomical telescopes, and with converting such instruments to terrestial telescopes by means of an erecting eyepiece. Only the basic properties are considered here. The question of resolving power is treated in Experiment B16; the properties of various eyepieces are discussed in detail both by Jacobs and Martin loc. cit.

THEORY

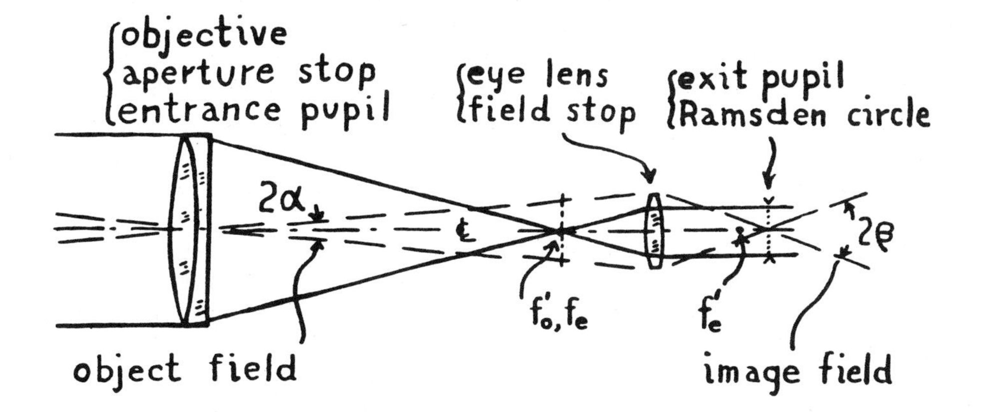

FIG. A16-1

Diagram of an astronomical telescope showing field angles and stops.

Figure A16-1 is a diagram of an astronomical telescope with a single lens eyepiece. The achromatic objective lens collects light from the object and forms an image of it in the secondary focal plane of the objective, f_o'. This image is viewed with a magnifier which forms a virtual image at minus infinity. Whereas the Galilean telescope (Fig. A15-1b) has a divergent lens as the eyepiece and thus produces an erect image, the convergent eyepiece of the astronomical telescope produces an inverted image. Other differences between the two telescopes are mentioned below in the section on stops and fields.

Magnification: The magnification of an optical system is commonly stated in either of two ways. In Experiment A1, lateral or linear magnification was defined as the ratio of image dimensions to object dimensions. Alternatively, one may define magnification as the ratio of the angle subtended by the image to that subtended by the eye as viewed by the observer. Since the object viewed with a telescope is usually at a very large distance, effectively at infinity, light from various portions of the object enters the objective in parallel beams. The final virtual image seen by the observer is conveniently placed at minus infinity, and thus the light leaves the telescope as a set of parallel beams. Thus it is meaningless to speak of image and object sizes in such a case, and therefore, we use angular magnification. In Fig. A16-1, the magnification is the ratio of $2\beta/2\alpha$.

More accurately, it is convenient to define magnification M in terms of the tangents of angles, as $M \equiv \tan\beta/\tan\alpha$. The difference between the tangents of these angles and their radian measure is generally small. As shown below, there are several ways to calculate this magnification.

Stops and fields: The aperture of a telescope determines how much light flux from the object reaches the image. For astronomical applications, the aperture is often a prime consideration, and it is then important to make sure that no light collected by the objective is wasted (as, for example, on the slit jaws of a spectrograph used in conjunction with the telescope). Thus in a properly designed telescope, the objective itself is the aperture stop, and the illumination of the image is proportional to the area of the objective divided by the square of its focal length, that is, the solid angle subtended by the objective at the image plane.

By definition the entrance pupil of an optical system is the image of the aperture stop formed by all the lenses which precede it. In the case of our telescope, the objective rim is also the aperture stop, and thus since no lenses precede it, the objective rim is also the entrance pupil. The exit pupil is defined as the image of the aperture stop formed by all the lenses which follow it. For the telescope of Fig. A16-1, the exit pupil (also called the Ramsden circle or the eye ring) is the image of the objective rim formed by the eye lens as indicated. (The exit pupil of a Galilean telescope lies between the lenses).

A third important characteristic of a telescope is its field of view, determined by the field stop. In the simple case of the lens and photographic plate -- either an ordinary camera or a photographic telescope, the field is clearly limited by the dimensions of the photographic plate so that it is the field stop. In the case of the telescope illustrated, the field stop is the rim of the eye lens. The tangent of the half-object field angle ($\tan\alpha$) is equal to the radius r of the field stop divided by the distance, ($f_o' + f_e$) to the entrance pupil. Similarly, the tangent of the half-image field angle ($\tan\beta$) is the same radius divided by the distance s' from the field stop to the exit pupil. The following relations hold:

$$\tan\alpha = \frac{r}{f_o' + f_e} \qquad \tan\beta = \frac{r}{s'}$$

$$s' = \frac{sf}{s-f} = \frac{(f_o' + f_e) f_e}{f_o' + f_e - f_e} = \frac{(f_o' + f_e) f_e}{f_o'}$$

$$M = \frac{\tan\theta}{\tan\alpha} = \frac{r}{(f_o' + f_e) f_e / f_o'} \times \frac{(f_o' + f_e)}{r} = \frac{f_o}{f_e}$$

Thus it is seen that the magnification can be expressed as (a) the ratio of the focal length of the objective to that of the eyepiece, (b) the ratio of the tangents of the image field angle to the object field angle, or (c) as the ratio of the diameters of the entrance and exit pupils. The latter relation is particularly convenient when an unknown compound eyepiece is used, for the two pupils are easily measured. Also, it permits a quick measurement of the focal length of the eyepiece itself if the focal length of the objective is known.

Field lenses: The field of view of the simple telescope of Fig. A16-1 is unnecessarily restricted and is also unsatisfactory in that the illumination falls off toward the edge of the field. Figure A16-2 illustrates the situation. In part (a) of the figure (the same optics as before), it is obvious that rays 1, 2, and 3 which illuminate the edge of the real image do not all reach the eye -- ray 1 is lost, and evidently only light which enters through the lower half of the objective reaches the eye, thus the illumination of the image as seen by the observer alls off to one half at the edge of the field.

Suppose however, a second lens is introduced between the objective and the eye lens as shown in part (b). This lens is (for now at least) located in the plane of the real image. It therefore has no effect on magnification, but it does direct all rays which pass through it into the eye lens, and it is therefore the field stop, and is called a field lens. Since it is closer to the entrance pupil than the eye lens (by the focal length of the latter) it subtends a larger angle at the entrance pupil thus giving a larger field of view than with the eye lens alone. Furthermore, the marginal rays through the objective (1 and 3) are both directed through the eye lens to the observer thereby making this larger field uniformly illuminated to the very edge. It is clear from the figure that there is a relation between the focal length of the field lens and its diameter which ensures that the extreme rays which fall on the field lens will also pass through the eye lens. (See question at the end of the experiment).

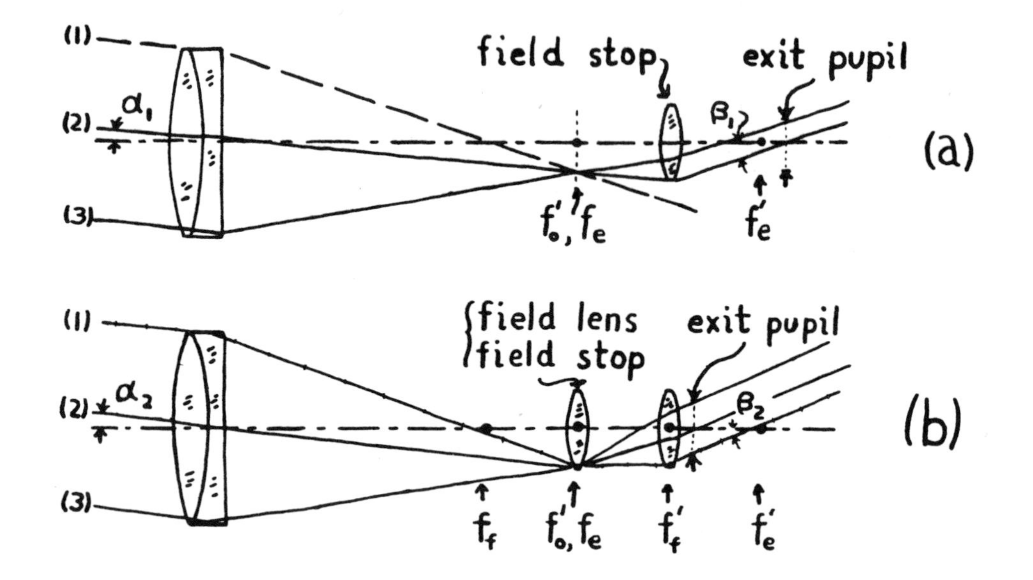

FIG. A16-2

Field of view of an astronomical telescope: (a) with simple eye lens and (b) with field lens added.

The positions of the various focal points of the lenses and the new position of the exit pupil are shown in the figure.

It should be noted that if a field lens is placed in the plane of a real image as shown in the figure, any scratch or dust on the lens will be in sharp focus in the observed image. In practice therefore, it is customary to displace the field lens slightly, either putting it in front of the real image as in the Huygens ocular, or behind it as in the Ramsden ocular. Both positions have their advantages. Since the field lens is no longer in the plane of the image, dirt and scratches are less apparent. More important, the displaced lens has some effect on the magnification and it may be used to help correct some of the aberrations of the eyepiece.

Terrestrial eyepiece: The inverted image of an astronomical telescope, while of no difficulty for astronomers, is obviously awkward for many applications. The image can be reinverted by the addition of an extra lens. Figure A16-3 (a) shows a very simple erecting system. The erecting lens, placed a distance twice its focal length from the first image, reforms the image an equal distance behind it. This erect image is now examined by the eye lens. The system obviously suffers from the same field limitations as the first telescope, and one practical solution is to add two field lenses resulting in an erecting or terrestrial eyepiece shown in Fig. A16-3 (b). Here the first two lenses effectively form a Ramsden eyepiece and the second pair a Huygens eyepiece.

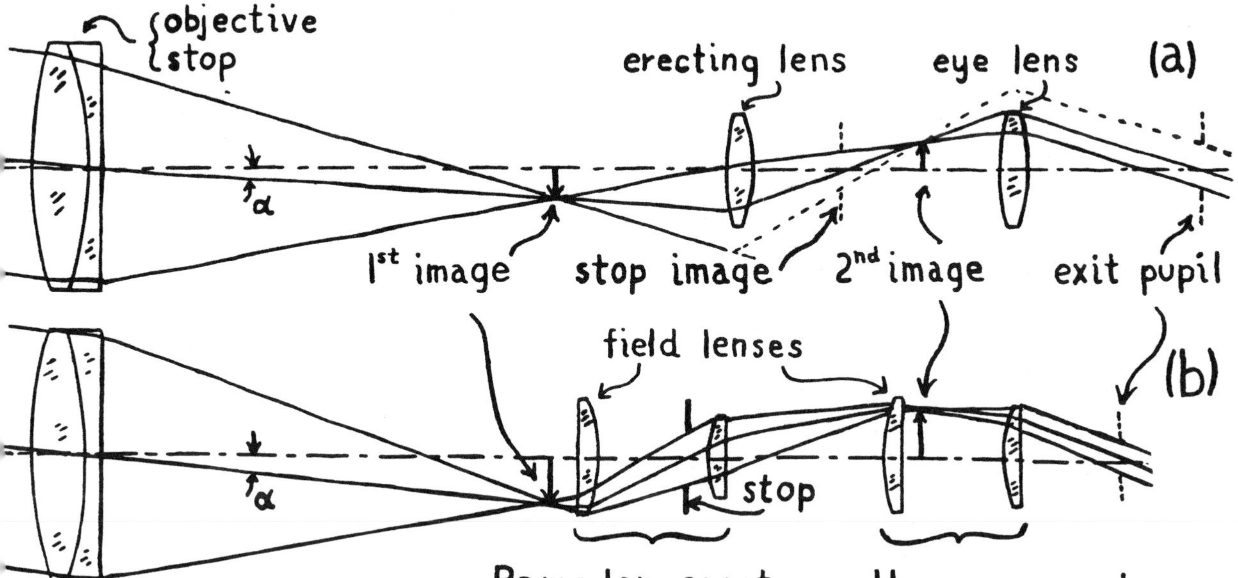

FIG. A16-3

Terrestrial eyepiece: (a) without field lenses, (b) with field lenses. The stop in the four lens system is used to eliminate stray light.

APPARATUS

1) Optical bench.

2) Achromatic objective lens, 1-2 in. dia., 10-15 in. focus.

3) Four matched plano-convex lenses, 1 in. dia., 1 1/2-2 in. focus.

4) Collimator lens, 10 in. or longer focus (need not be achromatic).

5) Point source.

6) Five lens mounts and sliders for the optical bench.

7) Bright illumination source for object (photoflood if possible).

8) Small ground glass screen, magnifier and millimeter scale.

9) Larger ground glass screen (2 in. x 2 in. or bigger).

10) (Optional) divergent lens of same focal length as in item 4, but negative.

11) Meter stick.

PROCEDURE

1) Simple astronomical telescope

Measure the focal length both of the objective and of the four matched plano-convex lenses. As an object, use a meter stick held at least 20 times the objective focal length distant from the optical bench. The numbers should be upside down. (Why?). It should be brightly illuminated, and a white background is helpful. Mount the objective lens at the end of the optical bench, correctly oriented for minimum spherical aberration. Eliminate most of the unwanted light from the object with a large piece of cardboard with a hole to admit light to the objective. Mount a ground glass screen on the bench and slide it to the image position of the objective. Next, mount one of the plano-convex lenses on the bench with its convex side toward the object, and slide it to put the screen in sharp focus. Now remove the screen and the object should be very nearly in perfect focus. Adjust the focus slightly if necessary. From the known focal lengths of the lenses, calculate the magnification. (If the object is not at infinity, is the magnification still the ratio of the focal lengths?) Cut a cardboard diaphragm for the objective 3 millimeters in diameter and place over the objective. The purpose is to limit the exit pupil to about 3 mm, a size which the eye can accept.

Remove the cardboard shield so that the observer can look at the meter stick with both eyes simultaneously -- one eye peering through the telescope and the other eye unaided. Two white strips of paper should be secured to the meter stick 25 or 50 cm apart. With the unaided eye these two strips are to be observed while at the same time reading the meter stick scale through the telescope. The two fields may be made to coincide reasonably well, though with some difficulty. Estimate the magnification of the telescope as well as practicable by noting how many cm of the scale seen through the instrument correspond to the separation of the paper strips. Is the agreement within experimental error?

Replace the cardboard shield and mount the small glass screen on the optical bench behind the eye lens (do not disturb the position of the lens). In front of the objective and close to it, mount the other ground glass screen. Illuminate this latter screen brightly and adjust the position of the rear screen until a small sharply defined ring of light is seen. This ring is the exit pupil. Use another plano-convex lens as a magnifier and with the millimeter scale, measure as accurately as possible the diameter of the exit pupil. Measure also the clear aperture of the objective (the stopped down size obviously). Compute the magnification from these measurements. Be sure to measure also the position of the exit pupil with respect to the eye lens.

Remove the ground glass from in front of the objective and direct into the telescope a collimated beam of light using the collimator lens and the point source of light. Note that the beam of light emerging from the eyepiece is now parallel and the diameter is constant for any position of the rear ground glass. Again measure the diameter and calculate the magnification.

Remove the collimator and ground glass screen. Observe once more the illuminated meter stick and after arranging the meter stick image to fall across the axis of the instrument, measure the angular object field of view, ($\tan\alpha$). Estimate the corresponding image field and again calculate the magnification. Is the result reasonable?

Measure the diameter of the eye lens, which is here the field stop, and calculate the object and image fields. Are the results in agreement with the measurements? Note the decreased illumination toward the edge of the field.

2) Field lens

In the image plane of the objective, place a second plano-convex lens (either orientation -- why?). Has the magnification been changed by the added lens? Study the field of view. Note that the illumination is much more uniform. Measure the object field angle, and compare with the calculated result. Note the appearance of any dust or scratches in sharp focus in the field of view.

3) Erecting lens

Slide the erstwhile field lens to the position where it forms an erect second image at unit magnification. Focus the telescope with the eye lens and note that the image is correctly erect (meter stick numbers now inverted). Note that the magnification is unchanged by the erector (or else it is not in the right place). Note also the much restricted field of view. Finally, add two more lenses in the image planes to form the terrestrial eyepiece of figure A16-3b. The field should be considerably improved. The field lenses may now be moved slightly away from the two image planes as indicated in the figure. Record your observations.

4) (Optional)

Use the negative eye lens to make a Galilean telescope and measure the field of view. Why is the field much more restricted than for the positive eyepiece?

RESULTS

Record all observations and measurements in such a way as to show the comparison between the various measurements.

Calculate the positions of the principal planes for the simple astronomical telescope and make a diagram to show these planes, the aperture and field stops, and the entrance and exit pupils of the instrument.

Briefly discuss the importance of the various characteristics of the telescopes you have studied.

QUESTION

Derive a relation between the maximum focal length of the field lens and its diameter assuming that the field lens is placed in the focal plane of the objective (Fig. A16-2(b)) and that the eye lens is of the same diameter. If the eye lens is smaller than the field lens, is the required focal length of the field lens changed? How?

EXPERIMENT A17. THE COMPOUND MICROSCOPE.

READING: Jenkins and White, Ch. 7-7.11, 10.7-10.10, 10.12 10.16; or Strong, Ch. 15.5, 15.7, 15.9, 15.11; also L. C. Martin, Technical Optics, (1950, Pitman), Vol II, Ch. 3; a useful older reference is A. C. Hardy and F. H. Perrin, The Principles of Optics, (1932, McGraw-Hill), Ch. 5, and Ch. 24.

This experiment involves the measurement of the magnification, numerical aperture, stops, and fields of view of a compound microscope and a comparison of these results with theory. The compound microscope is far too intricate an instrument to be thoroughly studied in any one experiment, and many important matters are omitted in this book. Some of the wave optic principles of image formation in microscopic instruments are considered in Experiments B21 and B22, but such matters are substage condensers, types of illumination, phase contrast, polarizing microscopes, projection microscopes, infrared and ultraviolet objectives and eyepieces cannot be examined here. Some of these matters are treated in the reading, but for a thorough study, the student must refer to specialized books and papers on the subject, some of which are listed in the references in Martin's book.

THEORY

The optical system of the compound microscope consists of an objective lens system and an eyepiece. The objective system may be very complicated (as many as ten lens elements are sometimes used) or it may be reasonably simple. The eyepiece generally contains at least two lenses. For convenience, however, we may represent the microscope in terms of the principal planes of the objective and those of the eyepiece as illustrated in Fig. A17-1. The objective forms an inverted, enlarged, real image of the object as shown. The eyepiece forms a virtual image of this real image customarily either at minus 25 cm or at minus infinity. In the figure, the eyepiece is focused to produce the final virtual image at minus infinity -- the real image is set at the primary focal point of the eyepiece. The distance between the secondary focal point of the objective and the primary focal point of the eyepiece is the optical tube length L, commonly fixed at 160 mm in modern microscopes.

The overall magnification of a compound microscope is the product of the lateral magnification of the objective and the angular magnification of the eyepiece. The two magnifications are expressed in terms of the focal lengths of the objective and of the eyepiece, and it is necessary to be able to measure these quantities. The methods described in Experiments A1 and A14 are of no help here (at least with standard components) for the secondary focus of the objective usually lies within the mechanical structure of the system and the primary focal point of a Huygens eyepiece (often used with microscopes) lies between the lenses. We require special methods (except for a simple model compound microscope).

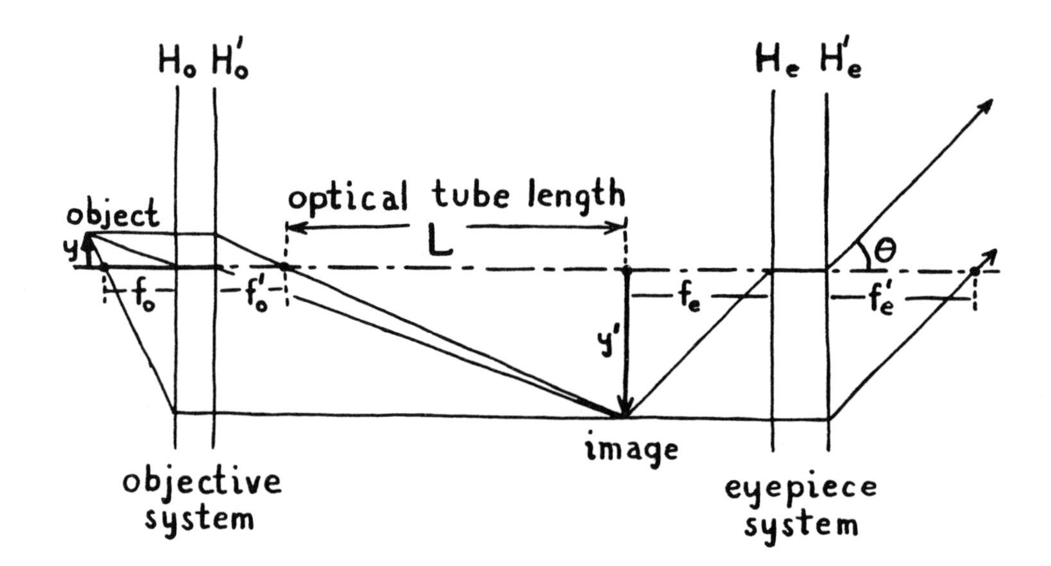

FIG. A17-1

Optical system of a compound microscope represented by its principal planes.

The focal length of the objective can be determined from two measurements of magnification and a distance. From Fig. A17-1, it is seen that the primary (lateral) magnification is given by:

$$M_o = |y'|/y = L/f_o'$$

If the magnification for two different values of L are measured, then it follows that:

$$M_1 = L_1/f_o' \qquad M_2 = L_2/f_o'$$

$$M_1 - M_2 = (L_1 - L_2)/f_o'$$

$$f_o' = (L_1 - L_2)/(M_1 - M_2)$$

That is, if the magnification for two values of L are found and the change in L measured, the focal length of the objective is determined. The principal planes can then be located approximately from the relations

$$L = f_o' M = X'$$

$$S' = L + f_o' = f_o'(1+M) \quad = \text{distance } H_o' \text{ to image}$$

$$S = X + f_o = f_o^2/X' + f_o = f_o^2/f_o' M + f_o = f_o(1 + 1/M)$$

$$= \text{distance } H_o \text{ to object}$$

The focal length of the eyepiece is easily found by combining it with a telescope objective lens of known or measured diameter D and focal length f_o. Figure A17-2 shows a telescope focused for infinity using the unknown eyepiece (represented by its principal planes). If the telescope is in focus, and the position and diameter, d, of the exit pupil are measured, the focal length and principal planes can be found.

$$f_e = f_o d/D$$

A distance f_e from the focal plane of the objective in the absence of the S,S' eyepiece determines the first principal plane, H_e. The second principal plane H_e' is located a distance S' from the exit pupil point where

$$S' = X' + f_e' = f_e^2/X + f_e' = f_e^2/f_o + f_e'.$$

The overall magnification of the microscope is then given by

$$M = M_o M_e = \frac{L}{f_o'} \cdot \frac{25\text{cm}}{f_e} \text{ or } \frac{L}{f_o'} \cdot \left(\frac{25\text{cm}}{f_e} + 1\right)$$

depending on the position of the final image -- at minus 25 cm or minus infinity. This overall magnification is readily measured as described below in the procedure.

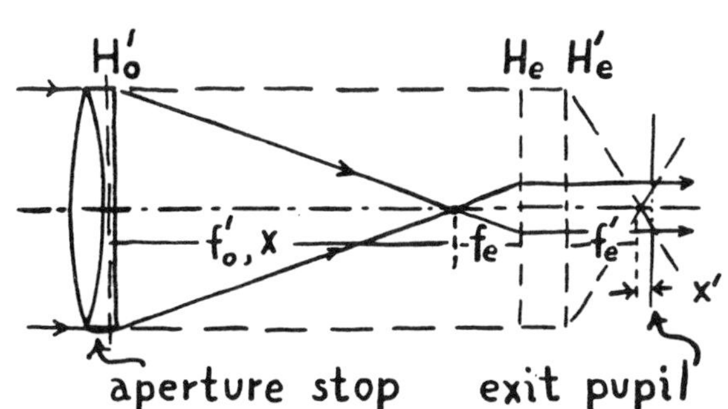

FIG. A17-2

Measurement of the focal length and principal planes of an eyepiece.

The resolving power of a microscope, its ability to distinguish detail, is dependent both upon the method of illuminating the object and upon the so called numerical aperture of the objective. The numerical aperture is the product of the refractive index of the medium between the objective and object and the sine of half the angle subtended by the objective at the object, $N \sin i$. For ordinary (dry) objectives, the index in unity, with oil immersion objectives, improved performance is obtained by immersing both the object and the objective in a drop of oil which has the same refractive index as the glass of the objective. In this experiment, we shall confine our attention to dry objectives.

APPARATUS

1) Optical bench.

2) Auto lamp light source.

3) 1 cm. (or 1/2 in.) reticle scale divided into 50 or 100 parts.

4) 2 pieces ground glass about 2 in. square.

5) Plano-convex lens, about 1 in. focal length, not more than 1 in. dia.

6) Plano-convex lens of about 2 in. focus.

7) Standard 1 in. (10X) microscope objective.

8) Standard 1 in. focal length eyepiece, preferably Huygenian.

9) Transparent millimeter scale.

10) Telescope objective lens 1 1/2 - 2 1/2 in. dia., 10-20 in. focal length.

11) Microscope slide, or small piece of glass 1 X 3 in.

PROCEDURE

A. Model compound microscope

In order to correlate experiment and theory more easily, we begin the experiment with a model microscope. First measure the focal length of each lens using any of the previous methods -- for simplicity, treat the lenses as thin lenses. Measure also the diameters of the lenses.

Set up the microscope on the optical bench arranging the components as illustrated in Fig. A17-3. Use the shorter focus plano-convex lens for the objective and the longer focus one for the eyepiece. Omit the microscope slide between the eyepiece and eye and view the object from position (a). Make the distance L equal to 10 cm first, and place the transparent millimeter scale in the desired image plane of the objective. Record the position of the scale for later use. Adjust the eye lens so that the mm scale is at the primary focal point of this lens. In front of the objective, mount the reticle and adjust it slightly so that it is in focus when observed through the eyepiece. The reticle is conveniently illuminated as shown with a piece of ground glass and a small lamp.

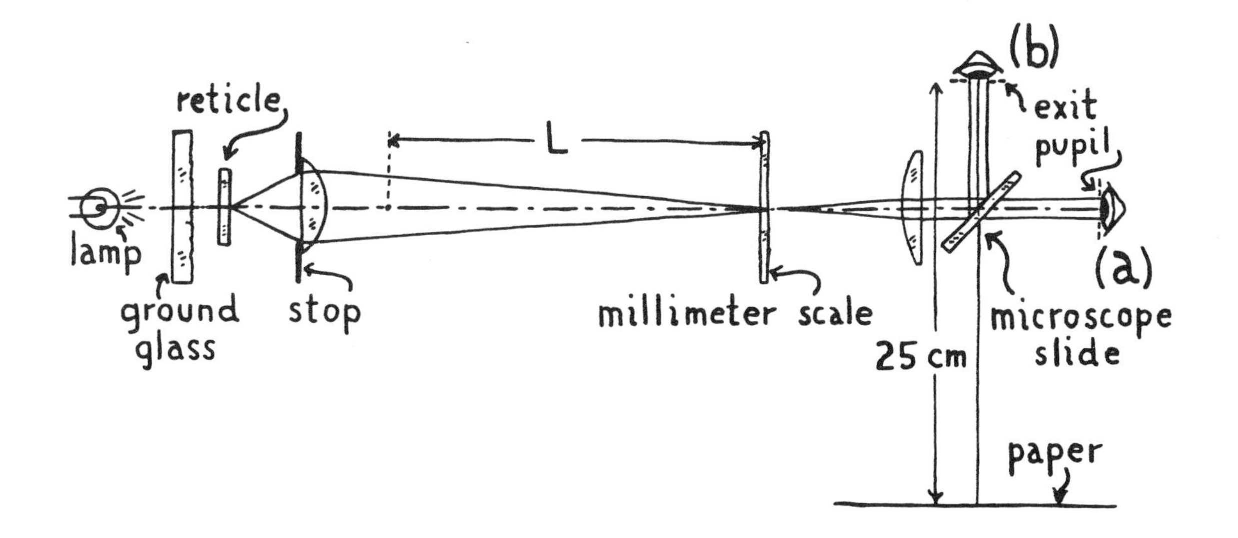

FIG. A17-3

Model compound microscope.

By comparison of the magnified image of the reticle scale with the millimeter scale, determine the magnification. Record the result. Move the mm scale and eyepiece so as to increase L to 20 cm., and record the new position of the scale. Record the change in L and measure the new magnification after adjusting the position of the reticle to bring it into focus once more.

From these measurements, calculate the focal length of the objective and compare with the previously determined value.

Set L equal to 16.0 cm and calculate the primary magnification. Focus the microscope by adjusting the object (reticle) until it is in focus. Use a fine ground glass screen to locate the exit pupil of the system. If its diameter is more than 3 mm, reduce it to about 2 1/2 mm by stopping down the objective lens with a piece of paper or cardboard with good circular hole.

Insert the microscope slide as in Fig. A17-3 to reflect part of the light from the microscope upwards to the new eye position, (b). Place a sheet of paper 25 cm from the new location of the exit pupil and refocus the eyepiece very slightly so that the millimeter scale, the reticle, and the paper sheet are all in focus. The exit pupil will, of course, move slightly when the eyepiece is refocused, and the position of the paper may be adjusted accordingly, though the change should be small.

Measure the magnification of the eyepiece by marking on the paper two points which correspond to one centimeter on the scale. Compute the magnification of the eyepiece and compare the result with the value $25/f_e + 1$. Determine also the overall magnification of the microscope by marking on the paper points corresponding to 1 mm of the reticle. Compare the measured overall magnification with the calculated overall magnification.

Remove the auxiliary glass slide. From the diameter of the objective lens (or its stopped down aperture) and the object and image distances involved, calculate the dimensions and position of the exit pupil. Compare with the measured values. From the positions of the entrance pupil and exit pupil and field stop, calculate the angular field of view for both the object field and the image field.

Remove the auxiliary glass slide and measure the object field of view by determining how much of the reticle scale can be seen. (If the reticle scale is too short, substitute a millimeter scale for the reticle). Estimate also the apparent image field of view, and from the tangents of the two half fields of view, calculate the overall magnification and compare with the calculated magnification.

Finally, determine the numerical aperture of the objective. Move the reticle and its illuminator and the eyepiece out of the way. About 6 inches in front of the objective, put a well illuminated centimeter scale and observe it through the objective at a distance of 25 cm or more. Keep the eye fixed and determine the angle subtended by the field, and from this the numerical aperture. Move the eye back several inches further and note any changes. What is being observed by the eye in this measurement?

B. Compound microscope

Mount the standard microscope objective on the optical bench and determine (a) the focal length, (b) the position of the principal planes, and (c) the numerical aperture. If practical, sketch the optical elements of the objective.

To determine the focal length of the standard eyepiece, first record or measure the diameter and focal length of the telescope objective. Focus the telescope for parallel light either using a distant object or by a collimation method. Put a ground glass screen immediately in front of the objective and illuminate it well. Locate the exit pupil, measure its position and size and determine the focal length of the eyepiece and its principal planes.

Calculate the overall magnification of the compound microscope assuming a value of L equal to 16 cm. and the virtual image at 25 cm.

Arrange the objective and the eyepiece either on the optical bench or in the microscope tube itself and use the illuminated reticle as the object. From the measured focal length of the objective and eyepiece, adjust the distances properly for L equal to 16 cm. and slide the reticle, illuminated as before, into focus. Measure the overall magnification as previously.

(OPTIONAL). Measure the field of view for the image and object fields. Compare with the theoretical values (after measuring the diameter of the field lens of the eyepiece). Does the field lens constitute the field stop in this case?

RESULTS

Record all measurements and calculations. What are the sources of error in the measurements? Determine the probable error in the various results.

Draw a diagram of the compound microscope with the standard components used, locate the principal planes and focal points. Locate the aperture stop, entrance pupil, exit pupil, and field stop.

QUESTIONS

If a microscope objective is designed for, say, an optical tube length of 16 cm, what differences would result from using a tube length twice as long?

Explain the significance of numerical aperture in terms of the rays accepted by the microscope. Suppose the object is immersed in water and that an oil immersion objective is used. Three refractive indices might be involved, N_{water}, N_{glass}, and N_{oil}. Which index is the proper one to use in determining N.A.?

Design an instrument to measure accurately (to 1%) distances of about 1/8th in. with the requirement that no part of the measuring instrument may be closer than 6 inches to the object. The problem might be encountered in measuring the dimension of some part within a piece of apparatus.

EXPERIMENT A18. SOME PROPERTIES OF THE HUMAN EYE.
(In collaboration with Leonard Matin*)

READING: Strong, pp. 313; 485-490 (H. W. Yates); and L. C. Martin, Technical Optics, (Vol. I 1948, Vol. II 1950, Pitman), Vol. I, Ch. 5, Vol. II, Ch. 4.

This group of demonstrations is intended to call attention to a few of the many special properties and limitations of the human eye of importance to the physical scientist or engineer. Three of these properties: color perception, resolving power, and polarization sensitivity are treated elsewhere -- in Experiments A19, B16, and C2 respectively.

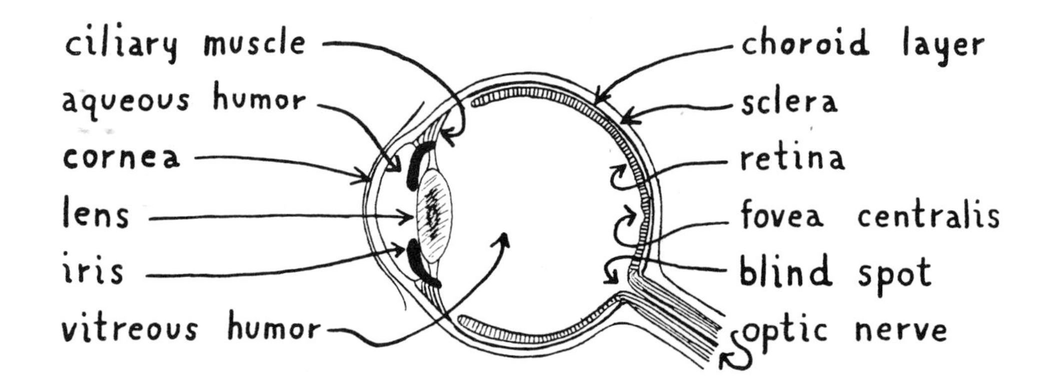

FIG. A18-1

Principal features of the human eye.

FIG. A18-2

Diagram to show location of the blind spot.

*Dept. of Psychology, Johns Hopkins University.

A. Blind spot

Figure A18-1 shows the principal features of the human eye. Where the optic nerve enters the eyeball, there is a blind spot having no photoreceptors. The existence and location of this blind spot can be easily demonstrated. Close the left eye and focus the right eye on the cross in Fig. A18-2. Move the eye slowly away from the page while observing the cross. At a distance of roughly ten inches, the black spot to the right of the cross will disappear. Farther away, the spot reappears. Measure the distance at which the spot vanishes and calculate the angular position of the blind spot with respect to the fovea centralis where the image of the cross is fixed.

B. Inversion of image

The image formed on the retina is inverted, nevertheless, objects appear erect to the observer. If an object is so close to the observer that it is within the primary focal distance of the eye, the object forms a blurred image on the retina and this image appears inverted.

Use a brightly illuminated pinhole an inch or two from the eye as a source and observe a pin point held between the pinhole and the eye. The image of the pin will appear inverted since the object is within the focal distance of the eye and thus too close for an inverted image to be formed.

C. Near point and accommodation

The normal eye can comfortably focus on objects from about ten inches to infinite distance away. With some strain the near point can be reduced to about five inches, but objects still closer are blurred. To achieve this focusing, the shape of the eye lens is changed by the action of the ciliary muscles so that the focal length is altered. This process of accommodation produces a focused image at the retina. By the expedient of reducing the aperture of the eye, objects only about an inch away can be seen with reasonable clarity. The near point is not changed, but the effect is to make the eye essentially a pinhole camera. If a card with the proper size pinhole (try various sizes of pinhole) is held close to the eye, it will be found possible to see objects only an inch away. It is, for example, possible to read the scale numbers and divisions on a reticle divided into tenths of a millimeter with comparative ease. If the object is held one inch away from the eye, the effective angular magnification is about ten times, and in this sense, the pinhole makes a very crude 10X magnifier.

Is the image erect or inverted? Why?

D. Spherical aberration

The refractive index of the eye lens is slightly greater at the center than at the margin and the spherical aberration of the eye is thus reduced. This aberration is not absent, however, as can be readily shown. Hold a card close to the eye and move it vertically across the path of the light rays entering the pupil from a distant horizontal straight edge such as the roof line of a building or even a cabinet across the room. As the card is moved upward the horizontal line of the object appears to shift upward as the card cuts off the paraxial rays leaving only the upper marginal rays.

Explain why the higher refractive index of the central part of the lens reduces spherical aberration.

E. Iris

A part of the adaptation of the eye to low and high light levels is achieved by varying the diameter of the iris aperture. At low illumination levels, its area may be ten times larger than at high levels. The change in pupil diameter is slow enough to be easily observed by looking directly at an incandescent bulb and then interposing a mirror between the bulb and the eye to see the pupil enlarge. The illumination on one eye affects the pupil diameter of the other. While observing the pupil of one eye, cover and uncover the other eye with your hand. Hold a millimeter scale quite close to the eye and a mirror about five inches away to observe both the pupil and the millimeter scale. Make measurements of the pupil diameter at both high and low light levels.

F. Rod and cone vision; dark adaptation

Vision involves both rods and cones. The rods, which are not sensitive to color, are widely distributed over the retina except in the center of the fovea where they are absent. The cone types, on the other hand, which enable us to sense color are concentrated chiefly in the foveal region. Curves of the relative distributions of rods and cones are given in Yates' appendix, Fig. I 10, p. 487 of Strong's book. The rods have greater sensitivity to light than the cones. Thus faint light sources, such as most stars, are best detected by averting the eyes slightly so that the images fall outside the retion of the fovea at a place where rods are plentiful. Such images appear colorless.

These properties are readily demonstrated in a darkened room. As a light source, use a 25 to 60 watt incandescent lamp operated from a Variac or rheostat at low voltage. Unless the room is completely dark, it is well to use some sort of cardboard shield such as a large box to keep out stray light. Begin with the lamp off and allow the eyes to become dark adapted by shielding the eyes from light for five minutes or so. After this period of adaptation, increase the lamp voltage very slowly from zero until the lamp can barely be seen with averted vision. We know the filament must be a dark red, but the light appears greyish-white. When the voltage is increased a bit more, the light can be seen by looking directly at it and the color is then properly orange-red.

Can dark adaptation be achieved by wearing red goggles? Can you explain why or why not?

G. Brightness constancy

Although a piece of coal may reflect some 100,000 times more light to the eye in bright sunlight than under a night sky, it appears black in both cases. Clearly then, the apparent brightness of an object does not depend only on the amount of light it reflects to the eye. The fact that the apparent brightness of objects does not change much with wide changes in illumination is called brightness constancy.

We must always judge the brightness of an object in a particular set of surroundings, and if the illumination changes, the light received from both the object and its surroundings changes in the same ratio. The importance of this fact for the preservation of constancy can be rather easily demonstrated.

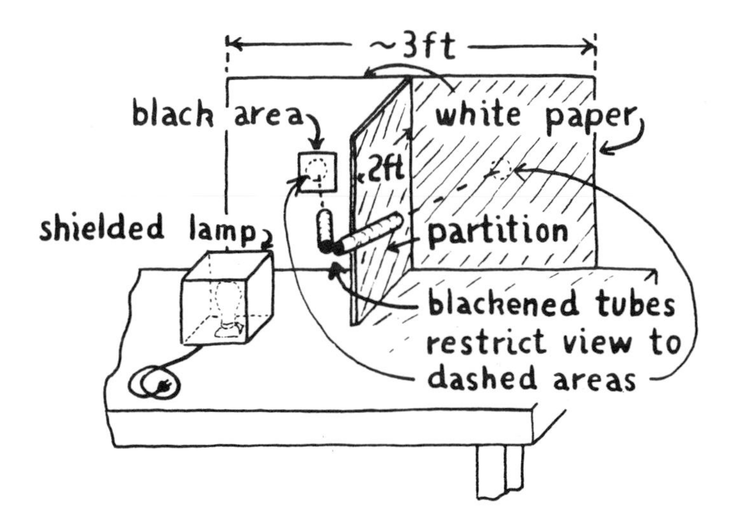

FIG. A18-3

Brightness constancy demonstration.

A large sheet of heavy white paper or cardboard (not glossy), perhaps 1 1/2 x 3 feet in size, is fastened to the wall just above a table top as shown in Fig. A18-3. Set up a partition two or three feet long to divide the background into two separated fields of view. At the center of the left field put a piece of non-glossy black paper a few inches square. Two paper tubes blackened on the inside are arranged so that through them the observer sees only the two areas (the left one black, the right one, white) indicated by dashes. The entire background on both sides is dimly illuminated by room lights. In addition, the left area is brightly illuminated with a shielded lamp so placed that the partition blocks its light from the right side. The amount of light falling on the left side is controlled by blocking off more or less light with a neutral density filter.

The observer first looks through both the restricting tubes, which he should be able to do with one eye, and matches the brightness of the two patches by adjusting the illumination on the left. Such a match can be made rather accurately -- (improved matching can be made by means commonly used in photometers). Obviously the illumination on the left must be far greater than on the right for this match. After the match is made, look at the two views unrestricted by the tubes. The black patch appears black; the white area on the right appears white. Yet from the equation made by looking through the tubes we know that the black area is actually reflecting as much light to the observer as the white of the other view.

H. Binocular vision

In the course of vertebrate evolution the two eyes have moved from the sides to the front of the head. This development has resulted in an increase in the extent to which the visual fields seen by the two eyes overlap each other until, in man, more than three quarters of what is seen by one eye is seen by the other also. The manner in which the two eyes in man cooperate to produce single vision with the attendant phenomenon of stereoscopic depth forms an interesting tale, several aspects of which we shall outline briefly.

Normally in order to fixate on a target with one eye, we turn the eye so that the target image falls on the center of the fovea of that eye. In binocular vision both eyes are so turned. This fixation target and the centers of rotation of the two eyes determine a circle called the Vieth-Müller horopter. (Actually the measured horopter is not always circular, but may be assumed to be so for present purposes). A visual stimulus arising from any point on the horopter produces images at corresponding locations on the two retinas as shown in Fig. A18-4. (Pairs of retinal points are called corresponding if they are equidistant from the centers of their respective retinas and at the same azimuth). A visual stimulus arising from any point in the plane of the horopter but not on it produces non-corresponding images. Thus in the figure, the fixation target P produces images P' and P" at the centers of the two foveas. Point B on the horopter produces corresponding images B' and B". Point A beyond the horopter, on the other hand, produces non-corresponding images A' and A". Similarly point C, inside the horopter produces non-corresponding images C' and C". Images A', B', and C' coincide but A" is nearer P" and C" farther from P" than the images for the other eye.

Prove that images B' and B" are equidistant from P' and P" respectively.

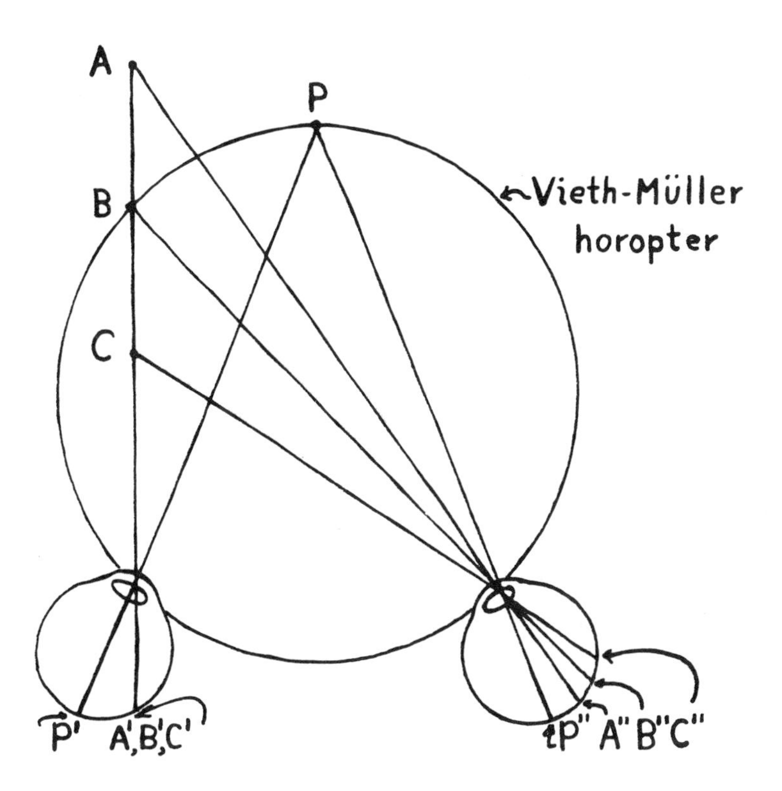

FIG. A18-4

The Vieth-Müller horopter.

The degree and sign of this non-correspondence (binocular retinal disparity) provides an important clue to the distance of an object from the observer relative to the fixation target. For objects sufficiently close to the horopter the images of both eyes fuse and a single image is seen. If, however, the object is far from the horopter, diplopia (or double vision) sets in since the degree of binocular disparity of the stimulation is too great to permit fusion to occur.

A stereoscope is usually thought of as a means of producing depth effects, but it is fundamentally a tool used to "unhook" the usual relations that stimulus objects normally set up on the two retinas. With this instrument two separate visual fields, one to each eye, can be manipulated independently. Martin (Vol. II, pp. 151-152) shows the optical system for three varieties of stereoscope. Actually, it is possible for many people with a bit of practice to view stereoscopically suitable pairs of pictures without a stereoscope. The following stereoscopic figures should be tried. (Those who cannot obtain satisfactory results could copy the figures -- except for the dot picture -- on cards and view them in any available stereoscope). In order to view these figures, it is suggested that the observer use a sheet of cardboard to block the right hand picture from the left eye and the left hand picture from the right eye. The pictures should be about equally illuminated and viewed at about ten inches. With a little effort, the pictures seen by the left and right eyes respectively, may be made to fuse. First, try to observe Fig. A18-5 in this way. If the figures are successfully fused, the dots should stand in front of or behind the square.

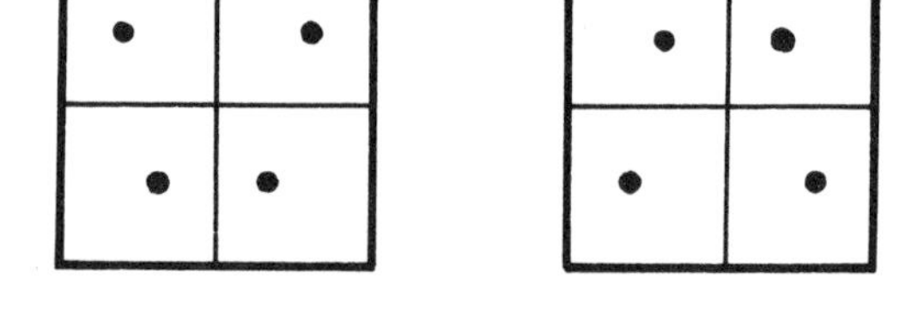

FIG. A18-5

Simple stereoscopic picture to view without a stereoscope.

It is interesting to note that when two visual fields have only non-congruent contours, almost any apparent relation can be set up between them. Thus, when patterns a and b of Fig. A18-6 are presented to the left and right eyes respectively, either as above without a stereoscope or else with a stereoscope, almost any possible arrangement can be seen. If the lateral separation of the two stimulus patterns is varied, one can see successively patterns c through g; if the patterns are displaced vertically, one can see patterns h through l. In this case, each eye is reasonably free to move about and fixate separately on many points of the figure. When, however, congruent contours are presented to the two eyes, such as two parallel lines (one to each eye), a single line will be seen for a wide range of lateral separations of the stimulus patterns. Congruent contours provide stimuli that produce a "compulsion to fusion" and the two eyes are no longer free to move separately; instead they tend to move to positions in the sockets so that corresponding retinal meridians are stimulated by congruent contours and result in single vision.

That correspondence is not necessarily perfect (binocular disparity zero) when fusion occurs, however, can be seen from the demonstration of Fig. A18-7. If cards like a and b are viewed stereoscopically by the left and right eyes respectively, the square provides a set of congruent contours, which produces a compulsion to fusion and for many lateral and vertical separations of the two cards a single square will be seen. However, the inner crossbars do not provide fusional stimuli since they are not congruent. By varying the separation of the two cards, the appearance of the binocular view may be changed so that patterns c through g or others may be seen. When the vertical halves of the crossbars do not appear to line up, as in f, a horizontal binocular disparity must exist; similarly when the horizontal halves of the crossbars do not appear to line up, as in g, a vertical binocular disparity must exist. The fact that a single square can be seen at the same time that some disparity is actually present indicates that where congruent contours exist, sizeable disparities can be masked by the strength of the fusional processes.

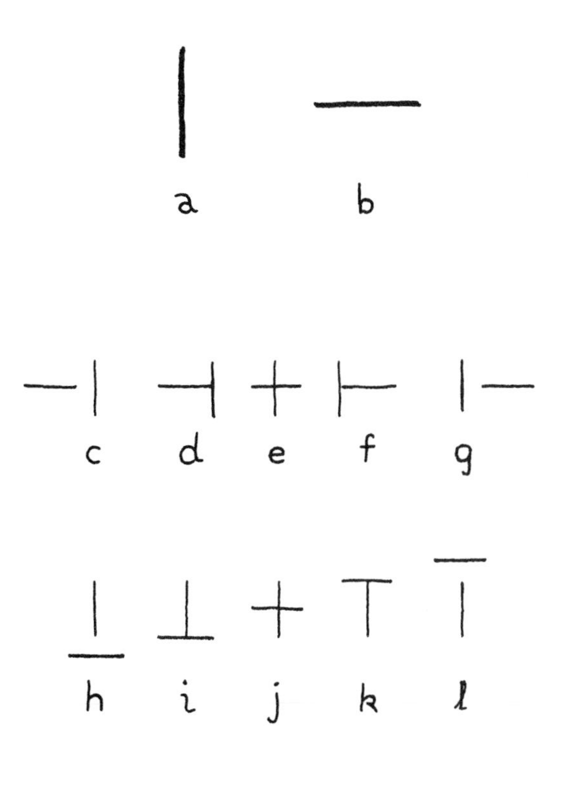

FIG. A18-6

Non-congruent contours. Patterns a and b are presented to the left and right eyes, respectively. Shifted laterally, a and b appear as c - g; shifted vertically they appear as h - l.

To observe how variations in binocular disparity may produce variations in depth present cards a_2 and b (Fig. A18-8) to the left and right eyes respectively and adjust them so that A' and B' appear fused. Then slip card a_1 in the left view so that A and B are also fused. Moving a_1 laterally to the left and right will then make the fused image of A and B appear to move toward and away from the observer while the fused image of A' and B' remains fixed.

Many more complex results can be produced with a stereoscope. One particularly interesting stereoscopic pair is shown in Fig. A18-9. The figure seems to be a meaningless jumble of dots. But under bright light, and, with effort and patience, about one person in five can see the three dimensional model. The picture was provided through the courtesy of the Westinghouse Air Arm Division.

The Pulfrich phenomenon illustrates an important property of the visual system. Attach a small white disk on a black background to a pendulum say 6 to 12 inches long. The pendulum must oscillate in a plane and should swing for a dozen or more cycles before coming to rest (a clock driven pendulum is ideal). Use bright illumination and observe the motion of the white dot from a direction normal to the plane of oscillation. The dot obviously executes simple harmonic motion in a plane. Now use a neutral density filter (two pieces of polaroid rotated with respect to each other make a good filter) in front of one eye. The white dot now appears to oscillate out of the plane of its motion in a nearly elliptical path; it appears to be closer to the observer than normal in one half of the cycle and farther than normal in the other half. The apparent deviation from the plane of oscillation increases with the density of the filter.

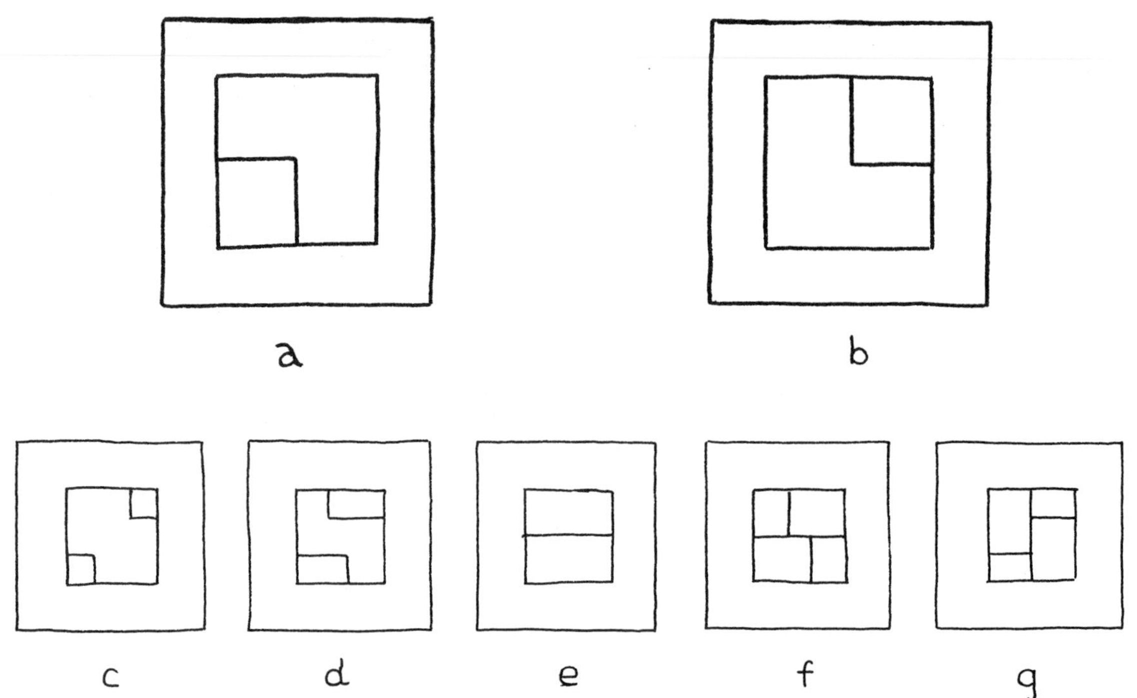

FIG. A18-7

Stimulus patterns a and b produce imperfect correspondence patterns c - g.

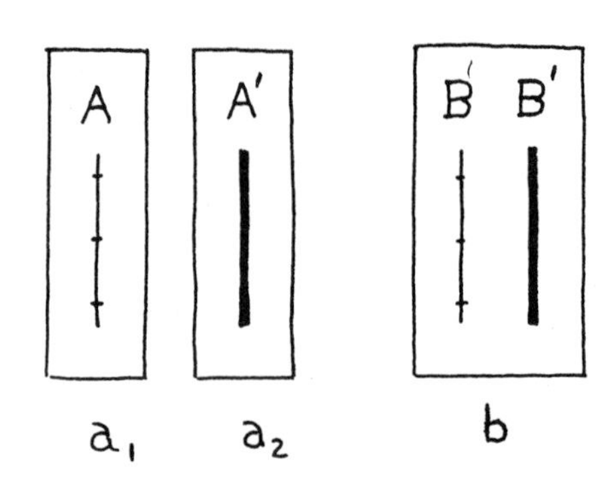

FIG. A18-8

Patterns to show depth variations due to binocular disparity.

This effect is explained as follows: the nervous response to the retinal stimulus takes some time to occur and this time increases with a decrease in the intensity of the stimulus. As the disk moves across the visual field the retinal stimulus arrives simultaneously at pairs of corresponding locations in the two eyes and moves successively to other pairs. Since the eye with the filter in front of it receives a lower intensity stimulus, however, the response aroused at a given location in that eye will occur later than the response aroused at the corresponding location in the other eye, and the time difference will be fixed for each pair of corresponding points. Thus, simultaneously occurring nervous responses in the two eyes are produced by nonsimultaneously presented stimuli, and since the disk is moving across the field with simple harmonic motion, the binocular disparity for simultaneously occurring responses in the two eyes will be the same simple harmonic function. Since the sensation aroused at a given moment depends on the pair of nervous responses at that moment, the apparent depth will then seem to vary from moment to moment as the disk moves across the field.

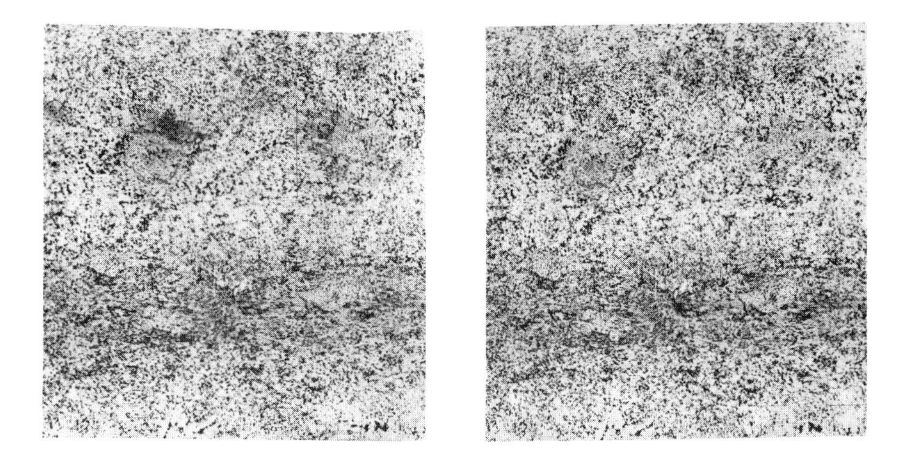

FIG. A18-9

Pattern recognition by stereoscopic viewing. See text for instructions. (Courtesy Westinghouse Air Arm Division).

I. Eye movements

Each of our eyes is suspended by six muscles whose coordinated contraction and relaxation enable us to turn our eyes in their sockets. When we fixate a particular stimulus point these muscles are in a certain state of contraction. Just as we cannot hold our arms straight out horizontally without their quivering, so our eye muscles are not perfectly steady either. The resulting small eye movements, called involuntary nystagmic movements, play a most important part in vision.

If we fixate a small spot of light, the retinal location at which the image falls changes continuously. Normally we are unaware of this fact. But if the spot is observed in otherwise complete darkness, it is found that after several seconds, the spot apparently begins to wander about randomly over rather large distances. This is the autokinetic effect. It was first noted by astronomers who observed it as a small wavering of stars in the night sky; recent work by psychologists has shown it to be a product of involuntary nystagmus.

How would you account for the immediate disappearance of autokinetic movement when room illumination is turned on?

It is possible to construct an optical system which will make the image of the spot of light fall on an unvarying position on the retina.

The observer wears a special contact lens with a small plane mirror imbedded on its side. The incident light beam reflected by this mirror (which turns it through twice the angle that the eye moves) is directed through a system of 0.5 angular magnification and then into the eye. When the image is thus stabilized on the retina, autokinetic movement ceases.

A further interesting result is obtained when the image of some object is stabilized on the retina. After a few seconds, the contours disappear and the observer sees only a uniform field of illumination. The probable explanation is that most of the neurons in the optic nerve transmit only changes in illumination; a very much smaller number signal steady illumination falling on the receptors to which they are attached. The nystagmic movements which normally change the position of the image on the retina thus shift contour lines from one set of receptors to another thereby continuing to generate a signal which the neurons of the optic nerve transmit.

EXPERIMENT A19. DEMONSTRATION OF COLOR BY CONTRAST.

By Frederic R. Stauffer

REFERENCE: E. H. Land, "Experiments in Color Vision," Scientific American, Vol. 200, pp. 84-99, May, 1959.

The perception of color by the human observer is a complicated and rather subjective process. Color is primarily identified with wavelength and the latter can be accurately determined by purely physical instruments, the eye entering only to the extent of reading scales, etc. However, when a light source is directly observed with the unaided eye as the detecting element and the mind as the data processing mechanism, we have a far more subjective situation for here the psychological factor enters. The color depends both on the source and on the observer.

Color or the sensation of color, may be perceived because of the abundance of that color, or because of the over-abundance of a complementary color. For example, green may be seen on a white card if (1) a green light is incident upon the card or if (2) white light and red-rich white light are used in a suitable manner. No green, other than that present in the white light is used. Yet green is clearly seen.

This demonstration is for the purpose of familiarizing the student with some of the simple things he sees each day, but perhaps does not notice. A more complete discussion, with application to three color photography, is given in the reference.

APPARATUS

1) Two flashlights.

2) White card or paper 8 1/2 x 11 in.

3) Red cellophane or a piece of red glass to cover one flashlight. Also green cellophane or glass.

4) A cylinder about 1 in. dia. and 2 or 3 in. high (flashlight battery is suitable).

PROCEDURE

Secure the red cellophane or glass to one flashlight. Then prop the card on a table top, arranging the flashlights and cylinder as shown in

Fig. A19-1. The distances and angles are not critical and should be adjusted for optimum effect during observation. Room illumination should be a minimum.

First, turn on the white flashlight. A sharp shadow of the cylinder is cast on the card, with the region surrounding the shadow being white. Turn off the white light and turn on the red light. The shadow, displaced from the first one, is, of course, surrounded by red light. Note that the region of the "red shadow" lies on the part of the card previously illuminated by white light, and the "white shadow" region is now illuminated by red light. Study the shadow regions and surrounding regions carefully with either one light or the other, not both, turned on. There is no abundance of green light to be seen.

Now turn on both lights and observe the shadow regions. The former "white shadow" is red, as expected, since the red light falls in that region. The former "red shadow", illuminated only by white light, appears green. By adjusting the lights, the shade of green can be varied.

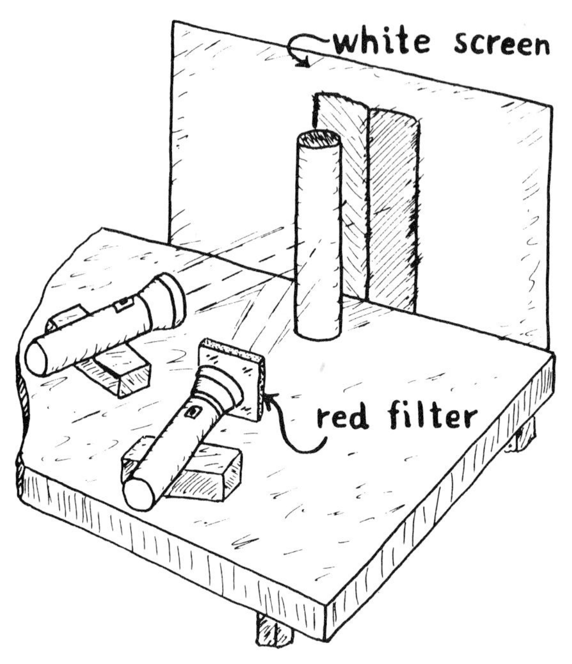

FIG. A19-1

Arrangement to observe color by contrast.

Try the same demonstration, substituting green cellophane or glass for the red.

A very striking demonstration of this same effect may be made by using somewhat more elaborate equipment. If two slide projectors are available, they could be used in the following manner:

First, make a slide, with an opaque patch in the middle. Then make a second slide, opaque everywhere except in the middle, where a transparent patch is made, identical to the opaque one on the first slide. Cover the projection lens of one projector with a red filter. Use the slide with the opaque patch in this projector. Using the other slide in the unfiltered projector, illuminate a screen with both projectors, fitting the white patch into the dark area caused by the opaque patch of the red projector. By adjusting intensities, the white patch area appears brilliantly green.

QUESTIONS

(1) With no abundance of green present, why does the "red shadow" appear green?

(2) What color is the "red shadow" region?

(3) How can you prove your answer to question (2)?

(4) If a color photograph were made of this demonstration, how would the shadow regions appear? Describe in some detail.

EXPERIMENT A20. DEMONSTRATION OF THE USE OF MOIRE FRINGES IN THE DETECTION OF ULTRA SMALL MIRROR ROTATIONS.

REFERENCES: R. V. Jones and J. C. S. Richards, Recording Optical Lever, Journal of Scientific Instruments, Vol 36, pp. 90-94 (1959); M. B. Stout, Basic Electrical Measurements, 2nd ed. (1960, Prentice Hall), Ch. 4.10, 4.32, 5.13.

This demonstration illustrates the nature of moire fringes, a purely ray optic phenomenon, and shows how they may be utilized in making measurements of extraordinary delicacy. In precision apparatus moire fringes can reveal mirror rotations of 10^{-10} radians or linear displacements of conceivably 10^{-12} cm -- smaller than the diameter of an atom! Stated another way, a telescope with a 5 millimeter diameter objective can, at best, distinguish stars as close together as 10^{-4} radians, yet a mirror of this size is used in conjunction with moire fringes to measure rotations of one millionth this angle. (There is, of course, no violation of wave optic principles, for here we measure an angle, we do not distinguish two sources subtending a very small angle). One very sensitive well known infrared detector, the Golay cell, makes use of moire fringes.

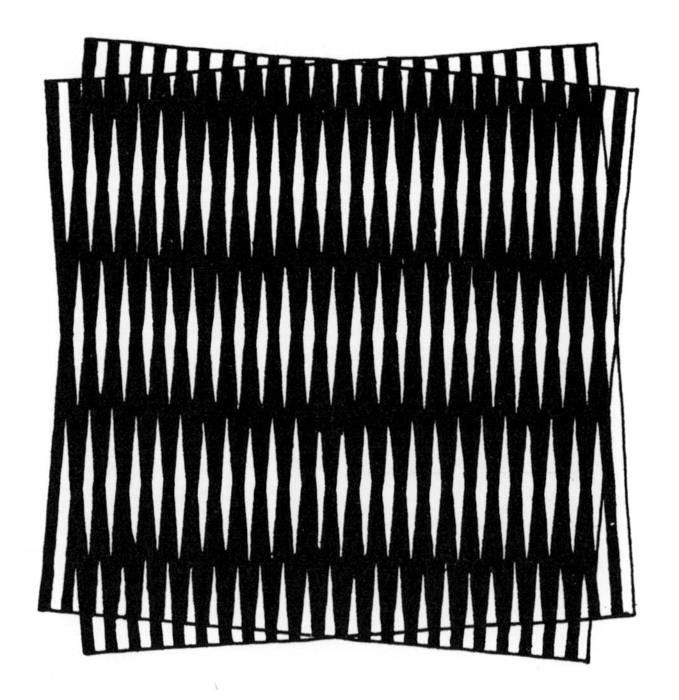

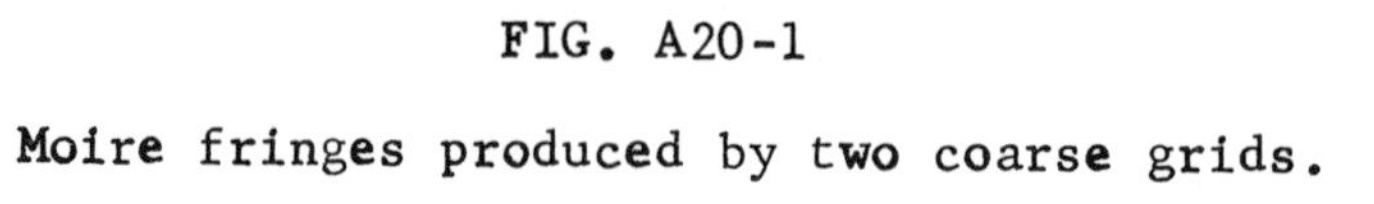

FIG. A20-1

Moire fringes produced by two coarse grids.

THEORY

Moire fringes are produced by the overlapping of two grid patterns or of one grid and the image of a grid. The grids themselves ideally consist of straight opaque stripes separated by transparent stripes of equal width sometimes called Ronchi rulings or Levey screens. They are made with high precision either by a photographic process or by a ruling process. Figure A20-1 shows the appearance of two such overlapping grids slightly tipped with respect to each other. The grids in the illustration are very coarse so that the structure of the broad, horizontal moire fringes is evident. The black moire fringes are separated by more or less transparent bands for which the transmission rises to a maximum of 50% (which is, of course, the transmission of a single grid). It is evident, then, that these fringes are a purely ray optic phenomenon having to do with the geometry of the grids. It is true that interference and diffraction may also be involved, especially when the grids have many lines per inch, but that aspect is not involved in the present discussion. The width of the moire bands depends upon the angle made between the two grids and the number of stripes per unit width of the grid. The smaller the angle and the less stripes per unit width, the wider the moire fringes will be. If one of the grids is translated in a direction at right angles to its stripes, the moire fringes move up or down. (Which way must the grids be moved in order for the fringes to move upward?)

Parenthetically, it is interesting to note that moire type fringes can often be seen between two fly screens, transparent organdy curtains, or various grilles. The fringes in such cases are often curved for the two grids are neither in the same plane nor parallel. If the grids have stripes running in both directions, there are two sets of moire fringes possible. The fringes are visible even though the opaque and transparent stripes are of unequal width -- only the contrast of the fringes is reduced. Moire fringes may even be seen between two parallel snow fences.

Suppose now that the two precision grids are accurately parallel. The transmission then obviously depends upon the "phase" between the two grids, that is, the relative degree of overlap of the two sets of stripes. If the black stripes are superposed, the transmission will be maximum, that of a single grid, or 50% on the average. If the black stripes of one grid overlie the transparent stripes of the other, the transmission is obviously zero. Figure A20-2(a) shows an end view of two superposed parallel grids. The upper grid is displaced a distance x with respect to the lower grid. The grid stripes have a periodic spacing of a, and parallel light incident from below is transmitted through the clear spaces of width b, common to the two grids. The transmission through the grids is $T = b/a = (a/2 - x)/a = 1/2 - x/a$ for values $0 \leq x \leq a/2$. For larger values of x, between a/2 and a, the transparent region is determined by the left edge of the upper opaque stripes and the right edge of the lower stripes. Here $T = b/a = (x - a/2)/a = x/a - 1/2$. For the complete period, $T = |1/2 - x/a|$ for $0 \leq x \leq a$. The transmission as a function of x is shown in Fig. A20-2(b).

A pair of precision grids may be used to measure distances of translation accurately. Suppose, for example, one wishes to make accurate measurements of the translation of the carriage of a lathe along its longitudinal ways. With the aid of the grids, a few simple optical parts, and some electronic components, distances of say ten inches can be measured with an error of less than 0.001 inch using grids with 100 lines per inch. Figure A20-3 shows a simplified diagram of how this measurement might be effected. The counter records the integral number of fringes moved and the meter the fraction of a fringe. Since two positions of the grids give the same transmission, one must note whether the photocell signal is increasing or decreasing during the fractional fringes at the beginning and end of the translation. This scheme with suitable refinements can be used for automatic machining systems.

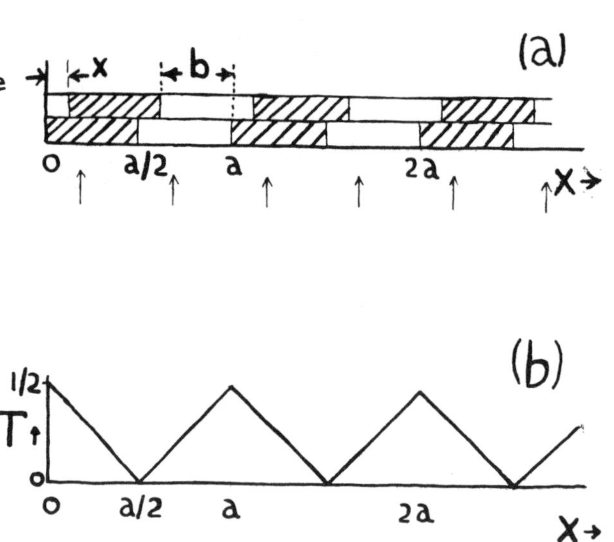

FIG. A20-2

(a) geometry to calculate the transmission of partly overlapping parallel grids; (b) the theoretical transmission.

So far we have assumed perfect grids; in practice, of course, errors occur. Small random errors in some of the lines have little effect. For instance, suppose one of the grids in Fig. A20-3 had one line displaced a/2 from its proper position but no other errors. If the smaller grid on the movable carriage had 100 lines per inch and were 2 inches long, the positional error would be 1/2% or 0.0005 at most. Such gross errors do not occur, but smaller ones are inevitable. Of course, the measurement of translation can be no more accurate than the average accuracy of the grids.

Small angular deflections can be measured. Suppose that an illuminated grid is placed at the focus of a lens behind which is a mirror. The light striking the mirror consists of a set of parallel beams one from each point of the object grid. The reflected light, passing again through the lens is imaged in the same plane as the grid itself, and is the same size. If the illuminated grid is moved to one side of the axis so that the image is formed on the other, then a second grid (or extension of the first) will or will not transmit light according to the phase of the image with respect to the second grid. A small rotation of the mirror will change light to dark. Figure A20-4 shows the basic principle involved.

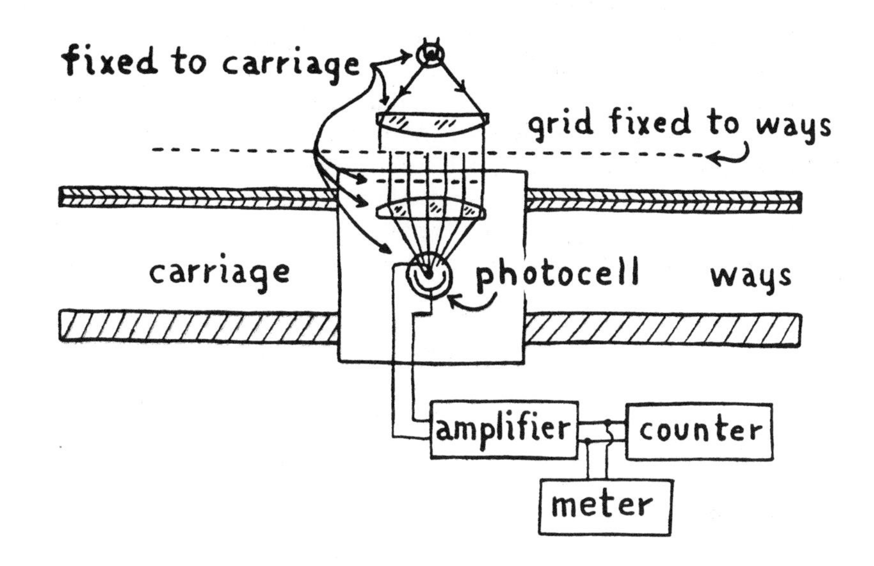

FIG. A20-3

Method for measurement of lathe carriage translation by moire fringe technique.

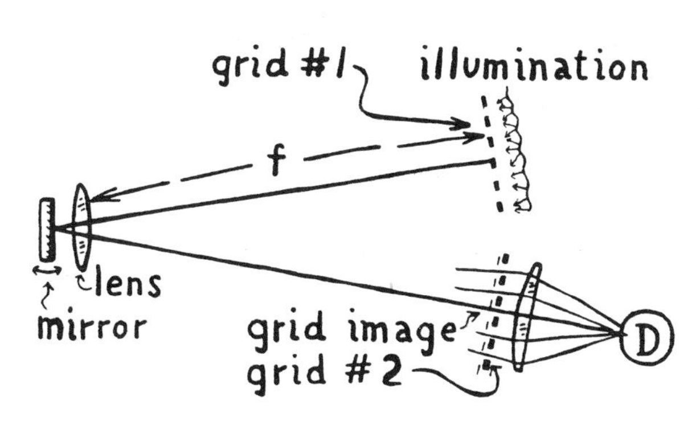

FIG. A20-4

Basic principle of small angle measurement.

If there are n lines per unit length on the grid and the focal length of the lens is f, a mirror rotation of $\Delta\theta = 1/2 \times 1/2n \times 1/f$ will cause the transmission to change from maximum to minimum or vice versa. The first factor of 1/2 occurs since the mirror doubles the angular deflection of the beam. The other two factors express the radian measure of the required displacement of the image. This scheme is very sensitive to small mirror rotations, but it is also very sensitive to small changes in illumination and detector sensitivity.

Considerable improvement in stability and also in sensitivity results from the use of two detectors arranged behind two grids which are out of phase so that if one grid transmits the other does not and vice versa. Figure A20-5 shows the complete system. Light from the source S is focused by lenses L_1 on M_1. (We ignore C_1 and C_2 for now). An image of the grid is formed by L_2 partly on the rest of grid 1 and partly on grid 2 which is out of phase with grid 1. Light passing out of the lower part of grid 1 is reflected by mirror M_2 through lenses L_3 to photocell P_2; light passing through grid 2 is reflected by M_3 through L_3 to photocell P_1. The two photocells form two arms of a Wheatstone bridge shown in the lower part of the figure. Mirror M_1 is attached to anything whose rotation is to be measured. The magnetostrictive device shown at the left of the figure is suitable. Properly set up the optical lever -- electrical bridge circuit is capable of giving full scale galvanometer deflection for very small magnetization of the iron. By direct measurement with conventional telescope and scale, the rotation of the mirror is all but unobservable.

APPARATUS

1) Beeswax and rosin mixture, stove to heat it, and Variac to control the heat.

2) Medicine dropper to apply wax.

3) Soft tackiwax.

4) Heavy aluminum (Dural) or steel base plate, 1/2 x 12 x 20 in. or larger.

5) Set of blocks for mounting optical components, either metal or wood, in the shape of rectangular prisms.

6) 6-8 volt auto lamp 21 cp. with small filament.

7) 2 plano-convex lenses, 1 1/4 in. dia., 3 in. focus.

8) 2 plano-convex lenses, 1 1/4 in. dia., 2 in. focus.

9) Achromatic doublet, 1 in. dia., 10 in. focus.

10) First surface mirrors (may be cut from one larger piece), 1/16 in. thick.

    A. 3/4 x 3/8 in.

    B. 3/4 x 3/4 in.

    C. Two 1 1/2 x 2 in.

11) Ronchi ruling, 50 to 150 lines per inch, for grid (a 4 x 4 inch piece diagonally ruled -- can be cut up). The minimum sizes required are approximately:

    A. 1 1/2 x 2 in.

    B. 1 1/2 x 3/4 in.

C. 1 1/4 x 3 in. (for testing).

The ruling in each case must be parallel or perpendicular to the edges, not diagonal.

12) 3 pieces of 1/8 - 3/16 in. thick plate glass.

A. Two compensators - 2" x 3".

B. One mount for grids - 4" x 2".

13) Two Clairex CL-3 photoconductive cells or the equivalent.

14) Sensitive galvanometer: period 1-3 sec., resistance 1-10 K , sensitivity about 0.001 microamp./mm.

15) Ayrton shunt for the galvanometer.

16) Two fixed resistors, 1/2 or 1 watt, 200 K and 250 K , (carbon all right -- other values are possible, of course).

17) Decade resistance box with steps of 10 KΩ, 1 KΩ, 100Ω, 10Ω, and possibly 1 ohm.

18) Batteries:

A. 7 1/2 volts of dry cells or 6 volt storage battery for auto lamp.

B. 90 volt B battery for bridge circuit.

C. 1 1/2 to 6 volts (d.c.) for magnetostrictive unit.

Note: It is not satisfactory to use an ac supply for the lamp -- fluctuations in line voltage upset the balance of the bridge.

19) Copper and iron wire of roughly the same size -- #18-24 gauge.

20) Coil of 100 to 500 ohm resistance from a solenoid or other device.

PROCEDURE

A very practical way to set up the demonstration is to secure the various optical components to metal or wood blocks of suitable size using a smoking hot mixture of beeswax and rosin applied with a medicine dropper. After the components are located on the base plate, they are secured either temporarily with tackiwax or more permanently with the hot beeswax and rosin mixture. Errors are easily corrected by using a razor blade to remove the waxed on components -- the hard wax may be melted up again.

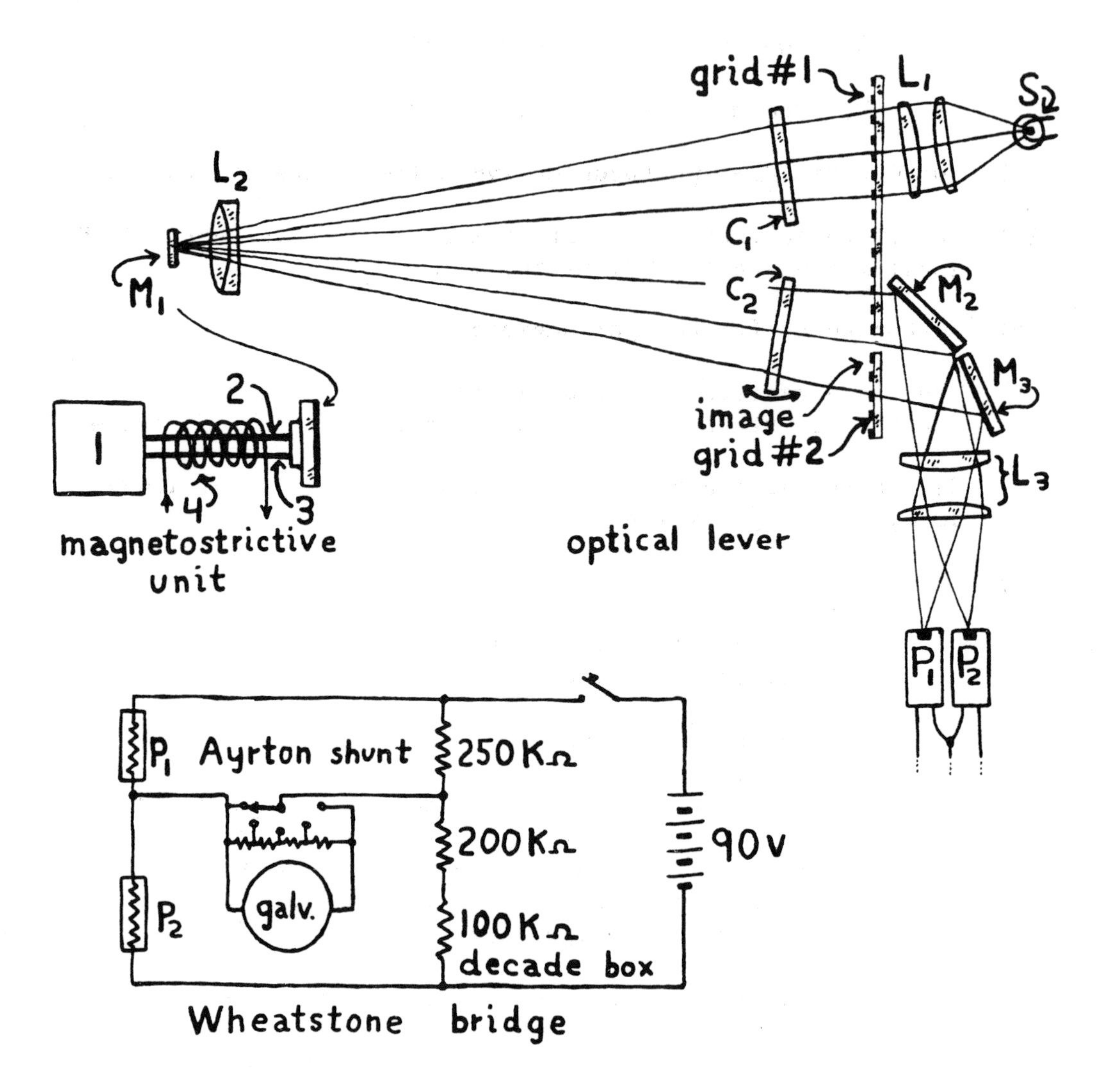

FIG. A20-5

Complete optical lever device.

The hot wax mixture is made in about equal proportions of beeswax and rosin so that when cool, it is just possible to make an impression in the wax with ones fingernail. The beeswax tempers the rosin so that the mixture is not brittle. Too much beeswax will make the mixture too soft. The wax should be applied smoking hot so that it will adhere well, but when very hot, the beeswax tends to distill off so the mixture becomes brittle. The addition of more beeswax retempers the mixture. The mixture burns, and it is well to have a cover to put out fires should they occur. About 20 cc of the mixture should be prepared, preferably melting it on a hot plate whose temperature may be controlled by a variac. (A Bunsen burner will also do).

I. Grids

The grids are mounted so that the stripes will be vertical when in use. The grids are waxed to one piece of plate glass with the photographic or ruled surface upward (not between the two glass pieces). First, mount the longer piece of grid on the glass base plate; carefully position it so that there is adequate room for the other grid and so that the base plate can be mounted on a block without injury to the grids. This first grid is carefully waxed in place by running a thin ribbon of hot beeswax-rosin mixture around the edges except for the common edge between the two grids. Next, lay the second grid carefully in place as close to the first grid as possible, but out of phase. The exact position of the second grid is determined by the use of the auxiliary piece of test grid laid over the other two, photographic emulsion face down for more intimate contact. By use of the moire fringe pattern, the correct phase relation is established as indicated in Fig. A20-6, even though it may be difficult to see the individual lines of the grids. The moire fringes should be straight over the whole surface, shifting from black to white at the junction of the grids. If there is a bend in the fringes at the junction, the stripes of the two grids (1 and 2) are not parallel. When the second grid is accurately positioned, wax it in place along three edges.

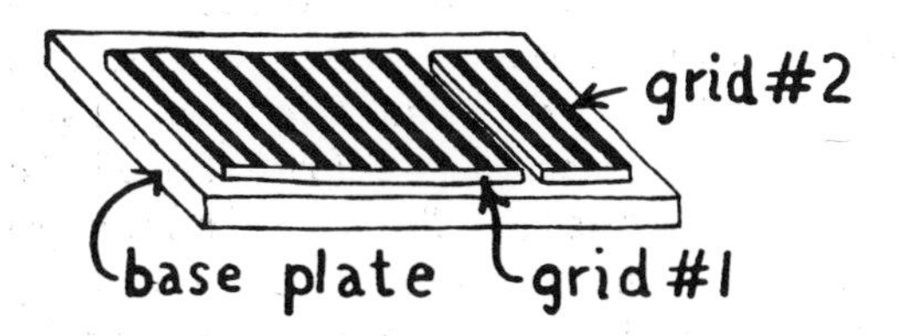

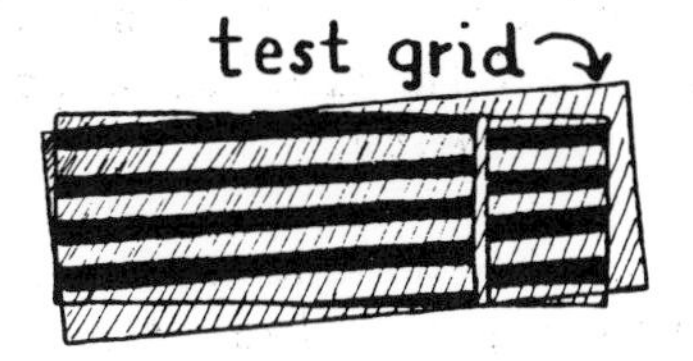

FIG. A20-6

Mounting and alignment of grids.

The grid unit is now waxed at the proper height to a block (which will later be fixed to the base plate). (The axis of the optical path should be arranged so that it is parallel to the surface of the base plate and about 2 inches above it). When waxing the unit to the block, keep in mind that the grid should be close to lens L_1 and mirror M_2 (Fig. A20-5). The emulsion surface should face lens L_2.

II. Preliminary set up

The demonstration should be set up on a sturdy table which will not shake when someone walks across the floor. Locate the heavy metal base plate so that the optical parts are readily accessible and the electrical components will also be within reach. The galvanometer should be placed where its scale is easily observed. Place the metal base plate on three corks, one midway along the left edge, and the other two in the corners at the right edge.

Wax the achromatic doublet L_2 to a support block so that its center will be at the chosen height of the optical beam axis. Keep in mind that mirror M_1 should be about half an inch behind L_2, so that the support block should be mostly to the right of the lens. Note the proper orientation of this lens. Wax the two plano-convex lenses of longer focal length to a mount to make L_1. (The focal length of the combination should be about 2-3 inches). The two lenses may be waxed to a ring spacer for better alignment. Next, mount the auto lamp at approximately the right height on a block with its filament in a vertical plane. The lamp should be secured to its mount by means other than wax, for when it gets hot the wax will obviously soften and the lamp will move. With soft wax, mount a small mirror (item 9B in list of apparatus) on a block to serve as M_1 for preliminary adjustments.

Wax L_2 on the left side of the base plate about 4 inches from the edge keeping the lens axis along the desired line (say parallel to the front edge of the base plate). The lens should be carefully positioned to avoid astigmatism in the imaging produced. Put M_1 about half an inch behind L_2 and locate the approximate focal plane of the lens using the lamp source and a white card to find the reflected image. Put the grid unit in approximately the proper place with the emulsion side toward L_2. Locate L_1 about half an inch behind the grid and about 3/4 inch off the axis of lens L_2. Find the proper location of the lamp S so that an image of the filament is formed on M_1. Tilt M_1 so that a reflected image of the grid falls symmetrically on the grid, both at the right height and so that the center of the illuminated circle of the image falls on the junction of grids 1 and 2. It is important that the optical system be as symmetrical as possible for good imagery. Adjust these components for optimum imagery using a white card behind grids 1 and 2 to check that minimum light is transmitted when the image coincides with either grid 1 or grid 2. It will probably be necessary to mask off part of the image of the grid because of deterioration at the edges. The mask should be centered on the boundary separating the two grids. When the positioning of these elements is as precise as possible, wax L_2, the grid structure, L_1, and the lamp unit to the base plate with the hard wax. (Do not wax M_1 in place). Tape the wires from S to the base plate at two places so that moving the ends of the wires will not disturb the lamp position at all.

III. Photocell unit and associated optics

Mount the two photoconductive cells side by side on a block at the beam height preferably using a clamp of some sort with screws rather than wax. Three wires should now be soldered to the photocells. Use stranded, color coded, hookup wire twisted into a cable (or special three conductor shielded cable if available). The cable should be long enough to reach from the cell position to the bridge circuit with perhaps a foot to spare. One stranded wire is soldered to the common junction of one lead from each photocell, and one to each of the other leads of the two cells.

Mount M_2 and M_3 on a block with soft wax so that M_2 may be as close as possible to the grid and that the block will not get in the way of the mounts for L_1 or L_3. Mount two remaining plano-convex lenses on another block (hard wax) to make L_3. The focal length of L_3 should be less than that of L_1 so that the final lamp images will be smaller than the lamp filament and will fall as nearly as possible on the sensitive surfaces of the photocells without waste. Locate the mirrors (M_2 and M_3), lens L_3, and the photocells so that images of the lamp filament fall on the two photocell sensitive areas. Mirrors M_2 and M_3 are adjusted so that the images are at the right height and separation for the detecting cells. Note that the two beams reflected by M_2 and M_3 cross. The purpose of this arrangement is to have the beams pass as nearly centrally through L_3 as practical. (Instead of M_2 and M_3, one can use a small angle prism to deviate one of the beams. The prism arrangement is described by Jones and Richards). After the images falling on the photocells are optimum, wax M_2 and M_3 in place with hard wax (do not remove the soft wax), and wax the mirror, lens, and photocell mounts to the base plate. Tape the photocell wires to the base plate at two different places so no motions of the wires disturb the cell positions.

IV. Light shielding

Blackened cardboard or wood blocks (flat black Krylon spray or poster paint, etc.) should be placed so as to shield the light path both from stray light and drafts. Leave enough area around M_1 so that the magnetostrictive device can be put in. Also leave a gap about 4 inches wide near C_2 so that this plate (see below) can be adjusted. A large sheet of heavy, blackened cardboard should be used for the top. Be sure that light from the lamp cannot directly reach the photocells and is not scattered to them. It is best to make at least part of the shielding removable for minor adjustments.

V. Wheatstone bridge circuit

Set up the Wheatstone bridge circuit as indicated in the diagram (Fig. A20-5). The Ayrton shunt (see Stout) is used to reduce the bridge sensitivity while keeping the galvanometer properly damped. A 90 volt B battery is used for the bridge source, though 180 volts would be suitable across the two photocells.

Since the image of the grid formed by L_2 has arbitrary phase with respect to the remainder of grid 1 and grid 2, a pair of compensating plates, C_1 and C_2 are used to restore the phase. These plates are simply located in the beam near the grid unit where the incoming and return beams do not overlap. Compensator C_1 may be waxed to the base plate normal to the beam direction through it and C_2 left free for adjustment to displace the beam slightly either direction.

Set the Ayrton shunt so that the sensitivity of the galvanometer is zero. Check the bridge circuit carefully to make certain that the wiring is correct and that all connections are tight. Connect the battery and turn on lamp S. Close the bridge circuit switch and set the decade resistance box at 50 KΩ or the approximate value required to make the two resistance arms of the bridge about equal to each other. Set the Ayrton shunt for minimum bridge sensitivity and balance the bridge reasonably well by adjusting the rotation of the compensator plate C_2 which displaces the image slightly either way. When the bridge is balanced (or nearly so), increase the sensitivity with the Ayrton shunt and obtain a better balance. At some galvanometer sensitivity it will be found that the required adjustment of the compensator is too delicate to make, and at this point, make the bridge balance adjustments with the decade resistance box. It may well be impossible to balance the bridge at full sensitivity, for the exact balance point will drift slowly. But full sensitivity is not needed. Now turn the shunt to zero sensitivity, disconnect the bridge battery, and turn off the galvanometer light (if any).

VI. Magnetostrictive unit

Determine the smallest useful mirror size for M_1. The size should be about 3/4 in. high x 3/8 in. wide (though a smaller one could be used with little loss of sensitivity).

Use a solenoid coil or wind a many turn coil for the production of the magnetic field. The unit shown schematically in Fig. A20-5 is mounted on a single block, number 1. The coil may be waxed in place or otherwise secured so that the wires can be mounted horizontally at about the right height along the axis of the coil. Wires 2 and 3, which support the mirror are about 2 1/2 to 3 inches long. One wire is copper and one soft iron. They are soldered at one end to a small block of copper about 1/4 in. square and 1/16 to 1/32 in. thick. The wires should be about a millimeter apart at the copper block. The other ends, passed through the coil, are fastened to the mounting block either with a clamp or with wax. The small mirror M_1 is waxed to the copper end piece with its long direction vertical. Two stranded wires are soldered to the coil terminals so that they may be connected to the magnetostriction drive battery later.

VII. Final set up

Remove the temporary mirror unit M_1 and replace it with the magnetostrictive unit. Position the magnetostrictive unit so that the mirror reflects the grid image back to the proper place, and wax the unit in position.

Connect lamp S to its battery source, activate the bridge, and balance it as before using the compensator as much as possible. With the bridge sensitivity as high as practical (the drift of the galvanometer should be less than 1 cm per 10 seconds) energize the magnetostrictive coil with one or more dry cells. The galvanometer deflection, which is almost instantaneous, should be at least half scale, probably off scale. With the author's unit, the deflection was perhaps twice full scale when 2 volts dc was applied to the 100 ohm coil. (The mirror deflection observed directly with a telescope and scale at a distance of one meter was not more than 0.1 mm at most).

* *

The demonstration could clearly be made quantitative if desired. One would have to calibrate the optical lever in terms of the signal corresponding to one full grid displacement using low sensitivity of the bridge and making sure the deflection was proportional to displacement -- that is assuming that the bridge was not too far off balance. The magnetostrictive device would require calibration -- the magnetic field within the iron wire would have to be determined, and the length and spacing of the wires ascertained so that the change in length of the iron wire could be determined from the measured mirror rotation. The calculations, in short, are not very straightforward.

A NOTE ON THE PROPERTIES OF INHOMOGENEOUS MEDIA.

REFERENCES: R. W. Wood, Physical Optics, 3rd ed. (1934, Macmillan), pp. 82-92; M. Minnaert, Light and Color, (1954, Dover); Jenkins and White, Ch. 22.13, pp. 506-507.

This note is intended to call attention to a few of the many phenomena associated with the propagation of electromagnetic (or acoustic waves) through an inhomogeneous medium. The diverse effects all result from variation in the refractive index of the medium, but the variations may be gradual and uniform or they may be abrupt and irregular. Accordingly, we consider several special cases.

I. Smooth, continuous change in refractive index.

The atmosphere, taken on a large scale, provides the most important example of a medium whose index changes continuously. (Here we ignore the many small scale variations). Rays propagated through a medium having spherical symmetry in the variation of refractive index (as the atmosphere) follow smooth curves which lie in planes passing through the center of symmetry. Bouguer's formula, derivable from electromagnetic theory, describes the path followed for a known variation in refractive index, $N = N(r)$. The relation is illustrated in Fig. 1. The formula describes, at least approximately, the paths of light rays and radio waves in the atmosphere. On the horizon the deviation of light due to atmospheric refraction amounts to about half a degree, the angular diameter of the sun. Thus at sunset, the sun is seen just above the horizon when it has actually just sunk below it.

Other examples of smooth changes in refractive index are found in the human eye lens whose index is greater at the center than at the margin (thus reducing the spherical aberration) and the Luneberg microwave scanning lens which is a dielectric sphere whose index varies radially. (In practice, these lenses are usually made with discontinuous changes in refractive index rather than a smooth change). Wood, pp. 88-90, describes a method of making gelatine lenses in the form of cylinders which have an index which varies radially from the center line.

II. Stratified layers.

It frequently happens that the density of air in our atmosphere is stratified because of temperature effects. Such layers produce unusual variations in refractive index and they are responsible for mirages in the desert, for what appears to be water on a hot dry road, for various distortions of the sun or moon near the horizon, and for extraordinary radar propagation sometimes observed. A laboratory demonstration of a mirage is

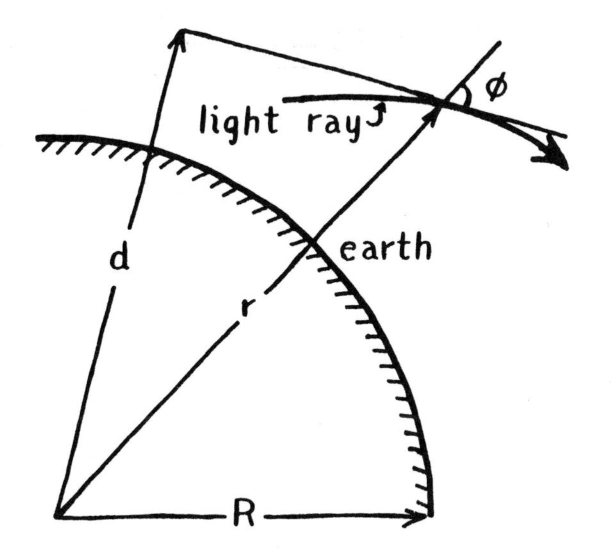

FIG. 1

Diagram to illustrate Bouguer's formula for the path of a light ray in the atmosphere: $N r \sin \phi = N d =$ constant where $N = N(r)$ is the refractive index for $r > R$ of a spherically symmetric medium.

described by Wood, pp. 87-88. This demonstration is readily assembled and clearly illustrates the nature of common mirages.

III. Small scale or random changes in refractive index.

In addition to the large scale variations in the refractive index of the atmosphere, there are small random changes associated with turbulence. These variations are responsible for the scintillation or twinkling of stars. A large portion of the turbulence apparently is associated with layers perhaps as high as 100,000 feet high. One can describe the effect in terms of weak lenses passing overhead. These weak lenses, high in the atmosphere, deviate the starlight so that the stars seem to dance about their normal positions. In addition, these lenses direct varying amounts of light from a star into a telescope objective and cause the image to go in and out of focus and to fluctuate in brightness.

A simple demonstration of this scintillation or twinkling of stars can be made with a pair of binoculars on a night when the atmosphere is rather unsteady. Focus the binoculars on a bright star, preferably low in the sky and move the objective end in a small circle so that the star appears to move in a circle in the field of view. The changes in brightness will be evidenced by changes in the ring which may even be broken up into a series of randomly distributed dots and dashes.

The next experiment will show how local variations in refractive index which result from density changes reveal airflow patterns around moving objects such as airplane wings.

IV. Scattering.

The relation between scattering and refractive index is explained in the reading. As shown there, forward scattering gives rise to the refractive index in the medium, and hence, is ultimately responsible for all the phenomena described above. Scattering in directions other than forward is also highly significant in the atmosphere, for it causes the blue color of the sky and the red of the sun at sunset. This type of scattering, perhaps in the troposphere, is of great utility in the propagation of short radio waves far beyond the horizon. (This scattering is different from ionospheric reflection or refraction by charged particles).

An interesting demonstration of "atmospheric" scattering is described by Jenkins and White, pp. 506-507.

EXPERIMENT A21. DEMONSTRATION OF SCHLIEREN AND DIRECT SHADOW.

This demonstration illustrates two ray optic methods by which very small changes in the refractive index of air may be made evident. A third method, closely related to these two, but depending on interference is mentioned briefly, but not demonstrated. The three methods, especially the two demonstrated, find important application in the study of airflow patterns in wind tunnels and are treated fully in books on aircraft design.

The three methods, direct shadow, Schlieren (German for streaks or striae), and interferometric all depend on the optical effects resulting from local variations of refractive index associated with density changes in air. Though they are all dependent on refractive index changes, the methods give different information* and supplement each other. The relationship between the methods can be illustrated by Fig. A21-1. A ray of light incident from the left would ordinarily strike the screen at point P_1. If a model is placed in the working area near the origin of coordinates, the airflow will be disturbed, and as a result there will be local changes in density and therefore in refractive index. The incident ray is deviated through a very small angle $\Delta\theta$, and will then strike the screen at the point P_2, a distance ΔQ from P_1. The direct shadow technique gives a measure of ΔQ ; the Schlieren method gives a measure of $\Delta\theta$; and the interferometric method a measure (in fringe shift) of the change in optical path, $\Delta(Nd)$. In the last expression, N is the refractive index, and d the geometric distance between an arbitrary point ahead of the working section and the screen; both quantities change, and it is the difference in the product which is measured in the interferometric method.

The interferometric method requires the use of high quality, expensive optical components. A less expensive type of interferometer is

---

*For those with more knowledge of calculus, the differences can be expressed in terms of partial derivatives: In the test region, the refractive index is a function of three coordinates (and perhaps time also). Thus $N - 1 = f(x,y,z)$. The optical path b is a function of x and y: $b(x,y) = \int_z (N-1)\,dz$. The interferometric method gives a fringe shift proportional to $b(x,y)$. The Schlieren contrast is proportional to $\partial b/\partial x$, and the direct shadow contrast is proportional to $(\partial^2 b/\partial x^2 + \partial^2 b/\partial y^2)$. The Schlieren method is independent of $\partial b/\partial y$ because a slit and knife edge parallel to the y direction is used. Of course, the apparatus or model can be rotated to interchange the x and y axes and thus obtain further information on this method.

described in Experiment B6. The interferometric method is not considered further.

Two arrangements for the direct shadow method are illustrated in Fig. A21-2. Light from a small bright flash of short duration passes through the test area, either as a divergent beam (a) or as a parallel beam (b), and falls on a distant screen. Changes in density and refractive index in the test area result in corresponding changes in illumination on the screen. Though the interpretation of the patterns obtained may not be of the simplest, the apparatus surely is.

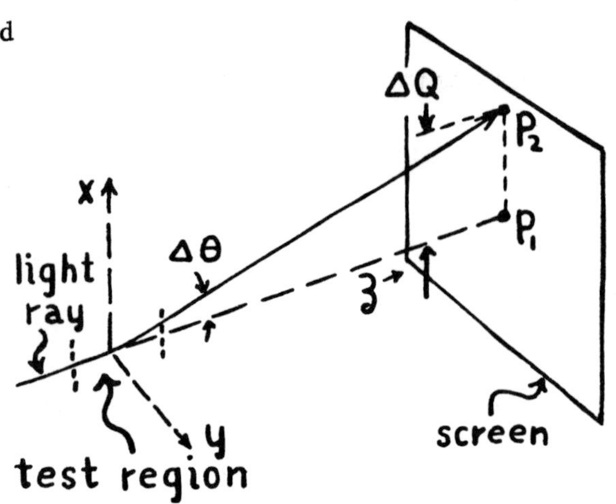

FIG. A21-1

Comparison of methods. Change of N in test region causes path difference $\Delta(Nd)$, deflection $\Delta\theta$, and displacement ΔQ.

The principle of the Schlieren method is illustrated in Fig. A21-3. With this technique, light from an illuminated slit is rendered parallel by a lens or mirror. After the parallel beam has traversed the test area, it is reconverged at a knife edge set to cut off about half of the rays otherwise transmitted to the screen. Rays deviated upwards -- as indicated in the figure -- pass over the knife edge and add to the illumination of the screen, whereas rays deviated downwards are blocked by the knife edge. Since the plane indicated by the vertical dotted line is in focus on the screen, an object placed at that point in the beam produces a sharp image on the screen. Changes in refractive index near the object, deviate the rays and cause changes in illumination at the corresponding point in the image. The image is thus surrounded by a shaded pattern indicating air density changes. This whole procedure is much like the Foucault knife edge test for mirrors (Experiment A12), but here it is the optical path that is tested, not the mirror surfaces.

Both the direct shadow and the Schlieren patterns can be demonstrated directly, or they may be photographed if a suitable short duration flash is available. With either arrangement, the sensitivity can be great enough to allow the air currents to be seen rising above one's hand because of its warmth. The air currents rising from a warm electric soldering gun give a spectacular demonstration of either smooth or turbulent air flow. With a little more elaboration, one could observe shock waves from a suitably shaped nozzle attached to the compressed air line. A small enclosed test section with glass or plexiglass windows might be very convenient for observing airflow over a model wing section, etc. One might also study the standing wave pattern of a reflected sound wave as produced in a rectangular pipe fitted with plexiglass walls.

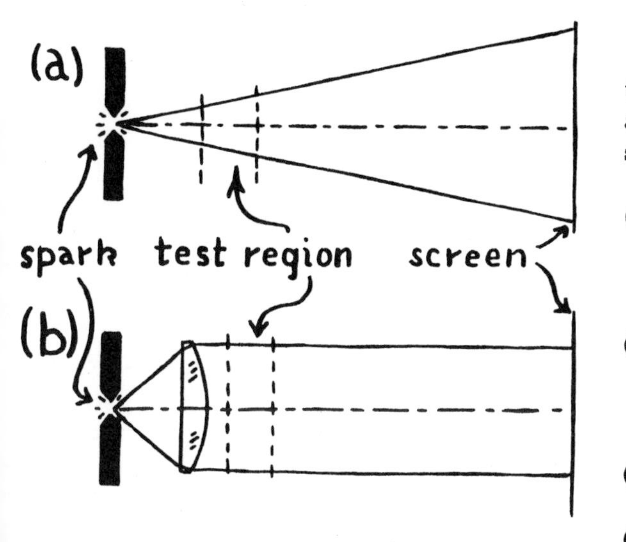

FIG. A21-2

Principle of the direct shadow technique, (a) divergent shadow, (b) parallel shadow.

APPARATUS

The required apparatus for the demonstrations is rather flexible, and the following list is merely suggestive.

(1) Two long focus mirrors or lenses 4 to 6 in. diameter. The focal lengths should be 50 cm. or more.

(2) A carbon arc or other high intensity source of small area and a 6 volt straight filament lamp.

(3) Adjustable slit.

(4) Razor blade and mount. The mount may be either a simple block of wood or it may be the unit used in Experiment A12.

If the patterns are to be photographed, one must also use some sort of high intensity flash tube. The tube will, of course, require some sort of electronic power supply. A few details are given in Figs. A21-4 and A21-6. Alternatively, it might well be possible to use ordinary photographic flash bulbs -- the brightest available. In addition, of course, one needs a camera -- a 4 x 5 in. view camera is excellent, though almost any camera could probably be used. Very fast film should be used. A pair of condenser lenses will be needed to focus the flash tube light on the slit for Schlieren pictures, and a simple lens to image a continuous source on the flash tube.

PROCEDURE

I. Direct shadow demonstration

For visual observation of the direct shadow technique, the procedure is evident from the figure. It is desirable to have the screen 20 feet or more away from the small bright source -- carbon arc or the equivalent. The illumination is better if a lens or mirror is used to make the light beam parallel. It is best to block off all the unwanted light in order to make the pattern readily visible.

For photographic purposes, the flash device is used, and if the room is completely darkened, the camera aimed at the area on the screen to be illuminated, may have its shutter left open for a few seconds prior to the flash. In this way, synchronization is not needed, and indeed with a flash tube, the synchronization problem might be difficult. Unless the flash has both very small dimensions (or a small hole is used to restrict its size) and is of very short duration (less than a millisecond), the photographs will show very little detail.

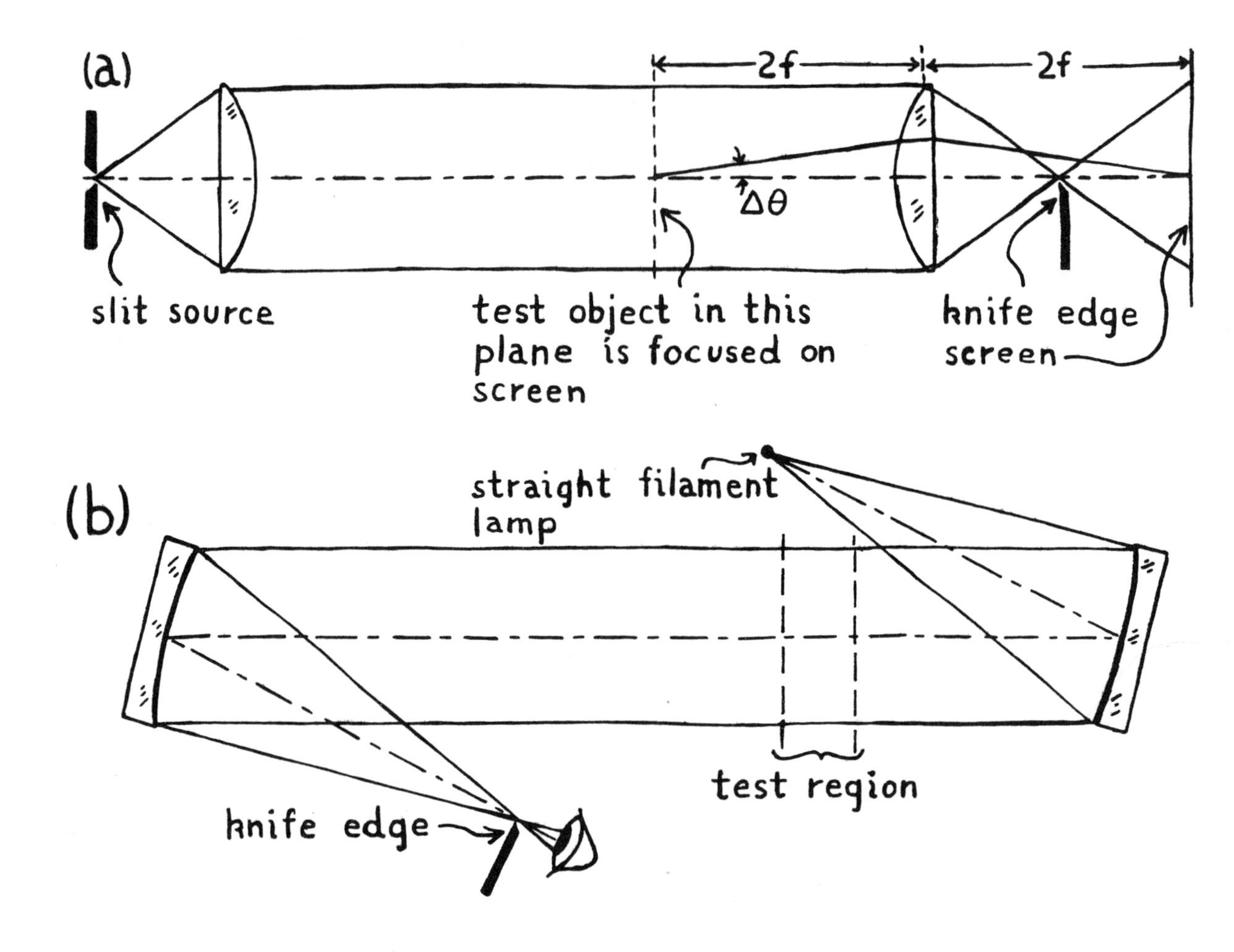

FIG. A21-3

Schlieren method; (a) with lenses, (b) with mirrors for visual observation. In (a) a positive deflection $\Delta\theta$ gives a bright spot on the screen; a negative $\Delta\theta$ gives a dark spot.

II. Schlieren demonstration

For visual observation of Schlieren patterns the optical system should be set up so that the light travels several meters from the test area to the knife edge. A straight filament lamp should be used as the source. It is placed at the focus of the mirror a little to one side of the projected beam. About three or four meters away, the second mirror is arranged to reflect the parallel beam to a focus on the other side of it. The patterns are directly observable at the knife edge or they may be projected on a ground glass screen. In the latter case, the position of the test area must be such that it is in focus on the screen. The necessary conditions are illustrated in Fig. A21-3.

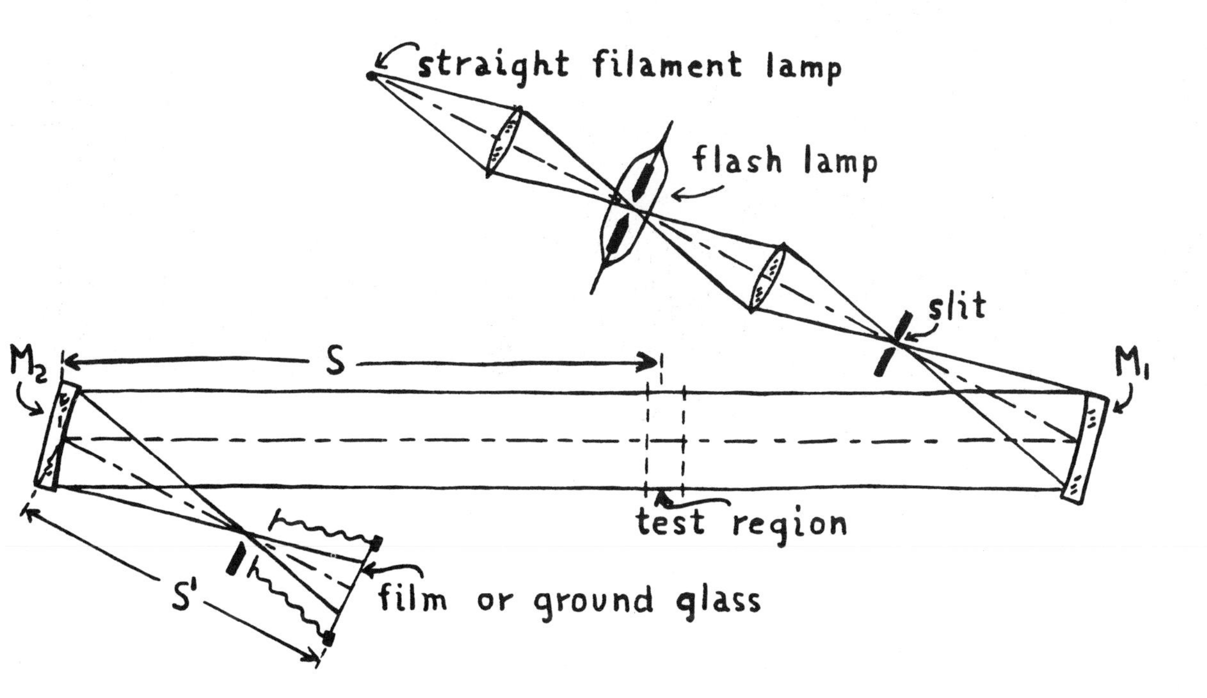

FIG. A21-4

Schlieren method adapted to photography.

A convenient arrangement of apparatus to photograph Schlieren patterns is shown in Fig. A21-4. The straight filament lamp is used for alignment purposes; it is focused at a spot between the electrodes of the flash lamp or at the spot where any other flash source would be put. The slit is needed to restrict the light rays to a narrow line. The lens board of the camera is removed if possible and the image focused on a ground glass plate at the position of the film holder. Black cloth or paper is used to cut out all unwanted illumination. The knife edge is carefully adjusted to give the optimum pattern, and the lamp then turned off.

At this point, a cardboard shutter should be held over the opening for the camera, and the film holder opened. A couple seconds before the flash takes place, the shutter is removed. After the flash, the shutter is, of course, replaced and the film holder covered.

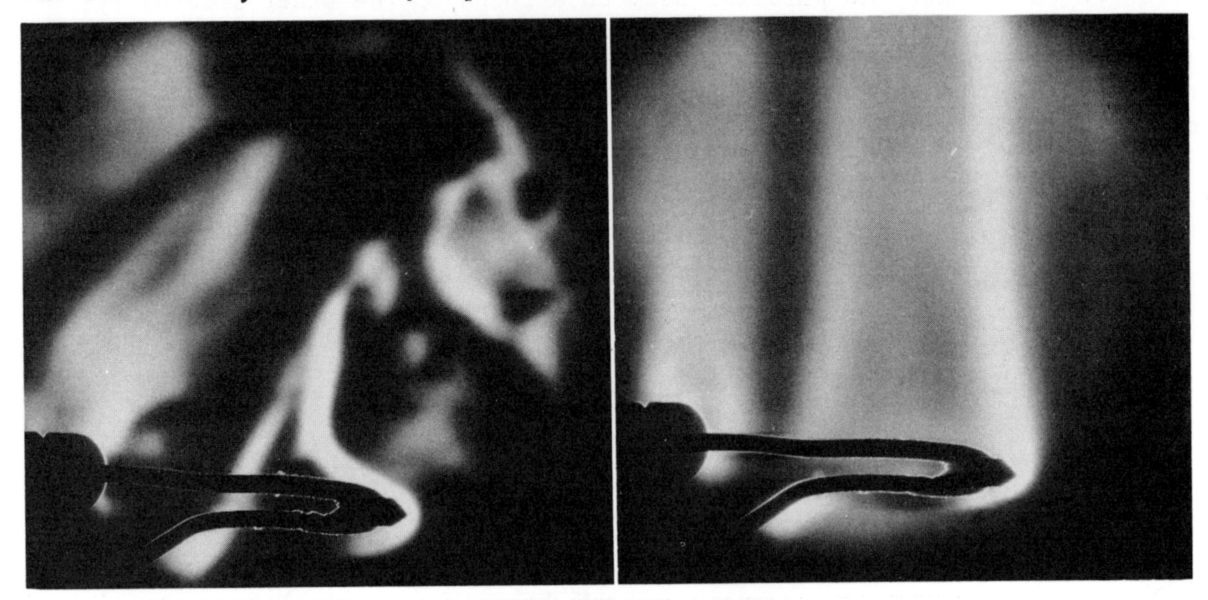

FIG. A21-5

Schlieren patterns for soldering gun:
(a) turbulent flow, (b) laminar flow.

As a preliminary test object for the system, a hot soldering iron or soldering gun is convenient. Figure A21-5 is a picture of the results obtained with a G.E. type FT 230 flash tube.

APPENDIX: Flash lamp circuits

A circuit suitable for operation of the G.E. type FT 230 flash tube is shown in Fig. A21-6. Peak currents of the order of several thousand amperes are involved in the discharge of the condensers through the flash tube, and consequently any appreciable resistance in this part of the circuit will be detrimental to the energy of the flash. It is, therefore, important to connect the flash lamp and the condensers with very heavy wire -- heavy braided ground strap wire is satisfactory. Even so, the length of these wires should be kept to a minimum. The remainder of the circuit may be put together with ordinary wire. Any direct current source which supplies the requisite high voltage, (2 1/2 kilovolts for the tube mentioned) is satisfactory. Detailed instructions on firing the tube come with it. A wire ring wound around the waist of the tube and triggered with the output of an old Ford coil can be used to trigger the tube.

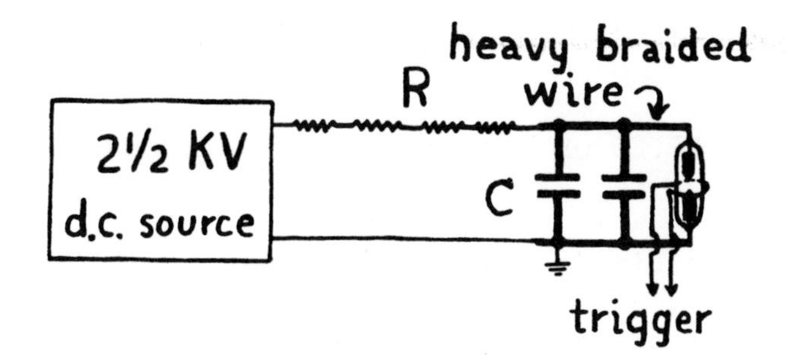

FIG. A21-6

Circuit for G.E. type FT 230 flash lamp.

R: several megohms -- use a series of 1 watt carbon resistors to avoid voltage breakdown.

C: 10-100 microfarad, 3000 volt DC condensers intended for flash tube purposes.

RC: time constant $\sim$1 minute

Trigger: use Ford coil spark generator to trigger flash gap (one turn of hook-up wire around lamp).

Caution: HIGH VOLTAGE DANGER

EXPERIMENT A22. THE SPEED OF LIGHT MEASURED BY A KERR CELL METHOD.

READING: Strong, Appendix H, esp. pp. 462-464; also (more detailed) C. H. Palmer, Jr. and G. S. Spratt, A Laboratory Experiment on the Velocity of Light, American Journal of Physics, Vol. 22, pp. 481-485 (1954).

In the experiment described in the reading, a Kerr electro-optic cell operated at approximately 20,000,000 cycles per second is used to measure the speed of light to within 1%. The experiment is patterned after Anderson's precision experiment but uses inexpensive optical parts and relatively simple electronic components. Some of the latter are built rather than purchased, so that setting up the experiment may take a fair amount of time. It will certainly require some knowledge of electronics. The Kerr cell requires both a high power, high voltage radio frequency supply and a 3 kilovolt dc adjustable bias supply. In addition, a special photomultiplier detector -- mixer is required.

The chief components required for the experiment (details are given in Am. J. Phys.) are as follows:

1) Three meter optical bench

2) Kerr cell (obtainable from Leybold)

3) Photomultiplier (931A)

4) Short wave communications receiver

5) Mixer unit for photomultiplier

6) Vacuum tube voltmeter

7) Cathode ray oscillograph

8) Two high voltage dc power supplies

9) High voltage rf power supply

10) 900 cps synchronous motor

11) Lenses, mirrors, Polaroid sheets

Section B

WAVE OPTICS

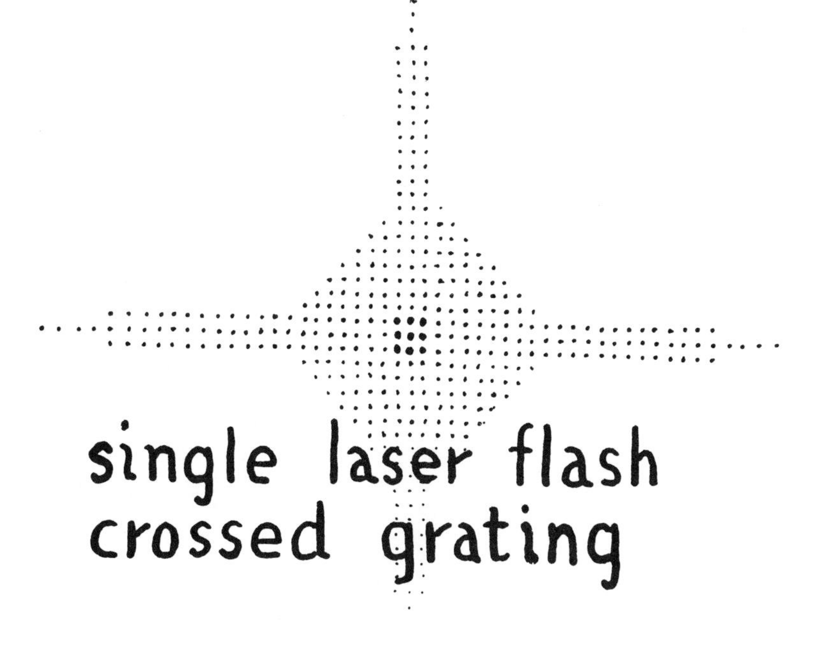

single laser flash
crossed grating

EXPERIMENT B1. YOUNG'S INTERFERENCE.

READING: Strong, Ch. 8-8.6; or Jenkins and White, Ch. 13-13.4, 13.7; also if possible, A. C. S. Van Heel, "High Precision Measurements with Simple Equipment," Journal of the Optical Society of America, Vol. 40, pp. 809-816 (1950).

This experiment on Young's method for producing interference will be used in several ways: (1) to determine the wavelength of light, (2) to show the nature of white light fringes and to illustrate their use in precision alignment, and (3) to determine roughly the refractive index of a flake of mica. In later experiments, it will be shown how Young's historic experiment can be modified to measure with high precision the refractive index of a gas and also how it may be used to measure the diameter of a star.

APPARATUS

1) One meter optical bench.

2) Mercury arc.

3) Filters for the mercury green, yellow (average), and blue lines: (For convenient filter combinations to isolate the mercury lines, see Appendix I at the end of the book).

4) White light source: auto lamp and transformer.

5) Adjustable slit.

6) Traveling microscope carriage and low power microscope with crosshair or reticle.

7) Double slit:
Inexpensive double slits of high quality are readily made from Ronchi rulings (also called Levi screens); these are precision coarse gratings consisting of black and clear spaces of equal width ruled on glass or photographed on glass plates. Either India ink or photographic opaque may be applied with a fine drawing pen or a ruling pen. The photographic surface of the Ronchi ruling must be clean for the ink or opaquing to adhere well. For this experiment, a Ronchi ruling with 50 to 100 lines per inch is suitable. Several slit combinations

should be made as illustrated in Fig. B1-1 by blackening one, two, three, and four adjacent transparent spaces. The blackening should be done under good lighting, on a convenient surface, and with a good magnifier. One clear space on each side of the blackened ones is left clear, and the outside area blackened to finish the slit combination. The precision, of course, is obtained by making use of the lines of the ruling, so that great care must be exercised in the opaquing to make sure that the slit edges are defined by the photographed lines and not the ink.

FIG. B1-1

Double slits made from Ronchi rulings.

This method can, of course, be used to obtain one slit or any number of slits. For the present purpose, however, it is preferable to use an adjustable single slit instead of one made from a Ronchi ruling.

The glass backing of the double slits does no harm in the experiment. Why?

PROCEDURE

(1) Measurement of wavelength

A. The experiment is set up as shown in Fig. B1-2. At one end of the optical bench, mount the mercury arc and the slit 15 or 20 cm away (to allow the white light source to be inserted later without moving the arc). At the other end of the bench, mount the traveling microscope carriage and microscope. The microscope objective should be removed, leaving only the eyepiece and crosshair or reticle. About halfway between the slit and the eyepiece, mount one of the double slits; mask out the unused double slits with paper or cardboard. Alignment of the system is readily achieved if the single slit is opened about a millimeter wide and the room lighting subdued; the beam from the arc is readily seen on a piece of white file card. Make sure the double slit is uniformly illuminated. Follow the beam after it passes through the double slit and position the apparatus to make the light fall on the field lens of the eyepiece (the lens away from the eye end). After these components are aligned, insert a converging condenser lens to focus the arc on the slit, insert the green filter near the slit (not near the arc where it may be cracked by the heat of the arc), and narrow the single slit so that it is barely open. Look for the fringes in the eyepiece and have someone adjust the tilt of the double slit and the slit width of the adjustable slit to obtain the most distinct fringes. Try also the other double slits. At least twenty distinct fringes should be obtained.

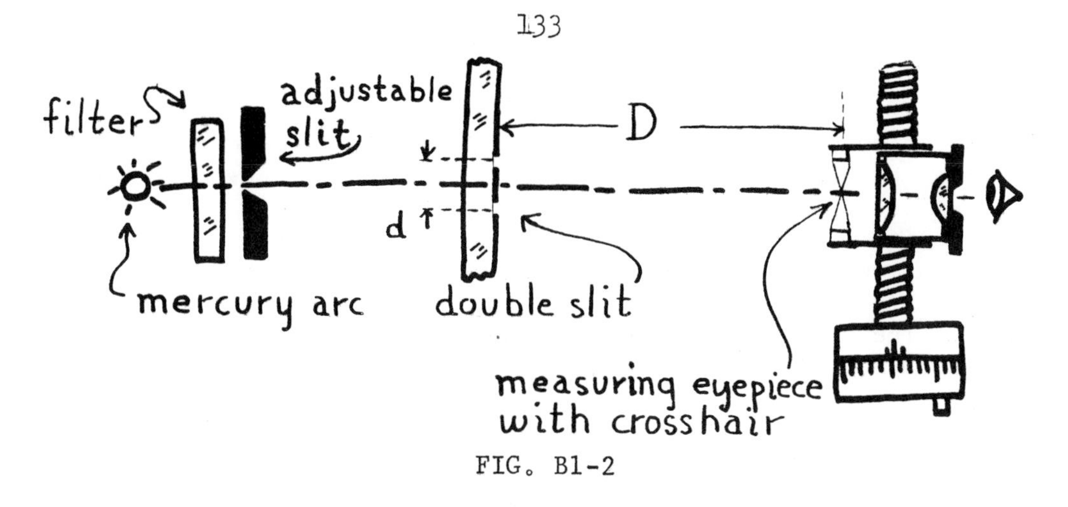

FIG. B1-2

Arrangement for studying
Young's interference.

B. Begin with the eyepiece at one side of the fringe pattern, a little beyond the first fringe to be measured. Crank the micrometer screw very slowly to bring the center of the first distinct fringe (bright or dark) to the center of the crosshair. Be careful to avoid backlash in the screw; if the mark is overshot, return well beyond the fringe center and slowly approach it from the original direction. Record the positions of the centers of all the distinct fringes. It is convenient to number the fringes arbitrarily beginning at one.

If time is available, replace the green filter with one to isolate the yellow lines, and repeat the fringe measurements; likewise use a blue line.

C. Measure the value of D, the distance from the double slit to the crosshair at the focal point of the eyepiece; record also the probable error of measurement.

D. Determine now the slit spacing of the double slit. Replace the microscope objective, and, after recording the position of the double slit for future reference, slide the slit carriage toward the microscope to the position where the slits are clearly in focus in the microscope field of view. The slits should be illuminated by rather diffuse white light from behind (i.e. from the arc side). Begin at one side and crank the microscope to bring the first of the slit edges under the crosshair, record the micrometer reading. Continue to the other edge of the slit and then to the two edges of the second slit, recording the results. Make three such series of measurements.

(2) White light fringes

A. Slide the double slit back to its original position halfway between the single slit and the eyepiece, (the position was recorded above). Remove the microscope objective again, and also the green filter. Observe the appearance of the fringes. (A very few measurements of the fringes would allow the approximate wavelength of white light to be determined). It is now easy to identify the central bright fringe. Between the mercury arc and the slit, substitute the auto lamp source and note the differences in the white light fringes. Record with care the colors of the first three fringes, beginning at the center (a rough sketch may be helpful).

B. The use of white light fringes in precision alignment is now to be demonstrated. As Van Heel explained in his paper, it is possible to line up three points very accurately using Young's interference fringes in white light. The three points may be separated by distances of a few feet or a few hundred feet. Van Heel gives several interesting examples. The required apparatus consists of an auto lamp source, a few cardboard slits (or Ronchi ruling slits for smaller distances), a low power eyepiece and crosshair. For the present demonstration, all the apparatus is already assembled. Observe the white light fringes critically again. Adjust the single slit width to the optimum value for seeing the first two fringes. Figure B1-3 shows the location of the "transition purple." On one side of the transition purple, the fringe is reddish, to the other side bluish; the eye is especially sensitive to this change in color (unless colorblind), and thus the transition provides a very sensitive reference line, appreciably better defined than the center of the central white fringe. Make several settings of the traveling microscope on the purple transition on one side of the central white fringe. Determine how accurately this setting may be made, and determine the angular uncertainty in the alignment of the three points. The appendix to this experiment describes the results of an experiment where the points were separated 9 meters.

(3) Refractive index of mica

Remove the white light source and use the unfiltered mercury arc source to see the colored fringes. Split off a flake of mica (see appendix at the end of the book) about 0.01 mm thick. The thickness must be uniform; measure it as accurately as possible with a good micrometer. Trim the edges of the flake to make a rectangular piece. Place the mica flake over half the length of one of the slits of the pair as shown in Fig. B1-4. If the slit height is correct, two interference patterns should be seen in the eyepiece, one of them the same as before and the other displaced to one side. The location of the central fringe should be apparent in both patterns. If the mica pattern is not clear, the mica may be too thick or not uniform. Measure the position of the central fringe in both patterns to determine the displacement. The refractive index is calculated from the relation $(N-1)t = d\Delta X/D$ where t is the flake thickness and X the displacement.

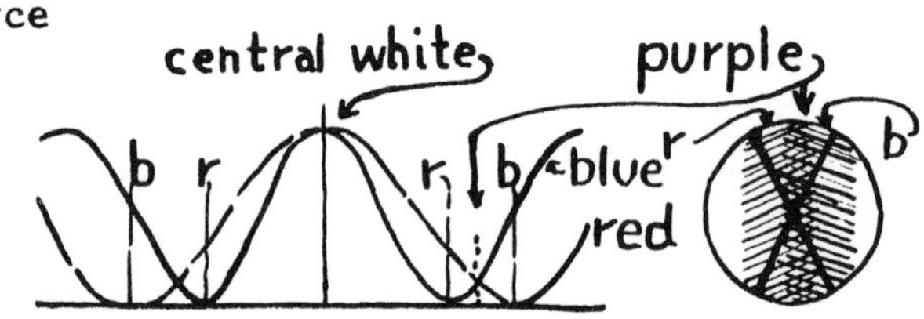

FIG. B1-3

Location of the transition purple.

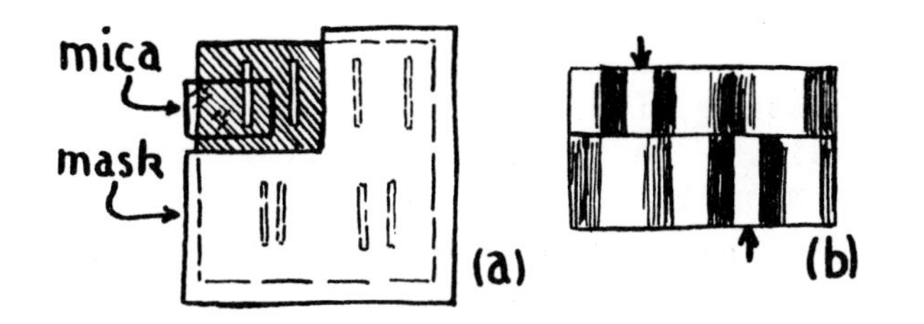

FIG. B1-4

Position of mica flake (a) and appearance of the field of view (b).

(4) (Optional) The Fresnel-Arago laws of interference

With the set up for green line fringes it is possible to verify the Fresnel-Arago laws of the interference of polarized light (Strong, pp. 178-9, or Jenkins and White, p. 559). The fringes must be made as bright as possible, and the room quite dark with a minimum of stray light. Several small rectangles of Polaroid are needed. First cover both the double slits with a piece of Polaroid. The fringes are still seen (though they may be very faint) and they are unchanged by any orientation of the Polaroid. Next, put one piece of Polaroid over one slit and another over the other so that they are crossed, that is, if the light coming through the slits is observed through a third Polaroid, one slit can be darkened while the other remains bright and a 90° rotation of the observing olaroid will interchange dark and light slits. In this condition, no interference is observed. The other Fresnel-Arago law could be verified with two additional Polaroids -- one over the single slit and one in front of the eyepiece, but the light is generally too faint to observe fringes.

RESULTS

Calculate the slit separation of the double slit. Plot a graph on rectangular coordinate paper of the fringe position vs. the (arbitrary) fringe number using the measurements made in part (1). Draw a smooth straight line through the data and calculate the wavelength of the mercury line or lines.

Discuss the results obtained in the demonstration of precision alignment.

Derive the relation for the refractive index of the mica. Calculate this index and evaluate the probable error in the result.

APPENDIX: An experiment in precision alignment

The three points which were to be aligned were separated 9 meters; the single slit was 4 meters from the double slit, and the eyepiece and crosshair mounted on a traveling microscope carriage were 5 meters on the other side of the double slit. The single slit was illuminated with a G.E. #1130 auto lamp (which has a small straight coil) and an achromatic condenser lens. The single slit was 3/4 mm wide and the double slits also 3/4 mm wide, separated by 3 1/3 mm. It was found that the mean deviation of the readings from the average was 0.017 mm, and the R.M.S. deviation in the result was 0.006 mm. Thus for a distance of 5 meters, the uncertainty in alignment (precision of pointing) corresponded to an angle $\alpha = 0.006 \text{ mm}/5 \times 10^3 \text{ mm} = 1.2$ microradians. In order to distinguish two stars separated by this angle, a telescope must have an objective half a meter in diameter. The comparison is not quite fair, however, for the angular error in pointing a telescope can be made somewhat less than its ability to resolve two points. The comparison is, however, suggestive of the precision attainable by simple apparatus.

The Van Heel method can be used to fix three points to make a right angle; in this case, a penta prism is used (see Experiment A4 for properties of penta prism), and the double slit placed at one surface of the prism. Very small rotations of a mirror can be accurately measured by masking out all but two small slits at each edge of the mirror. Circular slits or two pairs at right angles may be used with a point source to align three points in two dimensions.

EXPERIMENT B2. DEMONSTRATION OF MONOCHROMATIC AND ACHROMATIZED FRINGES WITH THE FRESNEL BIPRISM.

READING: Jenkins and White, Ch. 13.1, 13.5. An excellent discussion of various interference experiments is given in T. Preston, The Theory of Light, 5th ed. (1928, Macmillan).

This demonstration shows how interference fringes may be produced without resort to diffraction. Although the double slit experiment of Thomas Young showed true interference and yielded a measure of the wavelength of light, it was open to the objection that the fringes were in some way caused by the edges of the double slit. Fresnel showed that interference could also be produced without a double slit by causing two beams derived from one source to overlap by refraction with a biprism or by reflection from a pair of mirrors. The biprism method is commonly used to measure wavelength. In this experiment, however, the monochromatic fringes are to be observed qualitatively, for there is insufficient improvement over Young's method to warrant measurements. In addition, however, it is possible, by compounding the prism, to produce good achromatized fringes in white light.

THEORY

I. The simple biprism

Figure B2-1 shows the principle of the simple Fresnel biprism. Monochromatic light from the source passes through the slit S. One ray, passing through the upper part of the prism, is refracted through the angle δ to the point P. It appears to come from the point S". A second ray, passing through the lower part of the prism, also refracted through the angle δ , but in the other direction, reaches the same point P. It appears to come from the point S'. The prism angle α is about 1° and hence, $\delta = (N-1)\alpha$ (very nearly) where N is the refractive index of the glass. Thus the separation of the virtual sources, d, is given approximately by $d/2 = a\delta$ or $d = 2a\delta = 2a(N-1)\alpha$. Now, as in the Young's experiment, the optical path difference between S'P and S"P is related to the wavelength: $\Delta \equiv S'P - S''P = m\lambda$ for bright fringes where m is an integer. So one has approximately: $\Delta/d = x/(a+b)$ and hence

$$m\lambda = dx/(a+b) = 2a(N-1)\alpha x/(a+b) \quad \text{(bright fringes)}.$$

(An ingenious way to measure the prism angle is described by Preston and also by G. S. Monk, Light: Principles and Experiments, (1937, McGraw-Hill), p. 130).

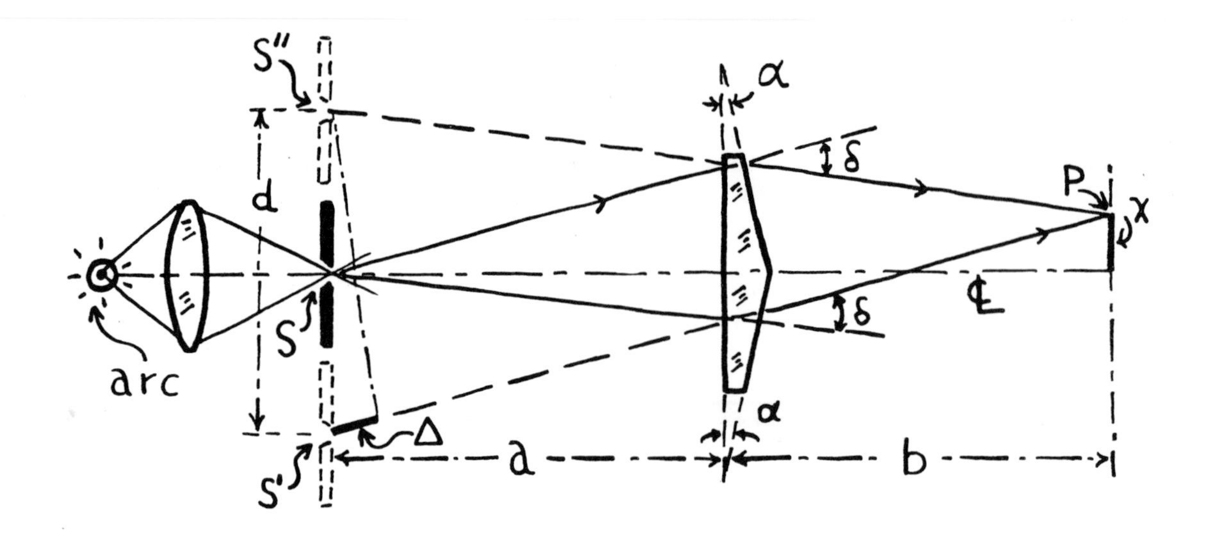

FIG. B2-1

Principle of the simple Fresnel biprism.

II. Achromatized fringes

To achieve achromatization of the fringes, the deviations for the mth fringe must be the same for any wavelength: $x = m\lambda(a+b)/d$. This necessitates $\lambda \propto d$, i.e. $\lambda \propto 2a(N-1)\alpha$. A combination of two prisms made of different materials is needed -- just as in the achromatization of a lens. One of these prisms must have positive deviation and the other negative. Therefore, the required condition is:

$$\frac{2a(N_b-1)\alpha + 2a(N_b'-1)\alpha'}{2a(N_r-1)\alpha + 2a(N_r'-1)\alpha'} = \frac{\lambda_b}{\lambda_r}$$

or:

$$\frac{(N_b-1)\alpha + (N_b'-1)\alpha'}{(N_r-1)\alpha + (N_r'-1)\alpha'} = \frac{\lambda_b}{\lambda_r}$$

where α and α' are the prism angles for the first and second prisms respectively, and N and N' their refractive indices. The subscripts r and b refer to two wavelengths.

Benzene is used to achromatize the glass component of the prism. Figure B2-2 shows how the biprism is constructed for achromatized fringes.

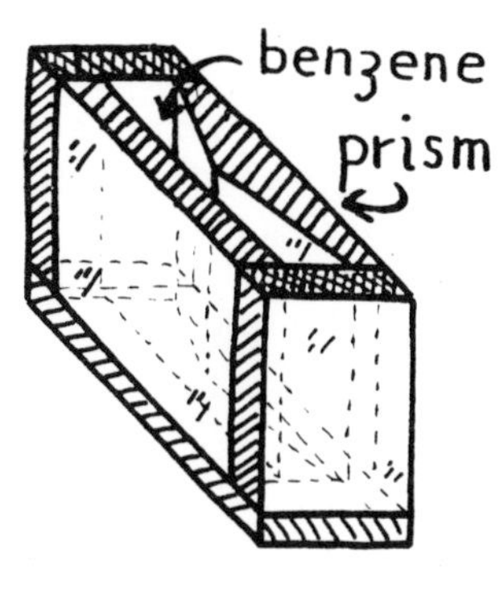

FIG. B2-2

Hollow prism for achromatized Fresnel biprism.

APPARATUS

(1) Optical bench.

(2) Mercury arc with green line filter.

(3) White light source: carbon arc preferable -- or auto lamp.

(4) Achromatic biprism (Leybold).

(5) Benzene (a few cc.).

(6) Ordinary biprism having about 1° refracting angle.

(7) Observing eyepiece.

(8) Condenser lens.

PROCEDURE

Set up the simple biprism as in the Young double slit experiment (Experiment B1), simply substituting the biprism for the double slit (see Fig. B2-1). Alignment of the vertex edge of the prism with the single slit is just as important as in the case of the double slit. With the mercury green line as the source, fifty or more fringes should be visible with the slit properly adjusted.

Substitute a white light source for the mercury arc. A carbon arc gives the best results, but either the unfiltered mercury arc light or an incandescent lamp may be used. The biprism should now show about six reasonably sharp colored fringes -- only the central one being white. Now substitute the hollow prism filled with benzene. At least a dozen black and white fringes should be very well defined and another dozen clearly measurable -- even if tinged with color.

APPENDIX: A more rigorous proof of the equation for fringe position is the following:

In the usual treatment of the Fresnel biprism, one calculates the position of two virtual sources from the refractive index and prism angle. These two virtual sources are then assumed to be entirely equivalent to the two slits in Young's interference experiment, and the correct formula is immediately written down. But there is a fallacy in the proof as indicated in Preston's book. Imagine the prism shown in Fig. B2-1 to be displaced slightly downward, at right angles to the axis ℄. Since the positions of the virtual sources depends only on constants of the prism, these sources remain fixed, and therefore, so does the interference pattern. This situation is contrary to fact; the error lies in disregarding the path of the rays through the prism.

A more accurate (and longer) treatment is based on the assumption that the deviation δ of the rays by the prism is the same on both sides, for it is to first order approximation equal to $(N-1)\alpha$. First, disregard the thickness of the glass. Figure B2-3 shows the geometry (highly exaggerated for clarity). The prism is assumed to deviate the two rays SA and SB, which arrive at point P, through the same angle δ, one ray upward and the other down. The geometric path for the two rays arriving at P is the same: Let the distances be designated as in the figure, then by the law of cosines $\overline{SP}^2 = q^2 + r^2 + 2qr\cos\delta = s^2 + t^2 + 2st\cos\delta$. Since δ is very small (less than 1°), $\cos\delta \approx 1$ so that $(q+r)^2 = (s+t)^2$, and the paths are the same to first order. Any path difference is then the result of differences in the thickness of glass traversed by the two rays.

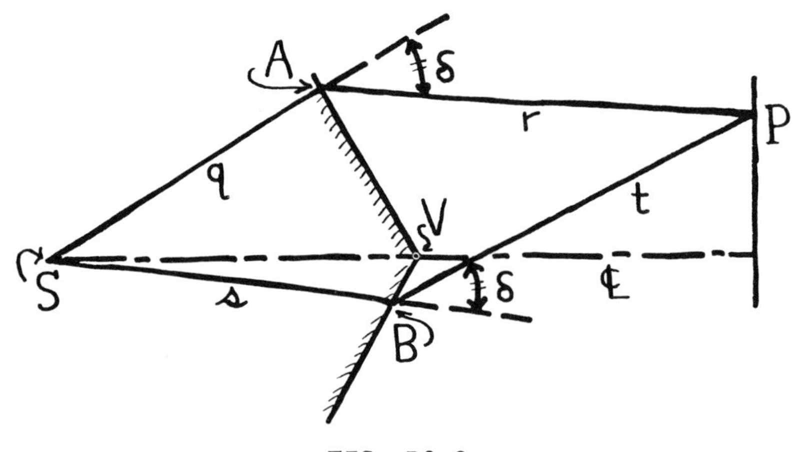

FIG. B2-3

Optical paths external to the biprism.

It remains to calculate the path difference in the glass prism between the two rays, 1 and 2 (Fig. B2-4), which meet at point P. Let θ_1 and θ_2 represent the angles made by the two rays with the axis before they strike the prism, and ϕ_1 and ϕ_2 the angles after they leave the prism. As before, δ is the total angle of deviation. The insert in the figure shows a much enlarged and exaggerated detail of the ray 1 in the prism. It is seen that $\delta = \theta_1 + \phi_1$, and similarly that $\delta = \theta_2 + \phi_2$. Ray 1 strikes the prism at a distance from the axis equal to $a\theta_1 = b\phi_1 + X$. For ray 2 the distance is $a\theta_2 = b\phi_2 - X$. If ϕ_1 and ϕ_2 are eliminated, one can write:

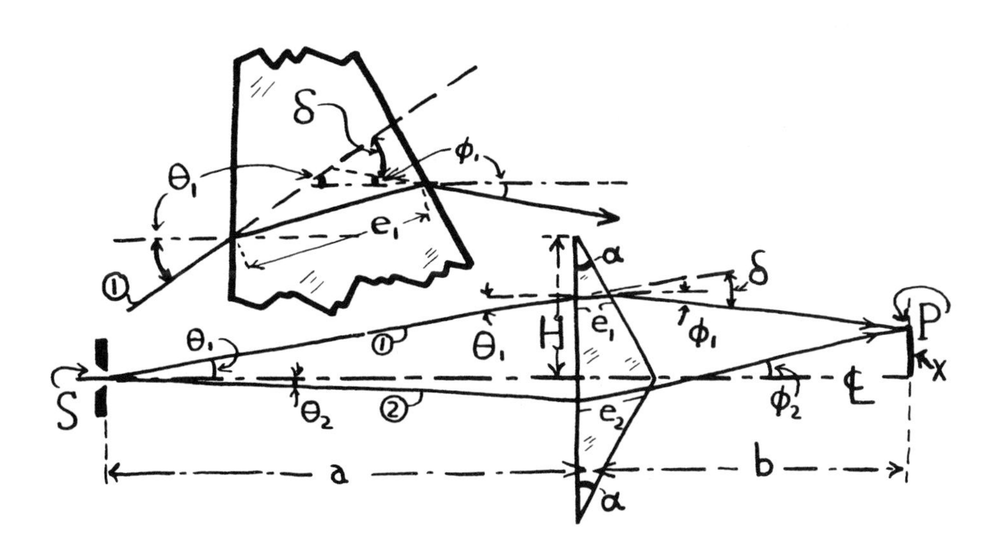

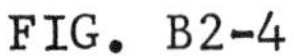
FIG. B2-4

Optical paths within the biprism.

$$a\theta_1 = b(\delta - \theta_1) + x \quad \text{or} \quad (a+b)\theta_1 = b\delta + x \quad \text{and}$$

$$a\theta_2 = b(\delta - \theta_2) - x \quad \text{or} \quad (a+b)\theta_2 = b\delta - x.$$

The thicknesses of glass, e_1 and e_2 traversed by the two rays are nearly proportional to the distances from the two vertices of the prism, that is, to $(H - a\theta_1)$ and $(H - a\theta_2)$. To a good approximation one has, therefore:

$$e_2 - e_1 = (H - a\theta_2) - (H - a\theta_1) = a(\theta_1 - \theta_2)$$

$$= a\left[\frac{b\delta + x}{a+b} - \frac{b\delta - x}{a+b}\right] = \frac{2ax}{a+b}$$

Thus bright fringes of order m result for

$$m\lambda = (N-1)\alpha\, 2ax/(a+b) = \frac{2ax\delta}{(a+b)}$$

as before. The answer is the same, but the approximations are now clearer.

EXPERIMENT B3. DEMONSTRATION OF LLOYD'S MIRROR FRINGES AND THEIR ACHROMATIZATION.

READING: Strong, Ch. 8.4-8.6; or Jenkins and White, Ch. 13.6; also read R. W. Wood, Physical Optics, 3rd ed. (1934, Macmillan), pp. 181-182.

Lloyd's mirror provides a simple and inexpensive method for obtaining excellent fringes. In this demonstration, it is used (1) to show several hundred monochromatic fringes, (2) to demonstrate the phase reversal upon (external) reflection, (3) to show the white light fringes, and (4) to produce high quality achromatized fringes using either a prism or grating. The experiment is a real challenge to one's skill in setting up a beautiful demonstration.

THEORY

Monochromatic fringes

Figure B3-1 shows the principle of Lloyd's mirror interference fringes. The slit S is illuminated by a monochromatic source (mercury arc, condenser lens, and green line filter). Ray #1 emerges from the slit and travels directly to point P. Ray #2 is reflected by the mirror and also reaches point P. This ray apparently comes from the virtual slit source S'. The geometry is the same as for Young's interference. The geometric path difference between ray #1 and ray #2 is $\Delta = dx/D$. In Young's experiment, if the path difference Δ is an integral number of wavelengths, constructive interference results. In Lloyd's experiment, on the other hand, this path difference gives destructive interference because there is a phase reversal upon reflection. Therefore, in Lloyd's experiment, the following relations hold:

$$(m + 1/2)\lambda = dx/D \qquad \text{bright fringes} \quad (1)$$

$$m\lambda = dx/D \qquad \text{dark fringes} \quad (2)$$

It follows that the fringe observed in the plane of the mirror ($x = 0$) must be dark -- there is no geometric path difference, only the phase reversal on reflection.

Another interesting consequence of this difference between the Young and Lloyd fringes is that the white light fringes for the two cases are complementary.

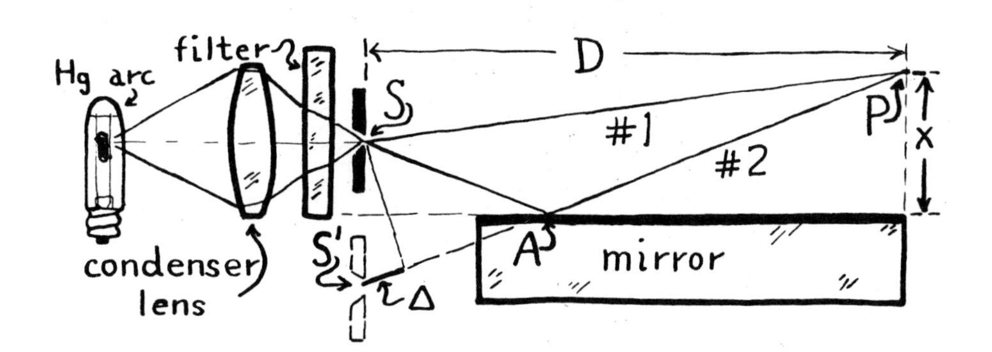

FIG. B3-1

Principle of Lloyd's mirror interference.

Achromatic fringes

As with the Fresnel biprism fringes, it is possible to achromatize the interference fringes of Lloyd's mirror arrangement. Equation (1) shows that the distance x of the mth fringe from the plane of the mirror is directly proportional to the wavelength. If means can be found to arrange the slit position S so that its distance d/2 from the plane of the mirror is proportional to wavelength, then the position of the mth fringe, x, will be independent of wavelength. That is, if $d \propto \lambda$, x will be independent of λ , and the fringes will be achromatized.

Figure B3-2 shows how a prism may be used to produce the desired effect. There is, of course, no real slit in the "source plane;" instead the real slit is located as shown. The light from a carbon arc which passes through the slit is collimated by lens L_2, dispersed by the prism, and focused by lens L_3 in the plane previously occupied by the real slit. In other words, one simply has a convenient spectroscope which forms a spectrum in the source plane. The position of the spectrum must be adjusted carefully with respect to the plane of the mirror in such a way that the distance requirement above is satisfied. This spectrum then provides the desired source for the Lloyd's mirror.

Although the prism method of achromatization will yield about 50 white light fringes of good quality, the grating method is superior because, unlike the prism, it produces a so called normal spectrum. With a grating spectrum, the wavelengths are spread out linearly with distance (to a very good approximation), with a prism, they are not; the dispersion of a prism is "irrational."

Figure B3-3 shows how a grating is arranged to produce the required spectral distribution for the source. A coarse grating (Ronchi ruling) is illuminated by a collimated beam of white light from the carbon arc. The grating disperses the light according to the simple grating equation for normal incidence; $M\lambda = \varepsilon \sin \theta$ where M is the order of

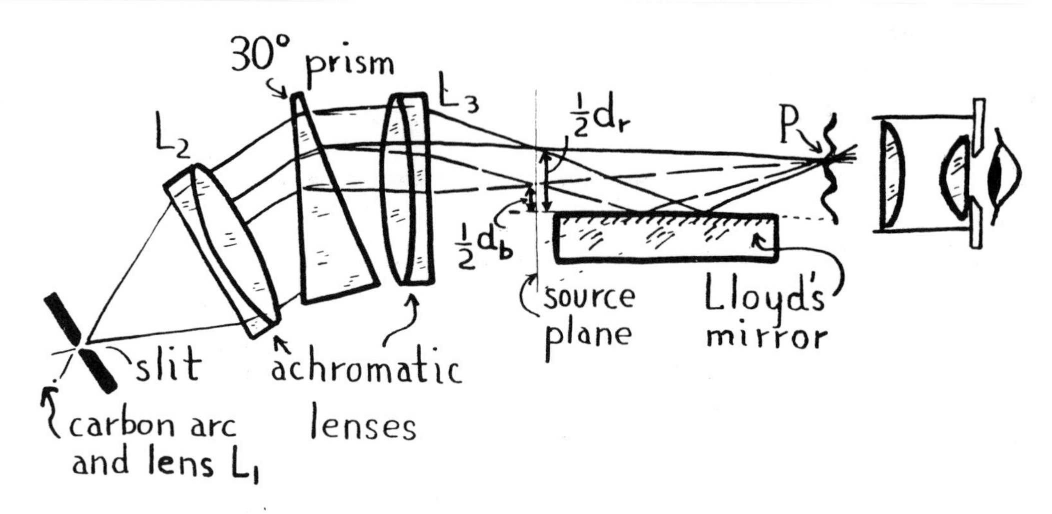

FIG. B3-2

Lloyd's mirror fringes achromatized with a prism.

the spectrum, ε the grating space, and θ the angle of diffraction as indicated in the figure. For the first order, it is seen that $\sin\theta = \lambda/\varepsilon = \frac{1}{2}d/f$. So here again $d \propto \lambda$, and the condition for achromatization can be fulfilled. Here however, as was pointed out above, the required condition is quite accurately fulfilled; with the prism, it was only approximated.

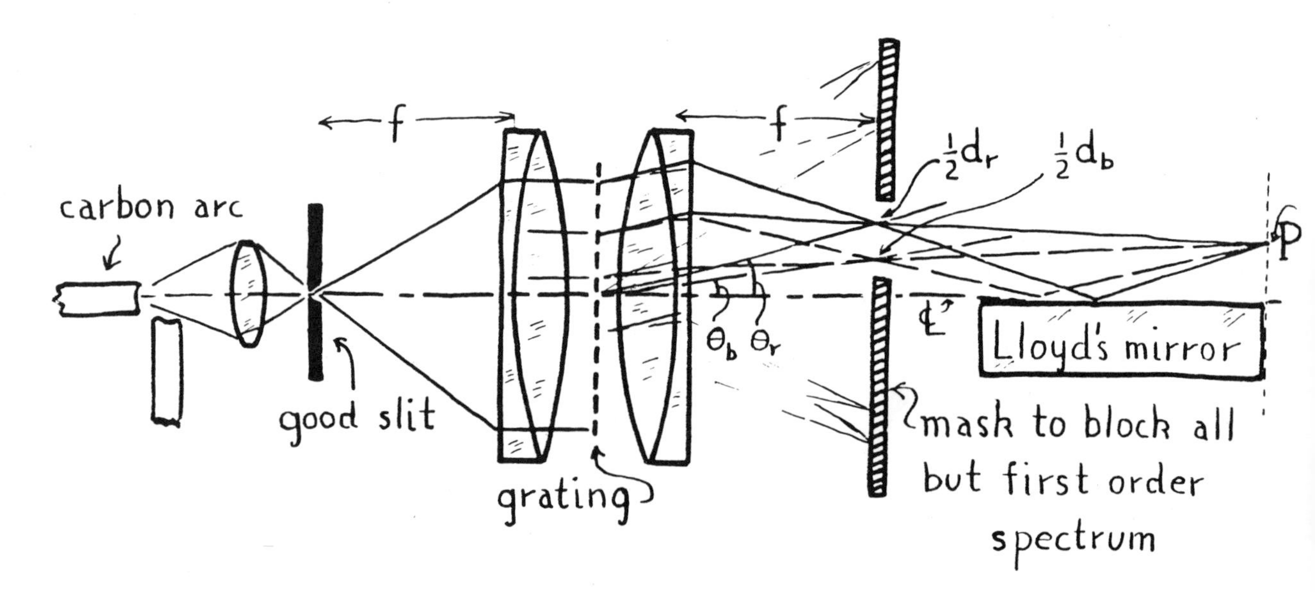

FIG. B3-3

Lloyd's mirror fringes achromatized with a grating.

APPARATUS

1) Optical bench.

2) Lloyd's mirror:

Although black glass is generally used for the mirror, a piece of good plate glass will do as well. The glass, black or clear, should be 3 to 5 in. long, 1 in. wide, and 1/4 to 1/2 in. thick. The back end of the mirror toward the eye should be fine ground to provide a convenient reference for seeing the phase reversal on reflection.

3) Mercury arc and green line filter.

4) Carbon arc source.

5) Traveling microscope of rather high power (100X).

6) 30° prism (to achromatize the fringes).

7) Fine Ronchi ruling (150 or 175 lines/in.) (also for achromatization).

8) Good optical slit.

9) Achromatic lenses, 5-10" focus, 1 1/2-2 1/2 in. dia.

10) Condenser lens, 2-6" focus, 2 in. dia.

PROCEDURE

1) Monochromatic fringes

On the optical bench, set up the Lloyd's mirror arrangement of Fig. B3-1. Use the mercury arc source with a condenser lens to illuminate the slit. The Lloyd mirror can be set either vertically or horizontally; the prime consideration is to be able to locate it very precisely. The slit should, of course, be parallel to the mirror surface. The separation of the slit and mirror is not important -- it may be 10 cm to a meter. The plane of the mirror must be adjusted so that it falls 1/4 to 1 mm to the side of the slit (giving twice this separation of the real and virtual slits, d). A smooth leveling screw is obviously advantageous, but not absolutely essential. The fringes should be observed with the traveling microscope which is focused on the ground end of the mirror. It is helpful to illuminate this end surface with white light. The fringes should be closely spaced if a large number (several hundred) of distinct ones are to be formed. The close spacing, of course, requires high magnification to clearly resolve them.

2) Phase reversal

Adjust the Lloyd mirror so that the fringes are broad (thus sacrificing the large number of fringes). Observe very critically, looking both at the illuminated end of the mirror and at the fringes nearby. The fact that the fringe at the very edge is black should be apparent.

3) White light fringes

Extinguish the illumination of the end of the mirror and substitute a white light source -- the carbon arc or an auto lamp source. Study critically the white light fringes, noting especially the colors of the two or three nearest the mirror surface. Compare these observations to those made of the white light fringes in Young's experiment.

4) Achromatized fringes

Set up the arrangement of Fig. B3-2 to observe achromatized fringes. First, produce a spectrum about a millimeter long. Good achromatic lenses are needed to get a satisfactory spectrum. Light incident on lens L_2 should be made parallel by autocollimation. The spectrum is first focused on a white card to determine the position of the source and indicate approximately where the mirror should be put. The plane of the mirror should be parallel to the slit, and thus, to the lines of red and blue in the spectrum; it should lie as far below the blue of the spectrum as the red is above the blue. Careful adjustment by trial and error should bring some fifty white light fringes into view. A certain amount of patience may be required to get the optimum adjustments of slit width, mirror location, etc. Estimate the number of fringes observed. If properly set up, the results are beautiful and very spectacular.

Try also the achromatization with the grating, using the arrangement of Fig. B3-3. First, line up the plane of the mirror and the slit. By autocollimation, make the beam of light from the carbon arc parallel, and then put the second lens in place to focus the light on a white card. The image should be a sharp line which falls on the axis ℄ . Insert the coarse Ronchi ruling which is chosen to give a first order spectrum about a millimeter long. Use a cardboard mask to block off the zero order spectrum and all other orders except the first order. If the slit width, the tilt of the mirror, the focusing of the lenses, the lines of the Ronchi ruling are all in good adjustment, one should be able to see about 80 white light fringes. Again estimate the number seen.

EXPERIMENT B4. THE MICHELSON INTERFEROMETER.

READING: Strong, Ch. 11.5-11.7; or Jenkins and White, Ch. 13.8-13.13.

In this experiment, the Michelson interferometer is used (1) to measure the wavelength of sodium light, (2) to measure the difference in wavelength between the two components of the D line, (3) to study the nature of the white light fringes, and (4) to make use of the white light fringes in a determination of the refractive index of a glass plate.

The Michelson interferometer is a versatile instrument of the utmost historical importance. One of these instruments, built with the greatest care and having extremely long paths (11 meters) was used by Michelson and Morley in 1887 to perform one of the most celebrated of all scientific experiments [1]. The negative result they obtained, in the words of Wood [2], " ... shook, to its very foundations, the wave theory of light, and gave rise to discussions which are not yet finished ...". Further experiments have failed to disprove Michelson and Morley's original results, and their observations formed the basis of Einstein's theory of relativity a few years later.

The interferometer was later used by Michelson to measure, in wavelengths of cadmium red, green, and blue light, the standard meter in Paris. It was also used to study the fine structure of spectrum lines, notably that of the red H_α line of hydrogen [3].

Although these experiments are not suitable for laboratory experiments, they illustrate the versatility of the instrument.

Figure B4-1 shows the basic design of the interferometer. Light from the sodium arc source is condensed by the lens L_1 and penetrates plate P_1 to the partly silvered surface nearest the observer. At this point, half the beam is reflected to M_1 and the other half is transmitted to M_2 (passing through the compensator plate P_2 on the way). Mirrors M_1 and M_2 reflect the light back to the half silvered plate, and half of each beam

[1] Michelson and Morley, Philosophical Magazine, Vol. 24, p. 449 (1887). A brief discussion is given in Jenkins and White, Ch. 19.15, and in R. W. Wood Physical Optics, 3rd ed. (1934, Macmillan).

[2] R. W. Wood, loc. cit. P817.

[3] Strong, pp. 172-173.

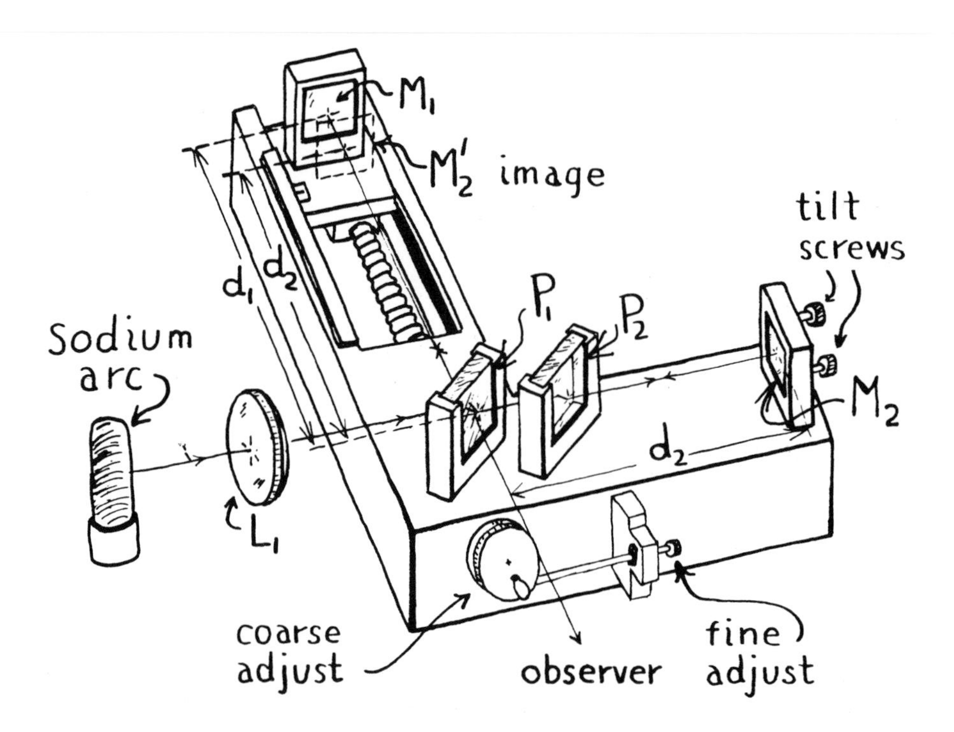

FIG. B4-1

The Michelson interferometer.

reaches the observer, the remainder being directed back to the source and lost. Mirror M_1, mounted on a carriage which slides in precision ways, can be translated toward or away from the observer by means of a precision screw turned with a crank. The screw itself can be turned through very small measured angles by the fine adjust worm which may be disengaged when desired. The displacements of the mirror M_1 can be measured very precisely in terms of the various calibrated scales suggested in the drawing.

Looking through plate P_1 toward mirror M_1, the observer sees both the real mirror M_1 and the reflected image of M_2 at M_2'. Mirror M_2 is adjustable with tilt screws so that its image can be made accurately parallel to M_1. It is the spacing of M_1 and M_2 or M_2' which determines the fringe pattern. It is clearly possible in this instrument to reduce the difference in path, $d_1 - d_2$ to zero or to make it negative, effectively moving one of the mirrors through the other (since one is an image).

The compensator plate is needed if white light fringes are to be obtained. It is to be noted that the light rays going to mirror M_1 traverse the plate P_1 three times before reaching the observer, whereas the rays going to mirror M_2 traverse it only once. In order to achieve exact equality of path, therefore, the compensator plate of exactly the same thickness as P_1 is added.

Not shown in the figure is the rotatable mount between P_2 and M_2 which is used in the measurement of the refractive index of a glass plate as described below.

THEORY

Measurement of wavelength difference

In the reading, it is shown that the wavelength of monochromatic light is determined in terms of the number of fringes disappearing into or appearing from the center of the pattern when circular fringes are used. The relation is

$$d_a - d_b = (m_a - m_b)\lambda/2 \qquad (1)$$

where d_a and d_b are two positions of mirror M_1, and m_a and m_b are the corresponding orders of interference (number of fringes), and of course λ is the wavelength.

The wavelength difference between two close lines such as the components of the sodium D line is determined from their average wavelength and the visibility of fringes. At certain positions of mirror M_1, it is found that the fringes are clear and sharp whereas at intermediate positions, they are very indistinct. The reason is that there are two sets of fringes which are not identical, and at some positions, the two sets are in step and the overall fringe pattern sharp, whereas at the intermediate positions the two sets overlap thus washing out the overall pattern. Figure B4-2 shows the situation schematically. The separation of the positions of maximum (or minimum) visibility of the fringe pattern determines the wavelength difference. Let the two wavelengths and their interference orders be distinguished by primes or no primes. Let d_a represent a point where both set of fringes are in step (maximum visibility). Then

$$d_a = m_a\lambda/2 = m'_a\lambda'/2\ , \quad \lambda > \lambda' .$$

Let M_1 be translated through a fringe visibility minimum to the next fringe visibility maximum, and call this position d_b. Then the longer wavelength will have given rise to $m_a - m_b$ fringes and the shorter wavelength to one more fringe than the longer one. That is

$$d_b = m_b\lambda/2 = m'_b\lambda'/2 .$$

Let $\Delta\lambda = \lambda - \lambda'$ and note that $m_a - m_b = m_a' - m_b' + 1$; eliminate the interference order and the result is

$$\Delta\lambda = \lambda\lambda'/2(d_a - d_b) . \qquad (2)$$

Refractive index of a glass plate

The refractive index of a glass plate can be measured by putting it in one of the two light beams in the interferometer and changing its effective thickness by rotation. The change in effective thickness, shown by fringe shifts, is directly related to the refractive index.

Consider the plate represented in Fig. B4-3. In the position normal to the beam (solid lines) the undeviated light ray travels a distance ce = t through the glass. If the plate is rotated through angle i to the dashed position, the light ray incident at point c is refracted to point g and then emerges parallel to the original solid ray, but displaced from it. The problem is to calculate the increase in optical path through the glass resulting from the rotation. Consider the optical path through the plate before rotation to the point f beyond (in the air) and then the path through the plate after rotation to the comparable point h. Before turning, the path is ce in glass plus ef in air; after turning the path is cg in glass plus gh in air. From the geometry these distances are respectively given by

$$ce = t$$

$$ef = cf - ce = t/\cos i - t$$

$$cg = t/\cos r$$

$$gh = fg \sin i = (fn - gn)\sin i$$

$$= (t \tan i - t \tan r)\sin i$$

Therefore the optical path before turning is

$$[d] = Nt + t/\cos i - t,$$

and after turning is

$$[d'] = Nt/\cos r + t \sin i\,(\tan i - \tan r).$$

The difference in optical path can be expressed as a shift of M fringes. Note that the light beam traverses the glass twice, so that $[d'] - [d] = M\lambda/2$ and hence $Nt/\cos r + t \sin i(\tan i - \tan r) - Nt - t/\cos i + t = M\lambda/2.$ To solve this rather involved expression for N, expand the second term and combine part of it with the first term and the rest with the term having cos i in the denominator, to obtain

$$\frac{N^2 - \sin^2 i}{N \cos r} - \frac{1 - \sin^2 i}{\cos i} - N + 1 = \frac{M\lambda}{2t}$$

Note that

$$N \cos r = \sqrt{N^2 - N^2 \sin^2 r} = \sqrt{N^2 - \sin^2 i},$$

so that one obtains

$$\sqrt{N^2 - \sin^2 i} = \cos i + N - 1 + M\lambda/2t.$$

Square both sides of the equation, then the N^2 terms cancel, and the term $N^2\lambda^2/4t^2$ is dropped (as it is of second order). After terms involving N are put on the left side of the equation and the others on the right, an expression in N is obtained. Finally, multiply the numerator and denominator of the fraction by 2t to obtain the result

$$N = \frac{(2t - M\lambda)(1 - \cos i)}{2t(1 - \cos i) + M\lambda} .$$

APPARATUS

1) Michelson interferometer.

2) Auxiliary glass plates and mounting to measure refractive index.

3) Sodium vapor light source.

4) White light source (25 watt incandescent frosted lamp).

5) Micrometer caliper (0-1 in. reading in 1/10000 in. or metric).

6) Mirror -- 1/4 x 1 in.

7) Galvanometer telescope and meter stick (should have good clear markings).

8) Condenser lens.

PROCEDURE

(1) Measurement of wavelength

A. Preliminary adjustment of the interferometer

Examine your instrument carefully. It will, of course, differ in some details from the one represented in Fig. B4-1. The optical surfaces must never be touched with one's hands for they may become

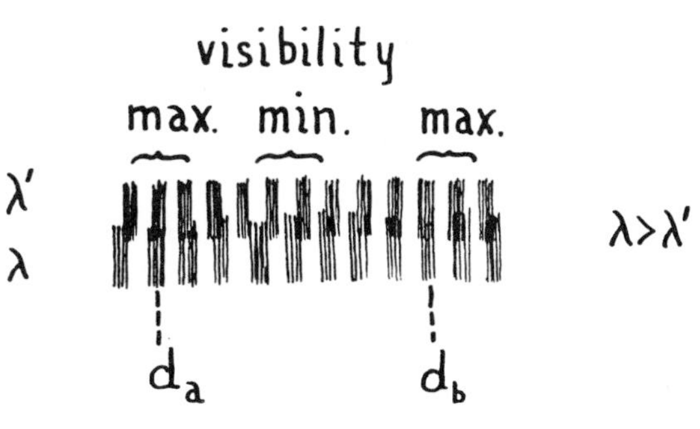

FIG. B4-2

Fringe visibility with two wavelengths.

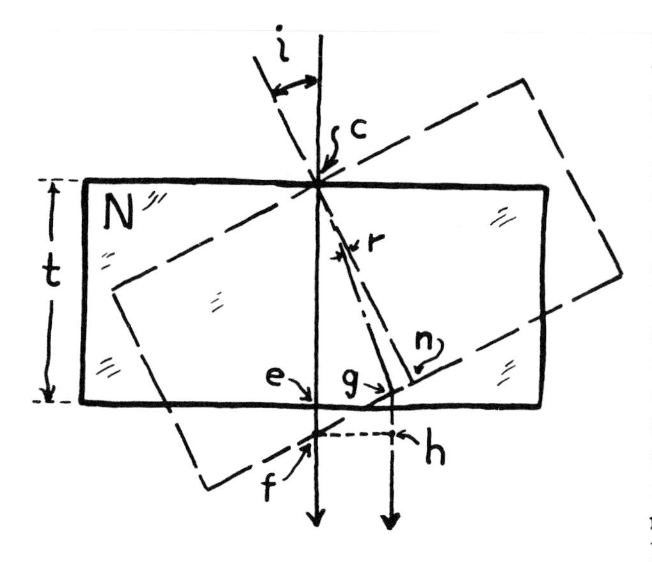

FIG. B4-3

Geometry for the refractive index measurement of a glass plate.

damaged or at the least, will require very careful cleaning. Note the primary drive screw, the crank mechanism, and the fine adjust worm and its disengaging mechanism. Never try to turn the primary screw crank unless the fine adjust worm is disengaged; also make sure that the latter is always either fully engaged or fully disengaged. Note and record the finest divisions (least count) of both the coarse and fine adjust mechanisms. Locate the mirror tilting screws which may be arranged in several ways -- all behind one mirror or one behind M_1 and one behind M_2.

Illuminate the instrument with sodium light using a condenser lens as shown in Fig. B4-1, to produce a wide, diffuse source to fill the field of view with light. Disengage the fine adjust screw, and with the aid of a millimeter scale (or simply a file card with a couple of pencil marks) make the paths d_1 and d_2 equal to within a millimeter. Between the lens and the half silvered mirror P_1, hold a file card with a pinhole in it so that only a small part of the field of view is illuminated. Four pinpoint images will, in general, be seen. One pair of these corresponds to the desired rays from M_1 and M_2 reflected off the half silvered side of $\mathbf{P}_1$; the other pair of images (undesired) corresponds to rays reflected from the back of P_1. Adjust the tilt screws for the mirrors (they may be arranged so that there is one tilt screw for M_1 and one for M_2) to bring into coincidence two pairs of images, i.e. reduce the number of images to two. When the card is removed, interference fringes should be seen. If they are not, check again the equality of the optical paths and the tilt of the mirrors.

When the fringes are found, adjust the tilt screws very carefully so that the fringes become circular and their common center lies at the center of the field of view. Now slowly translate M_1 with the coarse adjust screw in the direction that makes the fringes move inward and disappear at the center. As exact path equality is approached, the fringes become much broader, and, unless the instrument has optical surfaces of exceptional quality, the circular shape of the fringes will become badly distorted. Thus the point of exact path equality is somewhat uncertain. When the point is reached where the fringes begin to move outward from the center, the point of path equality has been passed. For present purposes, it is sufficient to be near the path equality point.

B. Measurements

Beginning near the point of path equality, move M_1 either direction in rather small steps with the coarse adjust until the fringes are as sharp as possible. Engage the fine adjust screw and remove the backlash by turning in the chosen direction until the fringes move smoothly into or out of the center.

Data to calculate the wavelength using equation (1) are now to be obtained. The interferometer reading (including fine adjust reading) is to be recorded for each tenth fringe up to the 190th. The data should be recorded in tabular form preferably as in the form shown:

Fringe number	Interferometer reading	Fringe number	Interferometer reading	Difference	Deviation
0th 10th 20th 90th		100th 110th 120th 190th			
			Mean		

(2) Measurement of wavelength difference

Equation (2) gives the wavelength difference of two nearly equal wavelengths in terms of the distance between two successive visibility maxima (or minima). The exact positions of the maxima or minima are, however, difficult to determine. The visibility of the fringes does not go to zero between the maxima unless the two spectral lines are of equal intensity. If they are identically bright, then the overlapping fringe systems completely wash each other out so that the field of view is uniformly illuminated. But, if one line is brighter than the other, the bright line fringes are not completely erased by the overlying pattern of the weaker line. The overall fringe pattern still goes through a minimum, but it is then more difficult to distinguish the minima or maxima of visibility. In the case of sodium, one of the D lines is about twice the brightness of the other so that the minima, though reasonably apparent, are not zero.

In order to obtain a good value of the wavelength difference, about 20 minima are to be observed, and, therefore, the result will have 1/20th of the error incurred in moving a distance for one minimum. Begin at a position where the path difference is rather small and move the mirror through successive minima merely observing the field without recording data. Return the moveable mirror back beyond the starting point and after removing the backlash, record the interferometer readings for at least 20 successive minima. Use the method of differences as in part (1) to obtain data for the wavelength difference.

(3) White light fringes

To find white light fringes, the observer needs considerable care and patience, for it is necessary for him to set the position of M_1 to within two or three wavelengths of the point of exact path equality.

Continue using the sodium source and move M_1 to a point about one-half turn of the primary screw away from exact path equality. It is worthwhile checking to make sure this position is found. It is convenient, though not imperative at this time, to tilt the mirrors to produce vertical localized fringes. About 5 to 10 fringes should be visible in the field of view. Now engage the fine adjust worm and begin to approach the point of path equality very slowly observing how the fringes move. Notice the amount of turning of the fine adjust dial to move five fringes past the center of the field. Substitute a white light source for the sodium. The white source should not be too bright -- a 15 to 25 watt frosted lamp is about right. The fine adjust worm is now moved stepwise (or continuously) while looking for the white light fringes. Move in steps corresponding to five fringes and use great care to avoid missing the fringes (depressingly easy to do). The time required to find the fringes depends upon both the skill of the observer and the quality of the instrument; it varies between 5 minutes and an hour. If they are found within 15 minutes on the first attempt, the observer may congratulate himself.

The instant any trace of color is seen in the field of view, stop turning at once. Now turn the worm with the utmost deliberation until the fringes are centered in the field of view. Note carefully the interferometer reading and the direction of approach (the reading is worthless if the point is approached from the other direction). Observe the fringes critically; make sure they are wide enough so that the colors may be clearly seen. Describe these colors beginning at the central fringe and compare them with the colors of the white light fringes of Young and Lloyd. Also pay attention to the central fringe -- is it white, grey, or black? Under what conditions would the central fringe be white? Grey? Black?

(4) Refractive index of a glass plate

For this part of the experiment, it is assumed that the interferometer is equipped with a pair of glass plates about 5 mm thick, cut from a single larger plate, and that there is provision for mounting them as indicated in Fig. B4-4. One plate, S_1, is mounted in a special rectangular frame which fits into a hole in the slide on which M_1 is mounted. On top of this frame is another hole into which a mirror, M_3, may be mounted for the purpose of measuring the angular position of S_1. The fixture into which plate S_2 fits has two slots. One slot is to hold S_2 and

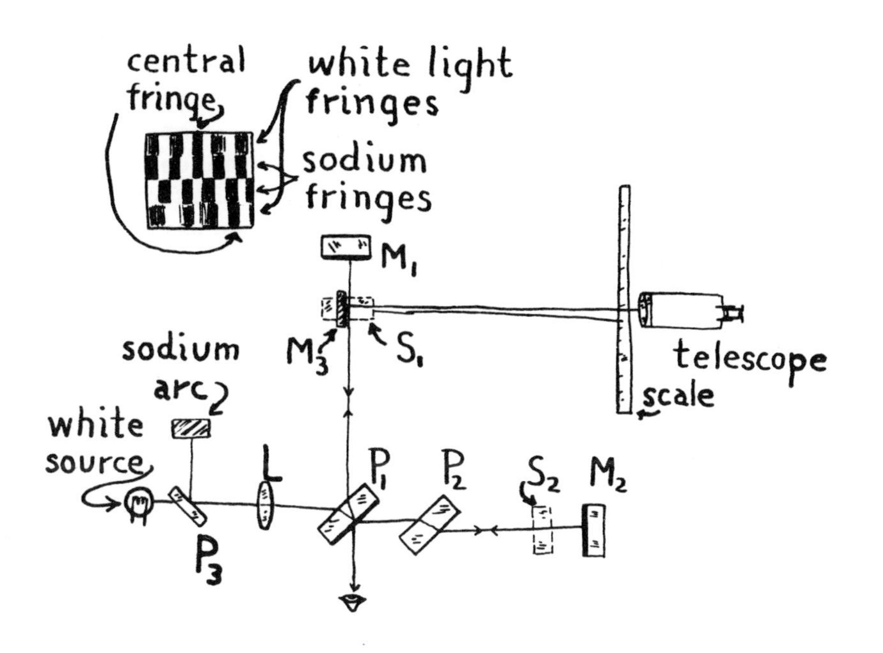

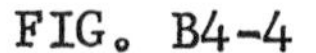

FIG. B4-4

The Michelson interferometer arranged to measure the refractive index of a glass plate. The insert shows the field of view with S_1 and S_2 in place.

the other is a slot into which the compensator plate P_2 may be placed in order to adjust S_2 exactly normal to the beam.

To begin with, the interferometer (without either S_1 or S_2 in position) is adjusted to produce white light fringes as in part (3). The mirrors M_1 and M_2 are adjusted to give about five vertical fringes centered in the field of view. The central fringe is to serve as an unambiguous reference line. The slide on which M_1 is mounted should not be moved after this step.

For this part of the experiment, it is convenient to illuminate the interferometer as indicated in Fig. B4-4. A narrow strip of mirror at P_3 is used to reflect sodium light into the central part of the field of view while white light passes over and under P_3 to illuminate the top and bottom of the field. Mirror P_3 may be a narrow strip of mirror perhaps a quarter of an inch wide cut from a larger piece, or the aluminum from a larger piece may be removed except for a narrow strip. The insert in the figure shows the appearance of the field with plates S_1 and S_2 in place, but not quite at right angles to each other. The zero order sodium fringe is then easily located by its coincidence with the central white light fringe.

Use a piece of lens paper to avoid scratches and finger prints, and carefully transfer the compensator plate P_2 to a parallel slot in the fixture for S_2 (the deep slot, not the shallow one). Rotate P_2 in the new position with coarse and fine adjustment until the central white light fringe is back in the center of the field of view. Replace P_2 in its original position. Install the two plates, one at S_2 using the shallow slot which makes the plate normal to the light beam, and one in the rectangular frame at S_1. Plate S_1 should also be normal to the beam of light passing through it. A slight rotation of the holder for S_1 should restore the white light fringes in the lower half of the field of view. It is assumed that white light fringes are also seen at the top of the field in the beam passing over the two plates S_1 and S_2. At this stage, a slight rotation of S_1 in either sense should displace the white light fringes in one direction only. Likewise, a rotation of S_2 either way should displace the white light fringes in the opposite direction. This condition can also be achieved rather easily without moving the compensator plate by alternately adjusting S_1 and S_2.

Arrange a telescope and scale as in the figure so that changes in the angular position of S_1 can be measured. Insert M_3 in position, being careful not to disturb S_1. Assuming that S_1 is still (or again) adjusted normal to the light beam, measure the reference position of M_3. Carefully rotate the frame with S_1 and the mirror 6 to 8° either way and measure the exact angle of rotation. A rotation of 6 to 8° will give a reasonable number (40 to 100) fringes to count. Smaller angles give too few fringes for accuracy, and larger angles far too many fringes to count easily.

After noting any fractional fringe displacement, use the fine adjust worm to rotate S_2 in either sense and count the number of fringes displaced with reference to the upper zero order sodium fringe. Count the fringes until the lower zero order sodium fringe (as determined by the lower white light fringes) again coincides with the upper zero order sodium fringe. As a check, rotate S_2 back in the opposite direction and count the number of fringes until they cease to move and are about to move in the reverse direction. As a final check, S_2 can be rotated still further in the same direction and fringes counted until coincidence of the central fringes is again observed.

Finally, remove plate S_1 from its holder and measure its thickness with a good micrometer. All the measurements needed to calculate the refractive index are in hand.

RESULTS

Calculate the wavelength of sodium light and the mean deviation of the result.

Calculate the difference in wavelength between the two components of the D line and the mean deviation.

Describe the white light fringes and explain the central fringe appearance.

Calculate the refractive index of the glass plate and the uncertainty in the result.

QUESTIONS

Is an extended source necessary? Could a point source be used? Discuss briefly.

EXPERIMENT B5. SIMPLE DEMONSTRATION OF INTERFERENCE WITH MICA.

READING: Strong, Ch. 11-11.2; or Jenkins and White, Ch. 14-14.1.

REFERENCE: R. W. Pohl, Einführung in die Optik, 4th and 5th ed., (1943, Springer, Berlin; photolithographed by Murray Printing Co., Cambridge, Mass.), p. 68.

R. W. Pohl has described in his book (loc. cit.) an elegant and simple means of demonstrating interference fringes of the Fizeau type. These fringes represent two beam interference between coherent sources separated in depth. The fringe position is given by the relation

$$2Nd\cos r = m\lambda \qquad \text{(dark bands)}$$

where the various quantities are indicated in Fig. B5-1. The order of interference is m and the wavelength λ. The theory of these fringes is given in the Reading.

APPARATUS

(1) High intensity mercury or sodium arc.

(2) Mica sheet:
The mica sheet should be three inches square or larger, 0.04 to 0.07 mm thick, and highly uniform. It may be purchased, or if a large sheet is available, it may be split using the technique described in an appendix to the book.

(3) Large screen or white wall.

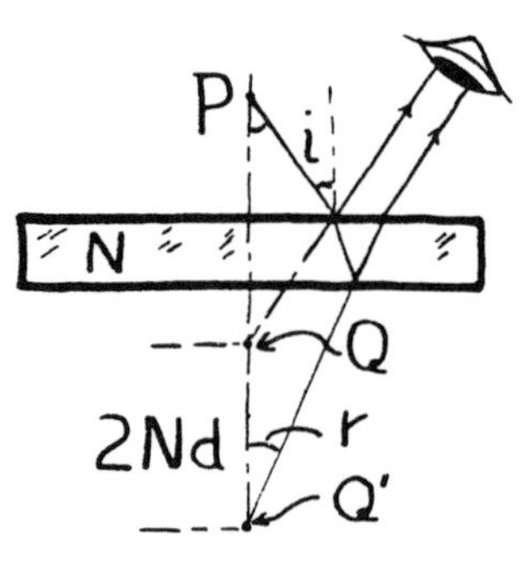

FIG. B5-1

Thin film interference.

PROCEDURE

The demonstration is set up as indicated in Fig. B5-2. The arc source is directed toward the audience, and placed several meters from the

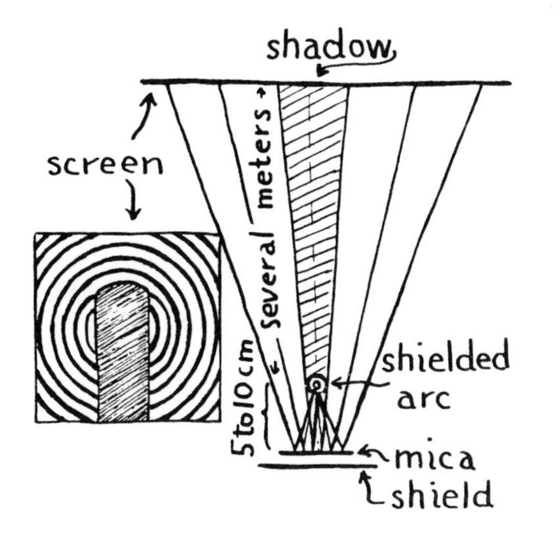

FIG. B5-2

Demonstration of mica sheet interference.

screen. The mica sheet is held 5 to 10 cm in front of the arc to reflect light back to the screen. It is desirable to put a light shield as indicated to keep the direct light out of the eyes of the audience. In addition, the shield over the arc should be replaced, if practical, by one of small diameter to avoid producing a large shadow on the screen which would otherwise cover the first two or three fringes. A large number of fringes up to a meter or more in diameter should be observed.

EXPERIMENT B6. DEMONSTRATION TRIANGLE PATH INTERFEROMETER.

REFERENCES: Strong, Appendix A (W. E. Williams) and Appendix B (J. Dyson); Hariharan and D. Sen, New Gauge Interferometer, Journal of the Optical Society of America, Vol. 49, pp. 232-234 (1959); Triangular Path Macro-Interferometer, ibid., pp. 1105-1106; P. Hariharan and R. G. Singh, Achromatic Fringes Formed in a Triangular Path Interferometer, ibid., p. 732.

The triangle path interferometer provides an interesting contrast to the Michelson interferometer. With the latter instrument, provided it is well built of precision parts, one may be able to find white light fringes in 5 to 30 minutes. With the triangle path interferometer, even built of plywood with crude adjustments and ordinary quality optical surfaces, it is possible to obtain white light fringes in a matter of seconds. Also, to find white light fringes with the Michelson, the use of a monochromatic source as an intermediate step is necessary; with the triangle path instrument, no monochromatic light is needed. The Michelson interferometer is better adapted to certain kinds of measurement (where the two paths must be widely separated) than the triangle path instrument, but the latter has also great capabilities. It may be used, when constructed of precision parts, to measure the length of end gauges as long as a meter in terms of wavelengths (provided a sufficiently coherent light source is available). Alternatively, if spherical mirrors are used, the triangle path instrument can be used for interferometric testing in wind tunnels at relatively low cost.

THEORY

Figure B6-1 shows a diagram of the triangle interferometer. Light from a diffuse source B, either white or monochromatic, is incident on a beam splitter B consisting of a partly silvered surface sandwiched between glass. The incident ray is divided by B into ray number 1 (solid line) transmitted clockwise as shown, and ray number 2 (dashed line) reflected in the equivalent counterclockwise path. In the figure, the rays have been sheared for clarity but actually for the symmetrical case shown, the rays coincide at all points -- there is neither shear nor path difference.

The more general case is illustrated in Fig. B6-2 which has been simplified by the omission of the glass on either side of the beam splitter. Here the successive images of the virtual source S_o (where the beam splitter divides the incident ray) produced by the various reflections are indicated. Mirror M_1 has been tipped and the amplitude divided rays from S_o appear to

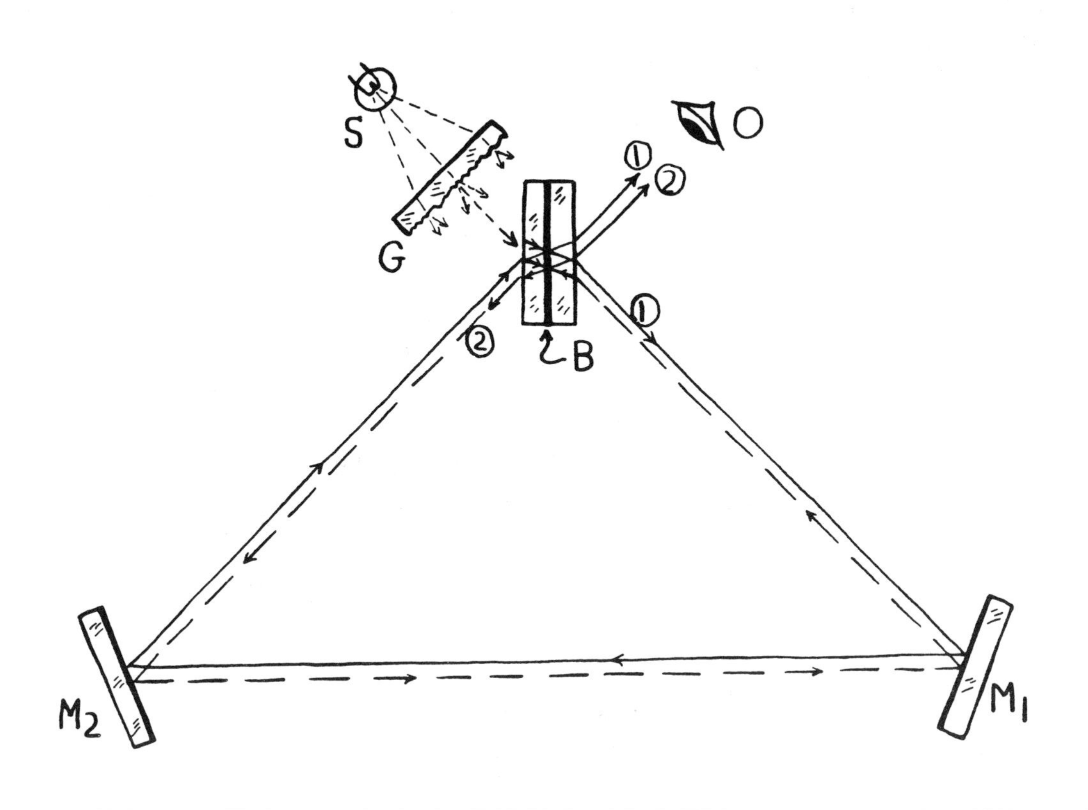

FIG. B6-1

Diagram of the triangle path interferometer.

come from S_1'' and S_2'''. These two virtual (coherent) sources are separated both laterally (sheared) and in depth. The latter path difference is responsible for the interference fringes. Simple considerations show that the emerging rays which originate from a common source point are parallel at an angle with the plane of B as indicated. Skew rays, out of the plane of the figure, are not shown, but their effect can be inferred from the diagram.

The beam splitter in Fig. B6-1 is symmetrical, and it gives rise to a black or white central fringe. An unsymmetrical beam splitter, coated on one side, gives rise to a colored central fringe. Why?

The interference fringes in the instrument come from pairs of virtual, coherent point sources, and therefore, as with other two beam interference arrangements, the fringes are colored. Why? A tipped thick glass plate introduced between mirrors M_1 and M_2 as indicated in Fig. B6-3, produces further shear in the optical path, and this may be added to or subtracted from the shear introduced by the mirrors alone. The tipped plate obviously affects red and blue light differently because of its dispersion, and it may be made to cancel the chromatism of the fringes so that these appear black and white. (A quantitative description is given in the last reference above).

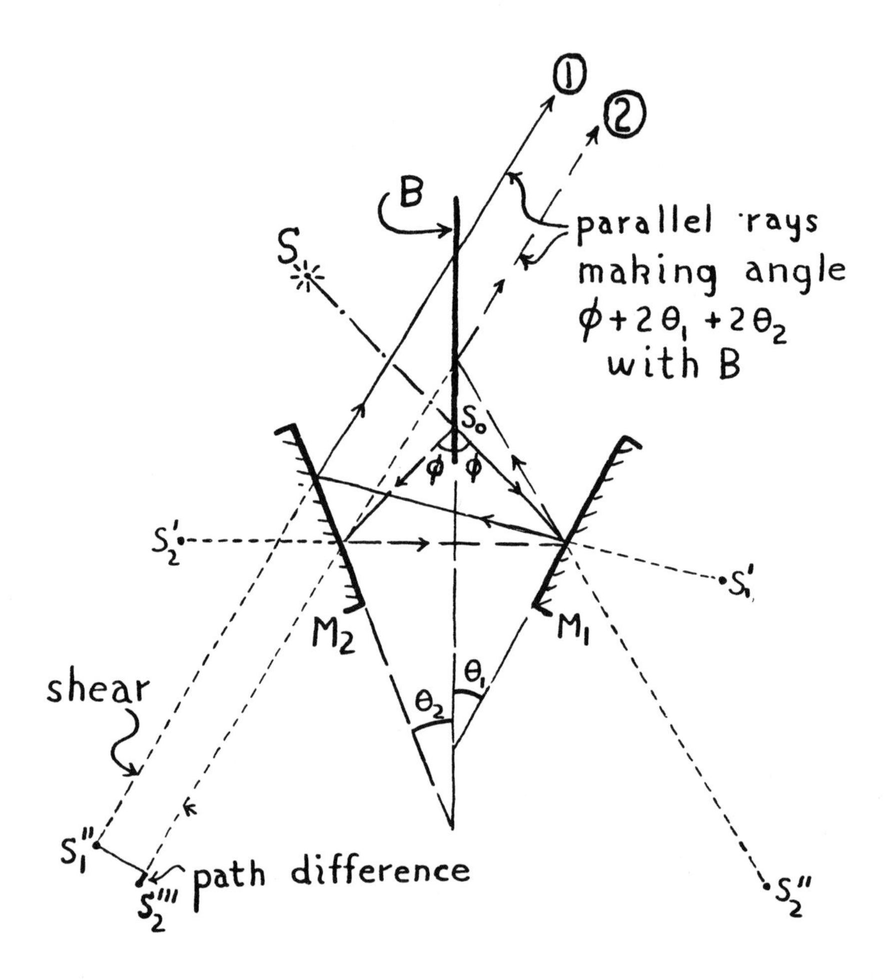

FIG. B6-2

Ray diagram to show virtual sources for interferometer (M_1 tipped).

APPARATUS

1) Interferometer

The interferometer for the demonstration is conveniently built of plywood. The base is a piece 1 inch thick and about a foot square. The mounts for the optical parts are made of 3/4 inch and 1/4 inch plywood as suggested in the sketches of Fig. B6-4. The mirrors are simply waxed to the adjustable plywood mounts as indicated. The mounts are equipped with adjusting screws (such as 8-32 threaded into the wood) and phosphor bronze springs. The base support has three supporting feet which are waxed to allow easy rotation about a screw pivot which holds the mount in place. The axis of rotation is approximately in the plane of the mirrors. The support for the beam splitter need not be adjustable, but should be provided with three feet for semi-kinematic design. The beam splitter mount should be arranged to hold either one or two plates so that both symmetrical and unsymmetrical beam splitters can be tried. The sketch in Fig. B6-4 illustrates a satisfactory mount. The angle of the beam splitter can be fixed at about 45° to the direction of the incident beam. It is perhaps best to lay out the parts symmetrically, making the exit beam emerge at right angles to the incident beam.

The mirrors are front silvered plate glass about 1 1/2 in. x 3 in. and 1/4 in. thick. The beam splitter can be made of 1/4 in. thick plates, obtainable from either Edmund Scientific Corp. or from Jaegers. There are several possibilities: some plates are available which reflect about 1/3 of the incident light, and one of these or a pair with their reflecting sides in contact may be used. Alternatively, so called one way mirrors (reflecting 80-90%) can be used either singly, or in combination with another piece of plate glass of the same thickness (need not be exactly the same). Fringes are obtainable with any of these combinations. Perhaps the 1/3 reflecting pairs are slightly superior in that the alignment is a bit easier.

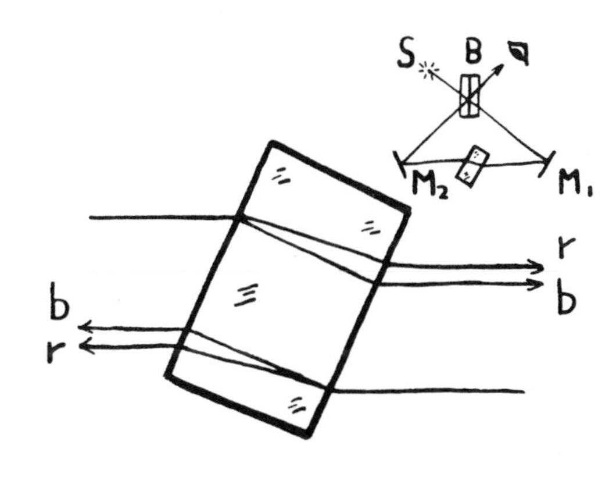

FIG. B6-3

Achromatization of fringes by tilted glass plate.

2) Auto lamp source.

3) Fine ground glass screen -- preferably ground so fine that the auto lamp filament is distinguishable through it (4 inch square).

4) Monochromatic source.

5) Thick glass plate about 2 inch square, 1/2 inch thick.

PROCEDURE

Set up the interferometer so that it is illuminated by the auto lamp which is placed about a foot away from the beam splitter. The ground glass is placed between the lamp and the beam splitter. Looking into the interferometer, one will see, if the mirrors are set at roughly the angles indicated in Fig. B6-1, several images of the source. If the glass is too coarsely ground for the filament to be seen, then a card with a small cross cut in it will provide the aligning method. There will be, in general, two sets of images, the number in each set depending upon whether the beam splitter is symmetrical or not. It is the brightest of each set which must be brought into coincidence by tilting or rotating either of the mirrors M_1 or M_2. When they are in coincidence, white light fringes should be seen; if they are not, adjust the mirrors very slightly until color is noted. The fringe system can be made circular or nearly straight depending on the adjustment of the mirrors. If one of the adjustable mirrors is not tilted or rotated, the fringe pattern will move off center and the fringes will become closer together and more nearly straight. In some cases, more than one set of fringes can be seen. This is particularly likely if the beam splitter is double (symmetrical), for then Fizeau type fringes are generally seen. If the beam splitter glass is made of Libbey-Owens-Ford twin grind glass, Brewster's fringes can be seen under the right conditions (see references in Experiment B8). Criss-cross fringes described briefly in the references may also be seen in some cases.

After observing the fringes in this manner, remove the ground glass screen and substitute an achromatic lens so as to produce collimated light incident on the beam splitter. Put the ground glass screen in the exit beam and observe the fringes on the glass. Finally, remove the ground glass and note that one can project the colored fringes on a white screen some distance away from the instrument.

If desired, one can align the instrument and obtain the white light fringes while the incident illumination is parallel. In this case, use a card with a small cross pattern cut in it and make the two brightest images as seen on the ground glass coincide.

Use either method of observation -- either with the ground glass in the exit beam and parallel light incident, or with the ground glass in the entering beam to diffuse the light incident. Rotate one of the mirrors so that the fringes are quite fine and vertical. Introduce the thick glass plate in the beam between mirrors M_1 and M_2 and tilt and rotate the glass plate slowly to restore the fringe pattern to its central position. One may obtain reasonably good achromatized fringes by proper adjustment of the plate, or still more colorful fringes by rotating the plate in the other direction and thus reversing the direction of shear introduced.

Remove the glass plate, obtain good vertical fringes with about 5

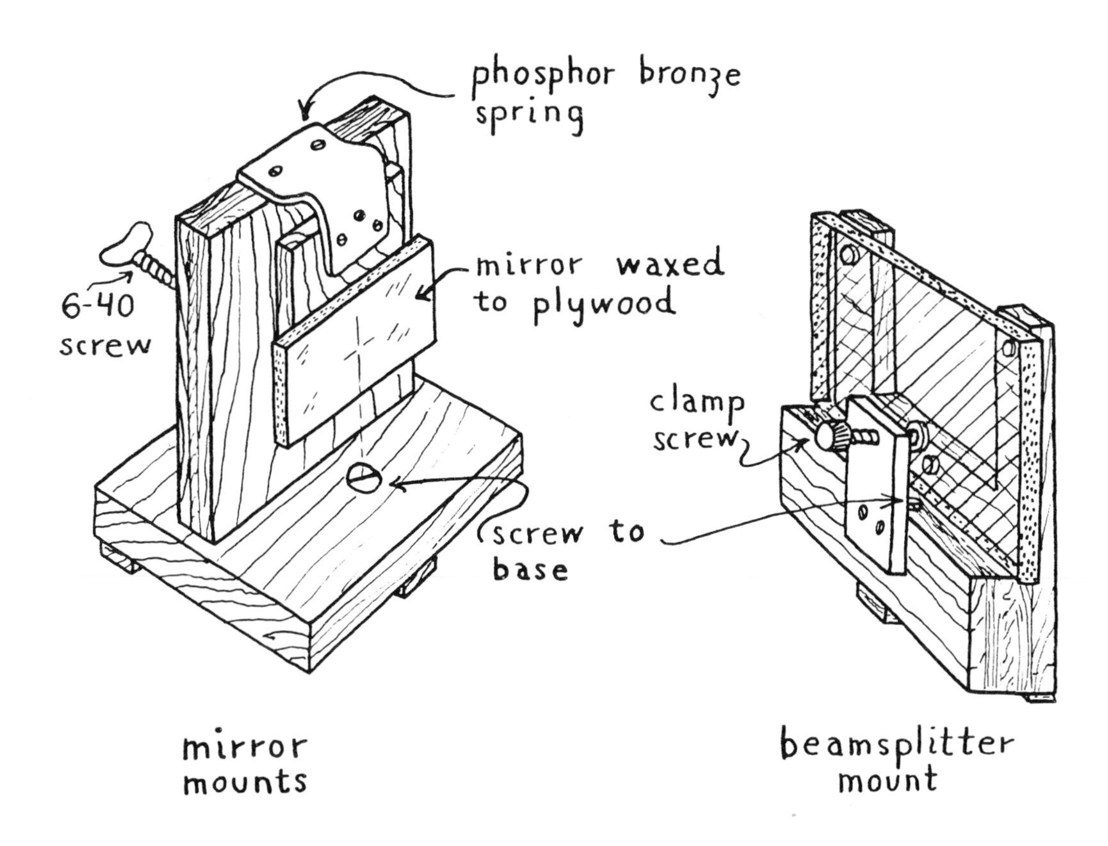

FIG. B6-4

Plywood mounts for the interferometer components.

fringes in the field of view, and substitute monochromatic light. If a 1 cm - 5 cm cell with plate glass ends or at least flat ends is available, fill it with water and insert in the beam in place of the glass plate. Put two or three drops of acetone in the water and observe the fringe pattern while disturbing the liquids to mix them together. The small changes in refractive index are readily apparent.

EXPERIMENT B7. CHANNELED SPECTRUM.

READING: Strong, Ch. 11-11.1, 11.4; or Jenkins and White, Ch. 14.16; or J. Valasek, <u>Introduction to Theoretical and Experimental Optics</u>, (1949, Wiley), pp. 388-391.

In this experiment, a channeled spectrum is used to calibrate a prism spectrometer. If white light is reflected from a thin dielectric film into a spectrometer, it is found that the spectrum is crossed by a number of black bands called FECO bands or fringes (Fringes of Equal Chromatic Order), and these interference bands may be used to calibrate a spectrometer. The procedure is especially useful in the infrared where few emission lines are available. (The film thickness is then directly measured).

THEORY

In the reading, it is shown that, if white light is reflected at normal incidence from a thin dielectric film, destructive interference results for wavelengths which satisfy the relation

$$m_1 \lambda_1 = 2Nt \qquad \text{dark band} \qquad (1)$$

where m_1 (the order of interference) is an integer, N is the refractive index of the film, and t its thickness. Obviously, other wavelengths give dark bands too when the interference order is different, that is

$$m_2 \lambda_2 = 2Nt \qquad (2)$$

$$m_3 \lambda_3 = 2Nt \qquad (3)$$

Let $M \equiv m_1 - m_2$ and $K \equiv m_1 - m_3$ be the integral differences in the orders of interference. Then

$$M = 2Nt\left(\frac{1}{\lambda_1} - \frac{1}{\lambda_2}\right) = m_1\lambda_1\left(\frac{1}{\lambda_1} - \frac{1}{\lambda_2}\right) = m_1\left(\frac{\lambda_2 - \lambda_1}{\lambda_2}\right)$$

so that

$$m_1 = M\lambda_2 / (\lambda_2 - \lambda_1) \qquad (4)$$

Thus, if the wavelengths corresponding to two black bands are known, and the number of bands between them is counted, i.e. M, it becomes possible to calculate m_1 from equation (4). Once the interference order for any one band is known, it is obviously possible to find the order for any other band simply by counting from the known band. (Does the interference order increase or decrease moving from red to blue?) It is thus evident that the wavelength for any band is calculable from m_1 and λ_1

$$\lambda_3 = m_1 \lambda_1 / m_3 = m_1 \lambda_1 / (m_1 - K) \tag{5}$$

There is one difficulty; the wavelengths for the two black band centers are unknown. The use of a line source such as a mercury arc to provide a comparison spectrum of known lines is indicated, but there remains a problem in that the mercury lines will not in general coincide with any of the band centers. The problem may be solved by using non-integral interference orders to determine both m_1 and λ_1. Figure B7-1 shows the situation. Two known wavelengths are used in order to find m_1, but now neither M nor m_1 will be integers. The centers of the black bands, of course, are still represented by integers, and the wavelength corresponding to any band is determined.

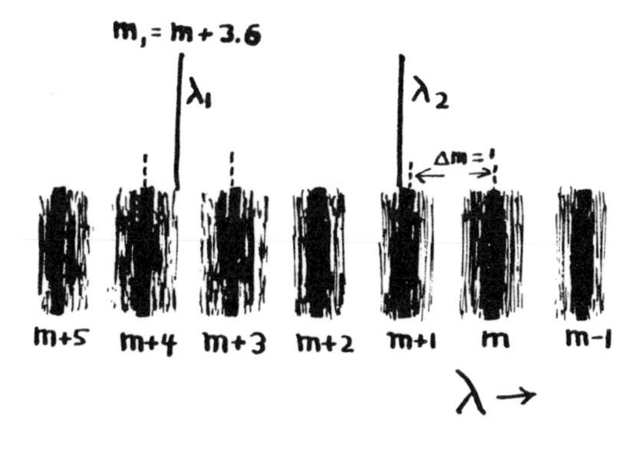

FIG. B7-1

Channeled spectrum and comparison spectrum.

APPARATUS

(1) Prism spectrometer:

Either a commercial prism instrument or a student spectrometer with a prism may be used for this experiment. A grating instrument is not desirable, since its calibration can be determined directly in terms of the grating equation.

(2) White light source (25 watt incandescent frosted lamp).

(3) Mercury or cadmium arc.

(4) Clear mica sheet an inch or so square.

(5) Micrometer caliper 0-1 in. or metric.

PROCEDURE

(1) Set up the apparatus

The first step is to set up the spectrometer and the light sources as shown in Fig. B7-2. The spectrometer must, of course, be in adjustment, (See Experiment A5). The mercury arc comparison source is generally so bright that with the spectrometer slit fairly wide open, it will overpower the observer's eye. It is therefore suggested that a ground glass screen be placed between the arc and the mica; here it serves the purpose of providing proper illumination for the collimator and also of scattering a good deal of excess light out of the way.

Prepare a thin flake of mica about 0.01 mm thick and at least one cm square from a thicker sheet. First bevel off one corner of the mica square at perhaps 45°. The splitting should be done under water or at least with a stream of water flowing over the mica. The razor blade or a pin is used to start a fracture in the mica and the water will allow the mica to split neatly along a cleaveage plane, especially if assisted a little with the razor blade or pin. Several trials will doubtless be required. Measure the thickness with a micrometer.

When a piece of about the right thickness is obtained and it appears to be of uniform thickness, dry it carefully with a piece of Kleenex and clamp it as shown in the figure so that it reflects white light into the spectrometer collimator. It may be desirable for the arc to be turned off during these preliminary stages.

Look through the spectrometer and observe the appearance of the FECO bands. If the bands are not straight and of reasonable contrast, the mica is not uniform in thickness. If more than about 30 bands can be seen, the mica is too thick; if less than 10, it is too thin.

When 10 to 30 good bands are observed, adjust the slit width for best visibility of the channeled spectrum. The mercury or cadmium arc should be turned on so that its spectrum is superposed on the channeled spectrum. A still neater arrangement is to illuminate one-half of the slit with the comparison spectrum, and the other half with the channeled spectrum, with perhaps a slight overlap. The white light source should be in such a position that the angle of incidence on the mica is as small as practical so that cos r in the mica will be nearly unity. (It is not necessary to obtain exactly normal incidence by means of a beam splitter, for example).

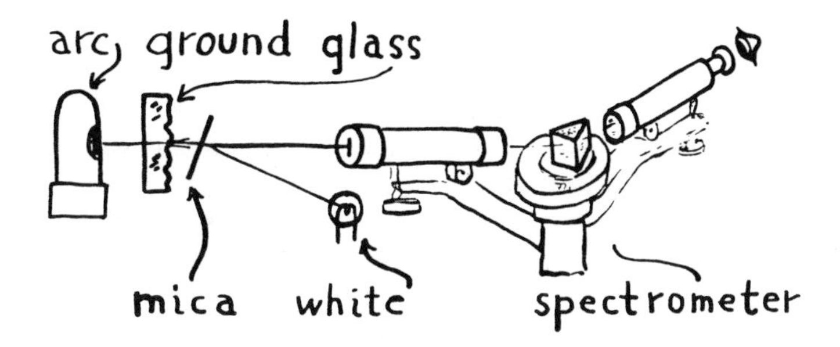

FIG. B7-2

Observation of a channeled spectrum.

(2) Calculate the thickness of the mica

Select two convenient known spectral lines of the arc -- say one yellow and one blue and count the number of black bands (including fractions) between the lines. Use equation (4) to calculate the interference order (non-integral) for one of the reference lines. Then use equation(1) to calculate the product Nt for the mica sample. Assume an approximate refractive index N = 1.58, and calculate the thickness. Make sure that this value agrees with the experimental error with the micrometer measurement.

(3) Calibration of the spectrometer

It is advisable to make a very careful sketch as in Fig. B7-1 of the spectrum showing the black bands and the comparison spectrum in the proper relationship. The (integral) interference order of all the black bands is now known, and the corresponding wavelengths may be calculated from Equation (5).

Measure the positions of about six bands distributed through the spectrum in terms of either the angle of deviation on the spectrometer or of the wavelength dial of a commercial instrument. Obviously with each recorded band position, it is essential to record the interference order so that the wavelength will be known.

(4) Unknown wavelength

Record the position of one or more unknown spectral lines designated by the instructor.

(5) Cellophane film

Substitute for the mica a piece of cellophane such as that used for cigarette packages (very nearly 1 mil thick). Describe the channeling of the spectrum observed.

RESULTS

Make a comparison of the thickness of the mica flake measured with the micrometer and calculated from the channeled spectrum. Indicate the sources of error. Which method is more accurate? Under what conditions?

Plot a calibration curve for the spectrometer indicating both the reference lines and their non-integral orders and the other calibration bands. If the instrument is already calibrated in wavelength, plot an error curve, i.e. a graph showing the correction to be applied to the calibrated wavelength dial.

Prove that the black bands of a channeled spectrum are separated by equal frequency differences.

Calculate the error made in assuming $\cos r = 1$ for a mica plate if the angle of incidence on the mica is 20°.

EXPERIMENT B8. DEMONSTRATION OF FIZEAU AND HAIDINGER BANDS.

READING: Strong, Ch. 11-11.3; or Jenkins and White, Ch. 14-14.6.

This experiment is primarily a demonstration of some of the properties of Fizeau and Haidinger bands. These bands have important applications in the quantitative testing of precision optical surfaces, and some of the methods will be indicated.

THEORY

Strong distinguishes three types of bands which result from two beam interference in dielectric plates: FECO bands, Fizeau bands, and Haidinger bands. Figure B8-1 shows the basis of this distinction. A ray of light (or wave front) is incident on a plane parallel plate at point P where it makes an angle i. Some of the incident amplitude is reflected as ray (1), the remainder is refracted at angle r and part of it again reflected as ray (2) which emerges parallel to ray (1). We ignore the part of ray (2) reflected back into the plate at the upper surface, for it has small amplitude in the case of low reflectivities we consider here. Also, we are not concerned here with the rays transmitted through the bottom of the plate. The optical path difference between ray (1) and ray (2) for any common plane AB normal to them (or parallel to a wavefront) is 2Nd cos r, and it gives rise to destructive interference if

$$m\lambda = 2Nd\cos r$$

where m is the integral order of interference for wavelength λ. The distinction between the three types of bands depends upon which factors in the equation are assumed to vary. If the right side of the equation is kept fixed and the wavelength allowed to vary, then for certain wavelengths, destructive interference results whereas for others, constructive interference according to the value of m. These interference bands are FECO bands which were illustrated in Experiment B7. On the other hand, if the wavelength is fixed, and d allowed to vary, the resulting interference bands are called Fizeau type. Finally, if the angle of refraction varies, the interference bands are Haidinger type.

The distinction between the three types of bands is not absolute, for more than one factor d, r, or λ may vary at the same time, and the types merge. In the Michelson interferometer, if the two paths are nearly identical, broad Fizeau type bands are observed; if the paths are quite different, Haidinger bands are observed. Clearly the two types must merge at some intermediate path difference where both d and r may change.

Newton's rings provide the classic illustration of thin film interference of the Fizeau type. They are observed near the point of contact of a convex spherical surface and a plane surface. In this case, both the film thickness (a function of the radial distance from the point of contact), and the angle r vary. But the band type is primarily Fizeau, for it is primarily the film thickness d that varies -- the effect of the variations in r is secondary. Newton's rings are often used to measure either the wavelength of the incident light in terms of the radius of curvature of the convex surface, or the converse. In this demonstration, they are to be observed as giving another illustration of Fizeau bands and showing the phase reversal on reflection. The relations between ring diameter and radius of curvature are given in the appendix to the experiment for those who may wish to make a quantitative study.

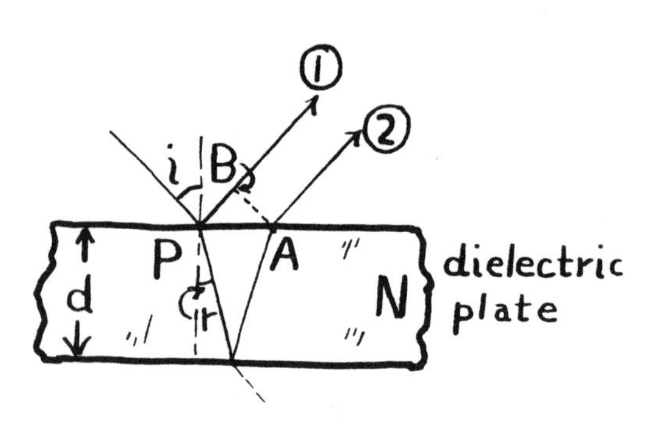

FIG. B8-1

Optical path difference for two rays produced by amplitude division of a wave front.

APPARATUS

(1) Sodium arc or mercury arc with green line filter.

(2) Low power microscope, preferably with mechanical stage drive.

(3) Two or more 2 x 2 in. squares of 1/4 in. thick plate glass, (Libbey-Owens-Ford twin grind plate).

(4) Small plano-convex lens of short focal length.

PROCEDURE

(1) Newton's rings

The demonstration should be set up using a low power microscope as shown in Fig. B8-2. First, clean a piece of plate glass and the plano-convex test lens with acetone to remove any grease or wax and then with a detergent solution. The pieces are then rinsed in clear water and dried with a soft towel or Kleenex. Finally, use lens paper or Kimwipes (Kimberly Clark Industrial Products Type 900-S) to remove any lint from the surfaces. Lay the flat plate with the convex surface of the lens over it on the microscope stage, using a sheet of lens paper between the surfaces.

Arrange a narrow thin piece of glass as shown to reflect diffuse light from the monochromatic source down normally on the field of view. Focus the microscope on the lens paper between the two glass surfaces; be careful to avoid letting the reflecting plate touch either the lens or the microscope objective. Next, adjust the substage condenser mirror to illuminate the field of from below, and then block this light with a piece of cardboard until later. Finally, pull the lens paper slowly from between the lens and glass plate while holding the glass parts. This procedure is a help in keeping dust particles from falling on the plate after it is cleaned.

Look for the Newton's rings at the point of contact of the two surfaces. If the radius of curvature of the lens is short, the pattern will be small and may be difficult to find; a stage adjusting mechanism on the microscope will help.

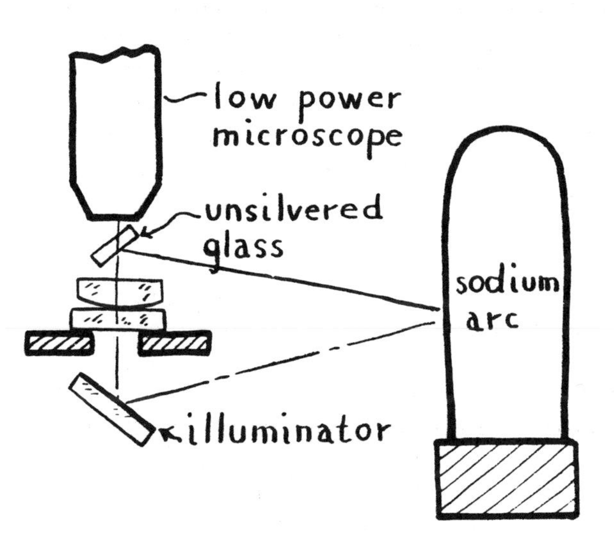

FIG. B8-2

Newton's rings by transmission and by reflection.

When the rings are found, center the pattern in the field of view and adjust the microscope focus slightly to obtain the sharpest rings. Note carefully the central portion. Is it dark or bright? Which should it be? If the central ring is not as it should be, the surfaces are not in optical contact. Repeat the cleaning.

Having observed the rings in reflected light, block the reflected beam with a cardboard shield and view the rings by transmitted light. Note the differences.

Replace the sodium arc source with a white light source and view the rings both in reflection and transmission. Compare these patterns to the Young and Lloyd white light patterns. Note that the Newton ring pattern arises from coherent sources separated in depth (amplitude division) whereas, the Young and Lloyd patterns arise from coherent sources laterally displaced (wave front division).

(2) Fizeau bands

Arrange the monochromatic light source as in Fig. B8-3 to observe the Fizeau fringes between the two pieces of plate glass. First, clean both pieces and put the sheet of lens paper between them as before. Fizeau fringes should now be seen by reflected light. Note whether the fringes are straight, curved, or irregular. Apply pressure to squeeze the plates together and note how the fringe pattern changes. From the appearance of the fringes, can any conclusions be drawn about the smoothness of the plate glass? Suppose that only one fringe were observed, would this fact necessarily prove that both plates were flat (to within a wavelength)? Suppose one plate were known to be flat within 1/100th wavelength, could any conclusion then be drawn about the figure of the other plate?

Use two thicknesses of very thin paper to separate the plates along one edge thus forming a wedge of air between them. Observe the Fizeau bands with a low power magnifier and make a mental note of their spacing. Now without disturbing the set-up, remove one thickness of paper so as to halve the wedge angle and note approximately how much the band spacing has changed. Also observe the effect on the bands when the angle of incidence is varied. Explain how the bands move in terms of the basic equation given in the theory.

Finally, remove the paper wedge altogether and obtain broad fringes once more. Observe the fringes in white light (the air film must be very thin, of course, to see them). Again using the monochromatic source and looking at the fringes, introduce a drop of water along one edge of the air film between the plates. As the capillary attraction pulls the water between the plates, observe carefully the fringes at the line of demarcation between the water and air. Explain why the fringes are less distinct and why they are closer together in the water.

If a master optical flat is available, it is possible to compare an unknown flat surface with it by means of Fizeau fringes and to determine whether the unknown surface is convex, concave, or flat. It is a question of observing the behavior of the fringes as the angle of incidence is varied. But suppose that no master flat is available? There remains an interesting possibility. Three flats may be made and tested against each other. If three surfaces, A, B, and C will make perfect contact with each other in any order or any position, they must all be flat. The procedure then, is to test A against B, B against C, and C against A. From observation of the Fizeau fringes, it is determined that A mated with B, say is 3 fringes convex, B against C is 2 fringes concave, and C against A is 1 fringe convex. If a plus sign represents convex, and minus concave, then three simultaneous equations result ($A + B = +3$, etc.) from which the three surfaces are calculated. (See A. G. Ingalls, E., Amateur Telescope Making, Book 1, (1933, Scientific American), pp. 54-56 (R. W. Porter). A more general discussion of optical testing is given by J. Strong, Procedures in Experimental Physics, (1939, Prentice-Hall), esp. pp. 63-66.

(3) Haidinger bands

Illuminate one of the glass plates using the same set up as for the Fizeau fringes (Fig. B8-3). The difference here is that the interference fringes must be viewed accurately normal to the plate, for, with a thick plate, small changes in angle correspond to large changes in path difference.

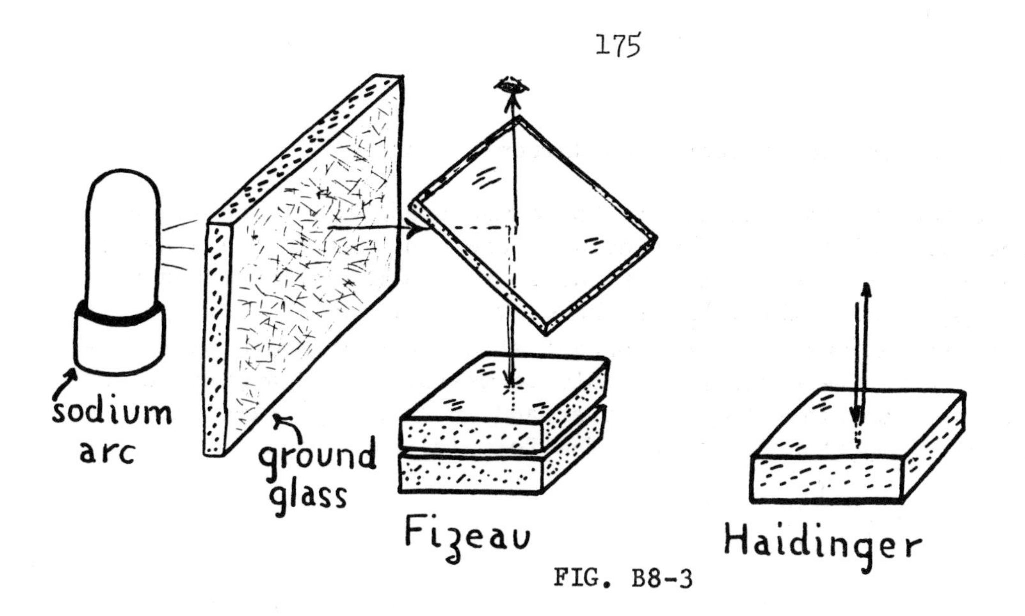

FIG. B8-3

Arrangement for Fizeau
and for Haidinger bands.

Unless the plate is very thin, the two sides must be accurately parallel in order to see the Haidinger fringes at all. A microscope slide will, in general, show them and so will a quarter inch thickness of Libbey-Owens-Ford "Twin grind" plate glass.

Observe the Haidinger fringes in the glass plate and note the change in fringe pattern as the plate is translated slowly in a transverse plane. It is readily possible in this way to determine any change in thickness of less than a quarter wavelength of sodium light. The technique is useful, for example, in the selection of the very uniform cell windows and compensator plates needed in the Rayleigh gas interferometer (next experiment).

Interference bands can also be obtained between various rays transmitted through two thick plates inclined at a very small angle. These bands, discovered by Sir David Brewster, are useful for certain precision adjustments of Fabry Perot interferometer systems. The bands are discussed by Jenkins and White in Ch. 14.11; additional information is given by T. Preston, The Theory of Light, 5th ed. (1928, Macmillan), pp. 224-226.

QUESTIONS

(1) What kinds of interference bands are obtained with soap bubbles in white light and monochromatic light?

(2) Thin film interference is commonly observed when white light is reflected from oil films floating on water. Is the interference constructive or destructive when the film is as thin as possible? Explain.

APPENDIX: Quantitative measurement with Newton's rings

If the Newton ring pattern is to be measured, the plane surface should be a good optical flat. In principle, one can measure the diameter of one ring and calculate the radius of curvature since the value radius is given by $R = D_m^2/4m\lambda$, where D_m is the diameter of the mth ring. This expression, however, assumes perfect contact at the center which is difficult to assure. It is therefore preferable to measure the diameters of at least two rings not too close to the center and in this way, eliminate the error introduced by imperfect contact. The formula then reads

$$R = \frac{D_m^2 - D_n^2}{4(m-n)\lambda} .$$

where D_m and D_n are the diameters of the mth and nth rings, and λ the wavelength.

EXPERIMENT B9. REFRACTIVE INDEX OF AIR WITH RAYLEIGH'S REFRACTOMETER*

REFERENCES: R. W. Wood, Physical Optics, 3rd ed. (1934, Macmillan), pp. 173-176; R. W. Ditchburn, Light, (1953, Interscience), Ch. 9.21-9.28.

Rayleigh's gas refractometer, which is based on Young's two slit interference phenomenon, is used in this experiment to determine the refractive index of air under standard conditions of pressure and temperature.

In Young's experiment, it was shown that the refractive index of a thin flake of mica could be measured by placing it in one of the two interfering beams. The results in that case were not very accurate because the mica thickness was uncertain and also because the fringe shift was not accurately measured. In the Rayleigh refractometer, both these difficulties are overcome.

Rayleigh's device used two gas cells, arranged so that one beam passed through one cell and the second beam through the other. In this experiment, only one cell is required.

THEORY

Figure B9-1 illustrates the principle of the method. If a Young type experiment is set up with a gas cell in one beam, it is only necessary to count the number of fringes passing the crosshair as the pressure in the cell varies from zero to atmospheric. If the cell is empty, the refractive index is unity and the optical path through the cell is simply L. When the cell is at atmospheric pressure, the optical path length is NL. The change in optical path, (N-1)L is equivalent to a fringe shift of Δm fringes of wavelength λ. It is not necessary to know the absolute order of the fringes, only the difference. In fact, the absolute order of the fringes cannot be known unless the paths are made identical at atmospheric pressure by introducing compensating plates of identical thickness in the reference path. Actually the compensation is required because the light source is not sufficiently coherent to produce fringes over the required path difference without compensation.

The theory of dispersion (Lorenz-Lorentz law) shows that the quantity $\delta \equiv N-1$ is directly proportional to the density ρ of a gas (as long as N is nearly unity). Thus for an ideal gas

$$\delta \propto \rho = PM/RT$$

*This experiment was devised by Dr. K. E. Erickson, formerly of the Laboratory of Astrophysics and Physical Meteorology.

where P is the pressure, M the molecular weight, R the gas constant, and T the absolute temperature. It then follows that for standard conditions, Po and To,

$$\frac{\delta_o}{\delta} = \frac{P_o T}{P T_o}$$

where $\delta_o = N_o - 1$ is the difference in refractive index for air under standard conditions. Thus

$$N_o - 1 = \delta_o = \frac{P_o T}{P T_o} \times \Delta m \times \frac{\lambda}{L}$$

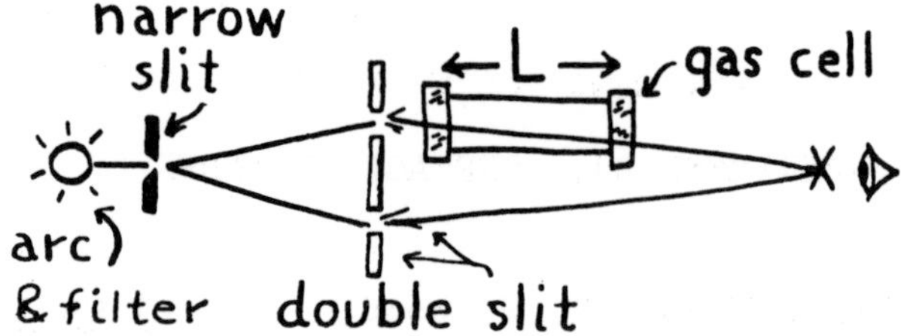

FIG. B9-1

The simplified Rayleigh gas refractometer.

APPARATUS

(1) High pressure mercury arc (AH3) with green line filter.

(2) Precision slit (only 1 mm length is used).

(3) Double slit made of cardboard and with slits 3 mm high, 1 1/2 mm wide and 10 mm between centers.

(4) Good quality lens at about 25 cm focus and with diameter 2 in. or more.

(5) Condenser lens: (See Fig. B9-2).

(6) Low power traveling microscope.

(7) Vacuum pump of moderate quality.

(8) Mercury manometer (0 to 1 atm. range) which can be read to within 1 mm.

(9) Controlled leak: it is best described by what it must do (see procedure).

(10) Glass tubing, connecting rubber vacuum tubing, and two stopcocks or pinch clamps.

(11) Gas cell: The cell is made of 1/2 in. diameter tubing, 30 to 40 cm. long and having a side arm for evacuating it. The ends of the tube are cut off as square as possible and carefully ground flat so that windows may be sealed on the ends. The cell windows and the compensator plates are made of microscope slide glass. Suitable microscope

slides of uniform thickness are selected by observation with Haidinger fringes as suggested in Experiment B8. One slide of good uniformity is cut up to make the two windows of the cell and one of the compensator plates. Another slide also of uniform thickness, but just about 1/1000 in. (25 microns) thicker than the first is cut to provide the second compensator plate. The extra thickness is advisable to compensate the optical path when the cell is at about half atmospheric pressure. Alternatively it's possible to use only one uniform slide and the extra compensation thickness required may be obtained by making a wedge of the compensator (see Experiment B4 for calculating the required angle of the wedge. Remember that here the beam traverses the compensators only once). The windows are cemented on the ends of the cell with beeswax and rosin mixture.

PROCEDURE

(1) Optical system

The apparatus is arranged as in Fig. B9-2. First line up the arc, the short slit opened to about a millimeter wide, and the lens L. The beam of light must pass through the center of the lens, which is located at a distance S equal to about 1 1/4f from the slit. The image of the slit, therefore, is formed at a distance S', about 5f, on the other side of the lens, giving a magnification of 4x. This image is made to fall on the crosshair of the low power traveling microscope. The double slit is now inserted, symmetrically placed, just ahead of the lens. The double slit will then produce diffraction patterns in the field of view, and these will overlap to give two beam Young interference. The condenser lens is next inserted between the arc and the single slit to achieve maximum illumination at the double slit. Put the gas cell in one beam near the lens as shown. In the other path, put the compensator plates -- either the combination of one thick and one regular plate as mentioned above, or the pair of equal thickness made into a wedge. Insert the green line filter. The single slit is now made as narrow as possible while still transmitting enough light for the fringes to be seen. It must be possible to see the fringes at all pressures in the cell.

(2) Vacuum system

Arrange the vacuum system with the three valves and the manometer as shown. It is very important to exercise care in operating the valves so that sudden changes in pressure do not occur -- otherwise the manometer may be broken and mercury spilled in the laboratory. (It might be a wise precaution to connect the manometer to the rest of the system with a piece of small bore tubing to prevent sudden changes in pressure from affecting the manometer too fast). The vacuum system need not be absolutely tight, but when sealed off, it must not leak more than 5 mm per minute.

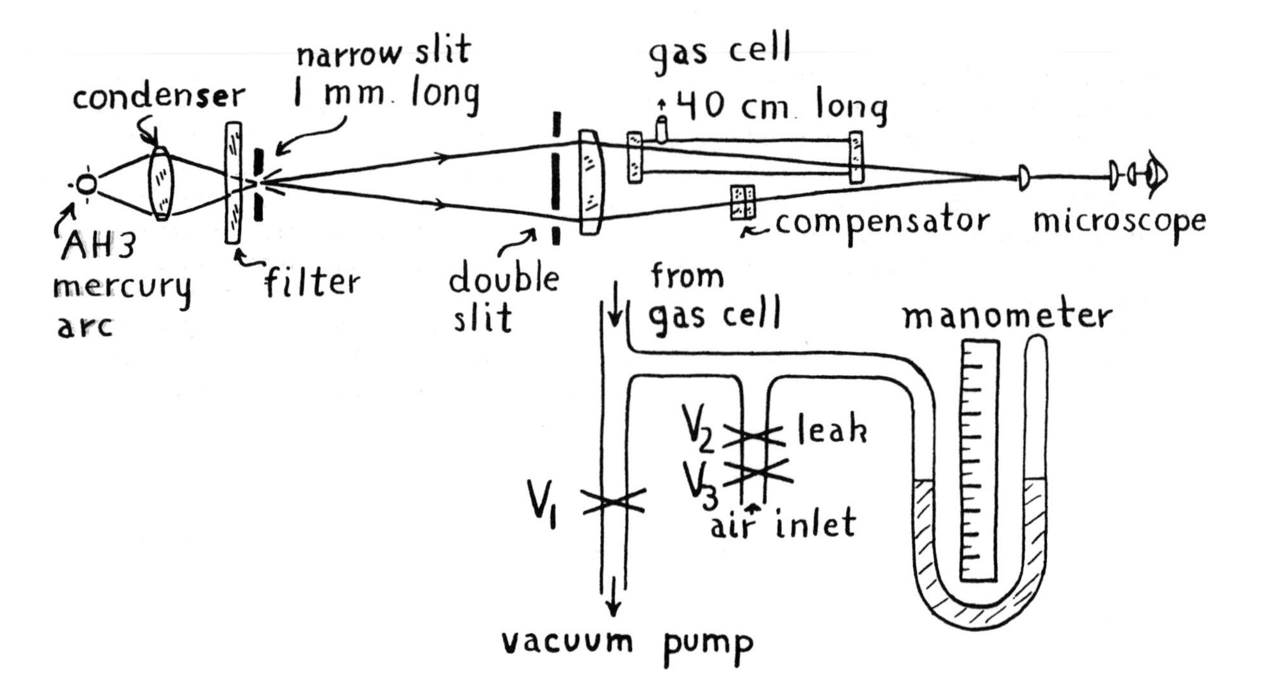

FIG. B9-2

Apparatus for the Rayleigh refractometer. Valves V_1 and V_3 are shut off pinch clamps or stopcocks, V_2 the controlled leak.

(3) Tests

Close all the valves and start the vacuum pump. The system (including the gas cell) is assumed to be at atmospheric pressure, and it is also assumed that green fringes are clearly seen in the microscope. Very slowly open V_1 and pump the cell down to about half atmospheric pressure, then close the valve and make sure the fringes are still seen -- they should be most distinct now, for both paths are supposed to be compensated at this pressure. Again open V_1 and pump down to a vacuum (less than 0.1 mm.), and again make sure that fringes are seen. If there is difficulty in seeing the fringes, then the adjustment of the compensator may not be right.

(4) Measurements

Evacuate the system until the fringes no longer move across the field of view. Close V_1, and, after checking that the controlled leak valve V_2 is closed, open V_3 carefully until it is wide open. Now open V_2 very slightly until the fringes begin to move and adjust the leak rate until about one fringe per second passes the crosshair -- an easily countable rate. Now, without disturbing V_2 or V_3, open V_1 to let the pump reevacuate the system (it must, of course, have enough pumping speed to override the leak).

When the fringes have again been brought to rest with the system evacuated and the leak in operation, set the crosshair on a dark fringe near the center of the pattern, and read the pressure on the manometer to the nearest half millimeter. Two alternative procedures may now be followed:

Procedure (A): Close V_1 and begin to count the fringes passing the crosshair. Count about 160 fringes and close V_3. After the fringes cease to move, record the exact number of fringes including the fraction of a fringe. Record the final pressure in the cell and the temperature.

Procedure (B): In this procedure, only about 20 fringes need be counted, and the pressure thereafter increased in steps without counting individual fringes. Begin as in procedure (A) with V_1 open and the controlled leak rate set. Read the initial pressure, close V_1, and count 19 fringes. Just as the twentieth dark fringe is passing by the crosshair, close V_2; record the pressure and the exact number of fringes (i.e. 20). The approximate value of fringes per millimeter pressure is now known.

Now, without bothering to count fringes, open V_3 and allow the pressure to rise to about double the previous value for 20 fringes, and while observing carefully, close V_3 just as a fringe is passing the crosshair. Record the pressure as before. The number of the fringe is readily found from the nearest integer to the value $m_2 = m_1 P_2/P_1$ where m_2 is the integral number of fringes for pressure P_2 and m_1 for P_1 (i.e. 20). Thus, a better approximation for fringes per millimeter is obtained.

Again double the pressure, stop just as a fringe passes the crosshair, and record the pressure. The integral number of fringes is again found using the improved approximation.

Continue doubling the pressure in steps, each time obtaining a better approximation, until a last step (not double the pressure) brings the final pressure to one atmosphere. This time, record the fractional fringe and determine as before the integral number of fringes. The final value is thus a number such as 213.6 fringes for 1 atmosphere pressure -- read in mm mercury. Be sure to record temperature also.

Measure the length of the cell, estimating to the nearest tenth of a millimeter. Leave V_1 closed, turn off the pump, and open V_3 to allow the system to remain at atmospheric pressure.

RESULTS

From the data, determine the exact fringe shift and the total pressure difference. Calculate the refractive index of the air under the laboratory conditions. Then correct these results to standard pressure and temperature. Determine the sources of error with care, and indicate the probable error. Is it legitimate to report for N_o a value such as

$$N_o = 1.0002791 \pm \text{a few parts in } 10^7?$$

EXPERIMENT B10. DEMONSTRATION OF THE STELLAR INTERFEROMETER.

READING: Strong, Ch. 8.8.

Stellar interferometers are used to measure both the angular separation of very close star pairs (double stars) and the diameters of single stars. In this demonstration, the Anderson interferometer is used to measure the diameter of an artificial star.

The difficulty in measuring the diameters of real stars is that they are so remote that, despite their enormous size, their disks in the focal plane of even the largest telescopes are smaller than the diffraction patterns produced by the telescope objective. As if this were not bad enough, the unsteadiness of the atmosphere, called seeing, is enough to blur the photographic images so that even the diffraction pattern is generally difficult to make out except in a small telescope. The stellar interferometer provides a successful solution to the problem -- at least for a few of the nearest and largest stars.

Two types of interferometer are described in the reading [1]. In both types, Young's white light fringes are caused to disappear (or nearly so) by properly adjusting a double slit to a rather critical value. In Michelson's arrangement, used for measurements of stellar diameters, an outrigger device carrying four mirrors, is attached to the telescope frame. The outer pair of mirrors slide on rails to provide the adjustable slit separation, and they reflect the light inwards to a second pair of fixed mirrors. The latter, in turn, reflect the light through two apertures in a cover over the telescope. The light then passes to the objective and eyepiece. The spacing of the fringes in Michelson's instrument is then determined by the separation of the inner pair of mirrors (or the apertures in the cover).

Anderson's interferometer is mechanically simpler than the Michelson since it has only one moving diaphragm containing two slits. The diaphragm allows light from only two small areas of the objective lens to

---

[1] A third type of interferometer recently invented by W. M. Sinton overcomes the chromatism of the fringes which handicap the Michelson and Anderson types. Sinton's instrument, which uses achromatized Lloyd's mirror fringes, is described in a paper entitled "An achromatic stellar interferometer," in the Astronomical Journal, Vol. 59, pp. 369-375, (1954).

reach the image plane where the interference pattern is observed. As it moves from the objective toward the observer, it selects areas of the lens which move farther and farther apart thus varying the effective double slit spacing. This interferometer is not suited to the measurement of the very small angles subtended by real stars (which require slit separations of twenty feet or more), but it is well adapted to the measurement of the much larger angular separation of many double stars.

In the laboratory, however, it is more convenient to use the Anderson interferometer than the Michelson to measure the diameter of an artificial star. The difficulty in the laboratory is to get a star small enough to measure, not the other way round, and the Anderson type is more than adequate for the job. To illustrate the point, it is easily shown that a 3 inch telescope should (under ideal conditions) resolve two stars having an angular separation of only 9×10^{-6} radians (1.8 seconds of arc). Therefore an artificial star at 10 meters distance should be only a few tenths of a millimeter in diameter for even a very small interferometer.

APPARATUS

(1) Artificial star:

The artificial star may be made from a tin can (bright inside) 2 or 3 in. in diameter. A quarter in. hole is drilled in the side and ventilating louvers cut in the top as shown in Fig. B10-1. A 6 volt, 21 cp. auto lamp provides the light. A hole about 1 mm dia. (#60 drill) is drilled in a piece of brass shim stock (heavy aluminum foil can also be used). The hole must be clean-cut and accurately round (triangular holes are all too easily drilled). The shim stock is secured over the quarter inch hole and is used to illuminate a microscope objective operated in reverse. A 20 to 40 mm focal length objective reduces the image by a factor of 4 to 8, thereby giving an artificial star uniformly round and bright of a fraction of a millimeter diameter. The microscope objective should be mounted at the standard distance (generally 16 cm) from the illuminated object in order that minimal aberration be introduced by the objective.

(2) Telescope objective:

The telescope requires a high quality well corrected achromatic doublet of 2 or 3 in. diameter and about 50 cm. focal length. High magnification is required, but if the focal length is too long, appreciable aberration will be introduced by the lens which will be working at the short object distance -- 5 to 10 meters -- available in the laboratory. The lens is supported on a special mounting on the optical bench which permits it to be very accurately tilted to make its optical axis coincide with that of the ways of the optical bench. A wood mount such as the one shown in Fig. B10-2 is satisfactory.

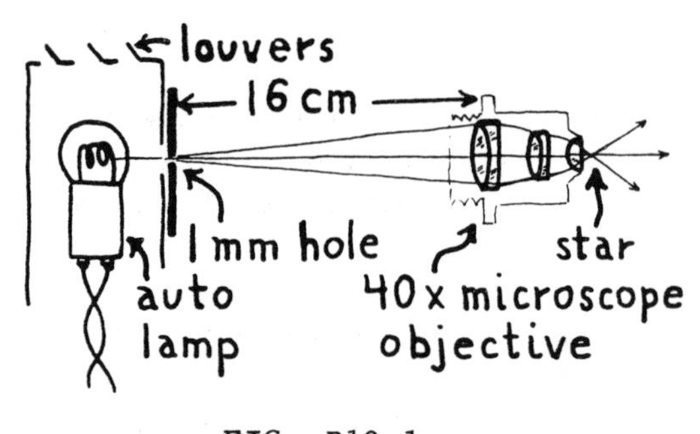

FIG. B10-1

Artificial star for the stellar interferometer.

(3) Eyepiece:

The telescope must have a high magnification and this is obtained by using a low power microscope as the eyepiece. The microscope is preferable to a very high power ocular because it has an accessible focal plane (and also it generally has better eye relief). It should have a magnification of about 40x.

(4) Double slit:

The double slit is made in the form of a pair of cat's eyes as shown in Fig. B10-2. Cat's eyes are preferable to a regular double slit because the illumination in the field of view is better distributed; the shape of the apertures apodizes the pattern vertically (Strong, p. 206). The slit separation should be about 5 or 6 mm, and the slits themselves about one mm wide and 4 or 5 mm high. (A special cat eye punch made in the shop was used to make a series of slit pairs for the original tests).

The double slit is mounted on a diaphragm which, in turn, is mounted on a slide on the optical bench. The position of the double slit should be adjustable transversly so that it may be accurately centered. It should be possible for the observer to slide the double slit diaphragm all the way from the primary image plane of the telescope to a point near the objective, and a system of strings and pulleys is suggested. The position of the diaphragm relative to the image plane must be measurable to within 1/2 cm.

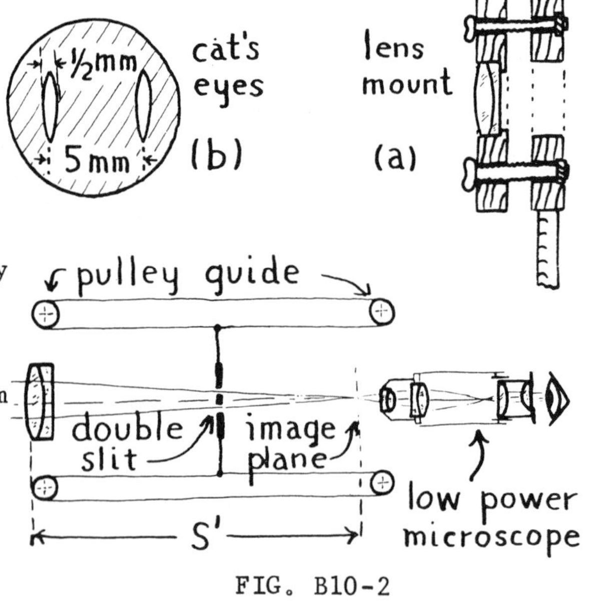

FIG. B10-2

The Anderson interferometer. Insert (a) shows an adjustable lens mount for the objective. Insert (b) shows the dimensions of the double slit diaphragm.

(5) Light shield:

Over the whole apparatus on the optical bench a framework is constructed to allow black cloth or paper to keep out most of the extraneous light, especially from the objective end of the instrument. Obviously the light shield must not interfere in any way with the motion of the diaphragm.

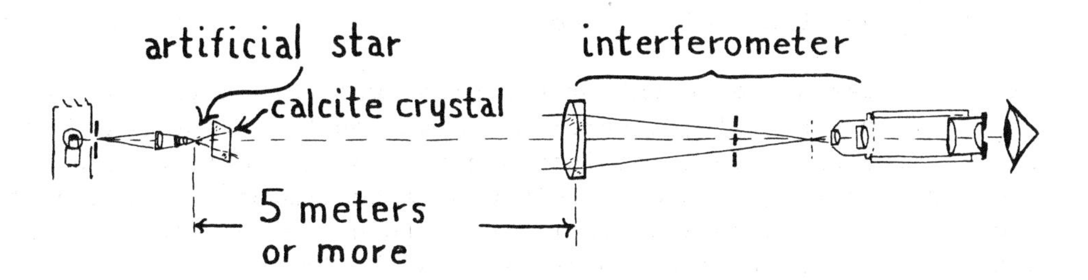

FIG. B10-3

Arrangement for Anderson interferometer. An artificial double star is made by using a calcite crystal as indicated.

PROCEDURE

Set up the apparatus as in Fig. B10-3. The separation of the telescope and the artificial star should be 5 meters or more. The star should be mounted at the same height from the floor as the telescope objective. Make sure that the star is properly set up so that it appears uniformly bright on either side of the telescope as well as above or below. Focus the telescope on the star with the double slit removed. The image should be clean-cut, round, and white. Small adjustments in the tilt of the objective lens and of the microscope make a considerable difference in the sharpness of the image. After the image has been made as good as possible, and very carefully centered in the field, clamp the microscope in place. Insert the double slit and slide it near the objective. Heavy black fringes should be seen in the central portion of the field. In general, there will be surrounding patches which will show appreciable color, but these may be disregarded for the measurements. As the double slit is moved closer to the observer, it is seen that the fringes get closer together (for the effective slit separation increases). The fringes also become fainter and fade.

In order to adjust the double slit properly, bring it to the point nearest the observer where light is still seen; this light comes from the outside edges of the lens. The pattern should be made as symmetrical as possible by adjusting the position of the double slit up or down or sideways in the diaphragm holder.

To make measurements, slide the diaphragm slowly from the objective to the near point (beyond which there is no light) several times while watching the fringe pattern critically. The fringes should be very distinct at the far point and gradually fade as the slits are brought closer. The bands will not vanish altogether, but should have a minimum visibility after which they will get sharper again at still closer distances. It is the minimum visibility position which is desired.

After a reasonable amount of practice, measure as well as possible the point of minimum visibility recording ten values to the nearest half centimeter. Average the values obtained and add any correction to determine the distance from the image plane -- this may be determined by sliding the diaphragm toward the microscope until it is in focus in the field of view. Also measure the distance from the objective to the image.

Make 1 or 2 measurements on the separation of the double slit. If practical, the measurements should be made with a traveling microscope as in the case of Young's interference (Experiment B1). In any case, read the positions of each edge of each slit taken across a line through the center perpendicular to the length of the slits as indicated in Fig. B10-2.

Calculate the angular diameter of the artificial star using the wavelength of white light (5500 Å) -- or the average wavelength seen through a red filter if one is used. Measure also the diameter of the hole in the shim (at the can) and calculate from the focal length and distance of the microscope objective the size of the artificial star. Finally, measure the distance from the star to the telescope objective and calculate the actual angle subtended by the star. The two angular values should agree to about 10 percent.

If difficulties are encountered in determining the fringe visibility minima, because of residual chromatism of the optical system, it may be helpful to use a red filter over the microscope eyepiece. (The filter should not be used earlier in the system for it must then be of exceptional optical quality).

In the event that it is desired to measure a double star, the use of a few millimeters of calcite crystal placed over a very small pinhole over the can might be tried.

Appendix: Sample measurements

Separation of cat's eyes, $a' = 6.51$ mm (average)

Image distance of telescope $s' = 77\ 1/2$ cm.

Distance of diaphragm from image plane for minimum visibility, D. (Average values).

D red =49 cm ($\lambda = 6 \times 10^{-4}$ mm)
D white=54 cm ($\lambda = 5.5 \times 10^{-4}$ mm)

Effective slit separation, a

$$a_{red} = 6.51 \times 77\ 1/2/49 = 10.3\ \text{mm}$$

$$a_{white} = 6.51 \times 77\ 1/2/54 = 9.33\ \text{mm}$$

angle subtended by star, $\alpha = 1.22\ \lambda/a$

$$\alpha_{red} = 1.22 \times 5 \times 15^{-4}/10.3 = 7.1 \times 10^{-5}\ \text{radian}$$

$$\alpha_{white} = 7.2 \times 10^{-5}\ \text{radian}$$

angle measured:

$$\text{star} = 0.735\ \text{mm}/8.9 \times 10^{3}\ \text{mm} = 8.2 \times 10^{-5}\ \text{radian}$$

The star in this case was larger than the recommended size, and the results not the best obtained, but perhaps typical of those that may be expected.

EXPERIMENT B11. DEMONSTRATION OF BREWSTER'S BANDS, HAIDINGER'S BANDS AND FABRY-PEROT BANDS.

READING: Jenkins and White, Ch. 14.2, 14.7-14.14; also Strong, Ch. 11.3, 12.5-12.7.

This demonstration illustrates some of the types of interference bands which can occur between two thick plates, partly silvered or unsilvered. Brewster's bands are two beam white light interference bands involving reflections from all four surfaces of the two plates. They are seen when the plates are inclined with respect to each other and they are different for silvered or unsilvered plates as will be shown. Haidinger bands are seen only with monochromatic light and involve reflections between the two surfaces of a thick plate with almost perfectly parallel surfaces, and incidence very near normal. Haidinger bands also involve two beam interference. Finally, the Fabry-Perot bands occuring between the inner surfaces of two silvered plates involve multiple reflections; they are seen only in highly monochromatic light and then only if the two silvered surfaces are very nearly parallel and are observed near normal incidence.

THEORY OF BREWSTER'S BANDS

The nature of Brewster's bands is illustrated in Fig. B11-1. Several parallel rays are incident on two plates of equal thickness t inclined to each other at an angle θ . These rays may be divided into several classes according to the kinds of reflections involved. Rays A, B, and C all traverse the space between the two plates just once, and they all emerge parallel to the incident direction. Ray A suffers no reflections while B and C are twice reflected in one or the other plate. Obviously ray A follows the shortest path through the system. Rays B and C, for the symmetrical situation shown, follow equal paths through the plates. The optical path of B through the lower plate is the same as the optical path of C through the upper plate.

Rays D, E, and F form another class. In each case, the rays traverse the space between the plates three times. Ray D, however, like A, traverses each plate but once. Rays E and F, like B and C, traverse one or the other plate three times. Note that because of the reflections between the (non-parallel) plates, the emerging rays are not parallel to the incident rays on the first plate. The optical paths of rays E and F may be equal (they will be so in the symmetrical situation).

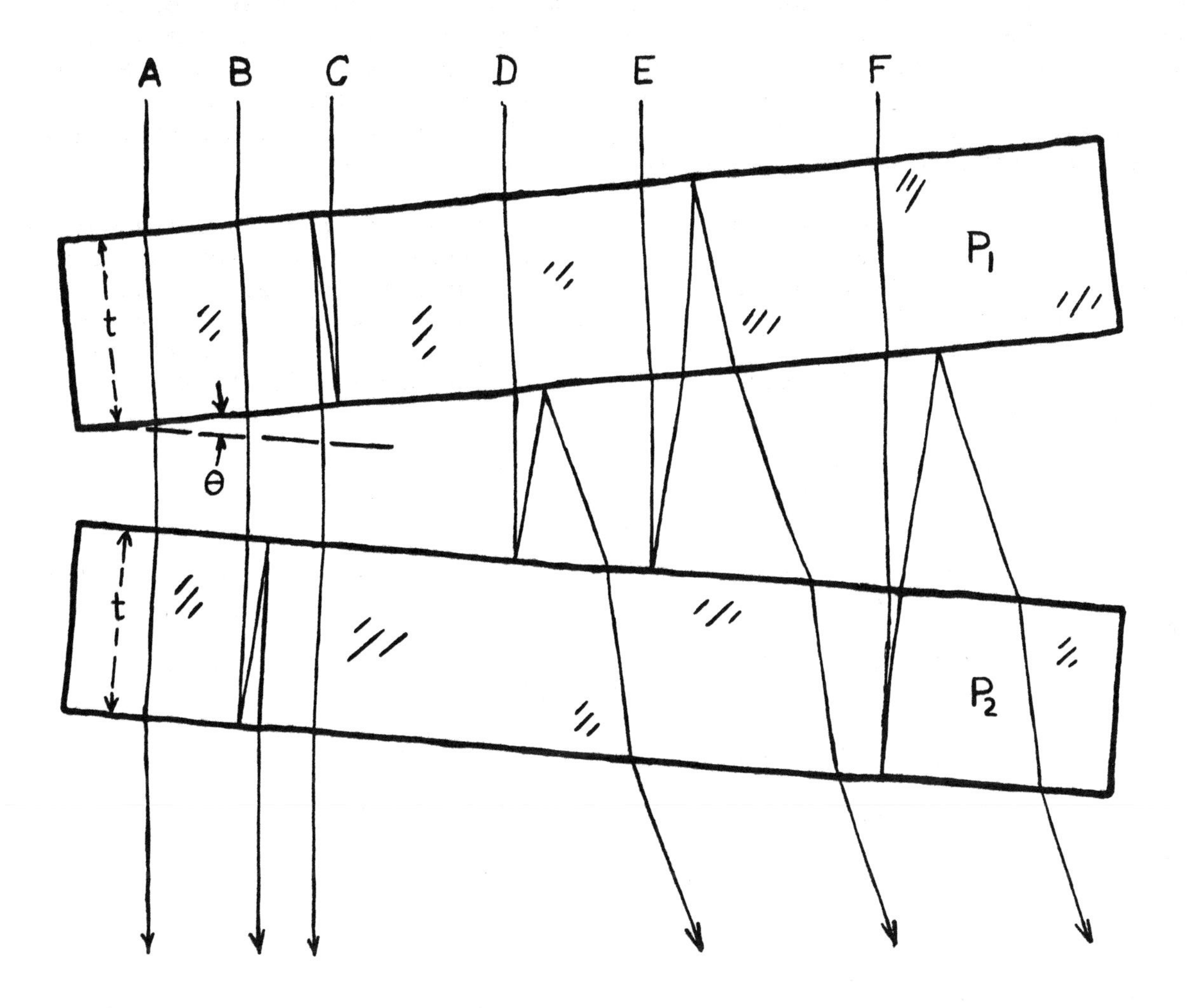

FIG. B11-1

Ray paths between two plates to show the nature of Brewster's bands.

The optical path difference between rays B and C and also between rays E and F depends on (1) the angle of incidence of the rays on the first plate and (2) the angle between the plates. Clearly the path length through the plates is least at normal incidence and increases either way off normal. For this reason, the central fringe occurs when the plates are traversed symmetrically.

If the glass plates are unsilvered, it is apparent that ray A is unreflected whereas both B and C are twice reflected. Assuming a reflectivity of about 4%, we note that the relative intensities of rays B or C to A are only about 0.16%. Thus Brewster's bands are not seen in the direct image. On the other hand, considering rays D, E, and F, we find that all three suffer two reflections and thus are about equally intense.

Actually, E and F suffer somewhat less than 8% additional reflection loss over and above that for ray D, but the difference is small. If then, an extended white light source is viewed through a pair of plates inclined to each other, the direct image shows no noticeable interference bands, but the image produced by the deviated rays, much weaker than the direct image, does show Brewster colored interference bands.

It should be mentioned that other classes of rays involving more reflections can also produce interference, but generally such rays are too weak to be easily observed, and need not be considered further. Jenkins and White mention the possibility of Brewster bands when one of the plates is twice as thick as the other -- specifically in two Fabry-Perot interferometers. Here the Brewster bands are of great use in adjusting the interferometers.

APPARATUS

1) Two 2 inch square pieces of 1/4 in. plate glass (good quality such as Libby-Owens-Ford Twin grind).

2) Two 2 inch square pieces of 1/4 in. "one way mirror", i.e. plate glass silvered about 90% reflecting.

3) 100 watt incandescent source (frosted).

4) High pressure mercury arc -- AH3.

5) Low pressure mercury, cadmium, or helium arc.

6) Tackiwax.

7) (Optional) Special mount described in the appendix to the experiment.

PROCEDURE

A. Brewster's bands

Clean all fingerprints from the two unsilvered glass plates and put a spacer of about 1/32 in. thick along one edge to form a wedge. Observe the incandescent source through the combination, the lamp being at a distance of about 5 to 10 feet. Tilt the pair of plates very slowly about the vertex of the wedge and the Brewster bands should be readily apparent in the deviated image of the lamp. If difficulty is encountered, one may use a 12 in. long cardboard tube with a cover having a 1/4 inch hole over the end to screen out most of the undesired room light.

Are the interference bands symmetrical? Is the central band dark or bright? How many bands can you see? How may the separation of the bands be changed? What is the optimum angular size of the light source?

Use the two silvered plates (cleaned with lens paper or Kleenex) placing them first with their silvered sides facing each other. Use the same wedge as before, and observe the white light source. Can interference bands be seen in the deviated beam? Can they be seen in the undeviated beam? Do not give up too easily, the bands are not nearly so obvious as with the unsilvered plates. Record the results and explain them.

Now put the silvered sides away from each other. Again look for Brewster's bands. Are there any differences from the previous case? Explain.

Finally, arrange the plates so that one silvered side is inside and one outside. Repeat the observations and explain the results.

B. Haidinger bands

In Experiment B8, the Haidinger bands were observed by reflection in a thick (unsilvered) plate by illuminating and observing at normal incidence from the same side (see Fig. B8-3). In this experiment, we are concerned with transmission. Use the low pressure arc source with a frosted glass screen to diffuse the light unless the source already is of large area. Hold one unsilvered plate at nearly normal incidence and by tilting it slightly about the normal, find the Haidinger bands by transmission. The contrast will be low compared to the reflected contrast in the earlier experiment -- just as Newton's rings have more contrast by reflection than by transmission. Next use of the partly silvered plates and find the Haidinger bands. How does the contrast of the rings appear in this case? Explain. Move the eye to observe the bands over various parts of the plates so that the variations in thickness are noted.

After the Haidinger bands are found, substitute the high pressure mercury source, but do not turn it on until ready to begin. Within a few seconds after the arc is turned on the Haidinger fringes should be clearly seen. Continue to observe them as long as possible. (A green filter may be used if desired). As soon as the arc warms up the fringes fade and vanish. How is this phenomenon explained?

C. Fabry-Perot bands

The two plates may be made accurately parallel either by the method about to be described, or with the slightly more elaborate scheme described in the appendix following.

Make three small balls of tackiwax about 3/16 in. diameter, and place one at each of two corners along one edge of one silvered plate on the silvered side. Place the third ball along the middle of the opposite edge, and put the other silvered plate against the tackiwax balls (silvered side to the wax) and press gently. Now look through the plates at a point source or a small source and by squeezing gently at the corners, bring into

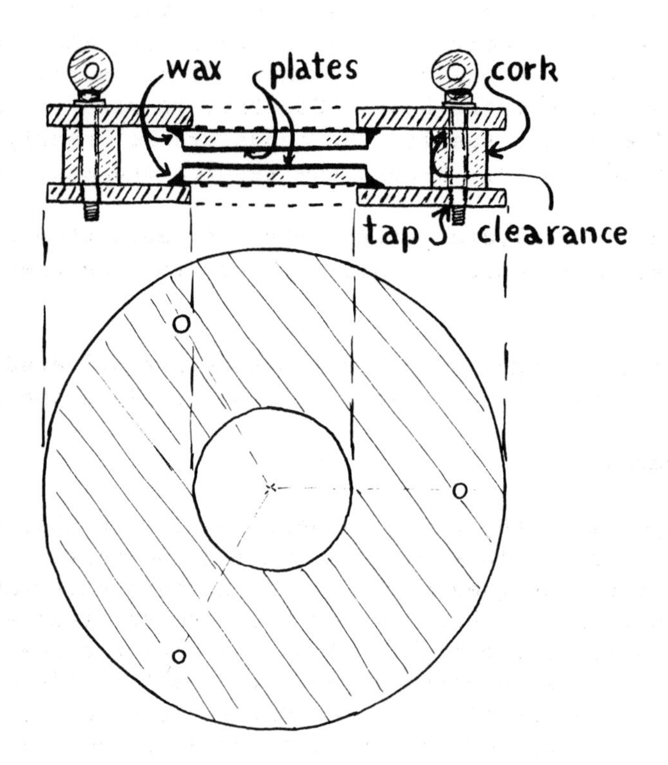

FIG. B11-2

A convenient adjustable mounting for demonstration of the various types of interference bands described in the experiment.

coincidence the multiple images seen. While this adjustment is being made, the reflection of the observer's eye should be seen and it should be made to coincide with the images. The last step, of course, insures that the observer is looking normally at the plates.

Look at the wide area monochromatic source and make very small adjustments in spacing until both the Haidinger bands in the plates and the Fabry-Perot bands are seen.

It will be found, of course, that the Fabry-Perot bands are highly unstable. The device described in the appendix results in good stability.

RESULTS

Record and explain the observations made in the demonstration. Explain what kinds of bands are observed in a Fabry-Perot interferometer made with parallel plates.

Why are the plates of a properly made Fabry-Perot interferometer made slightly prismatic?

APPENDIX

A convenient mounting method for the plates in this demonstration is to use two aluminum rings machined on a lathe as illustrated in Fig. B11-2. The plates, clear or silvered, are waxed between the rings as shown.

EXPERIMENT B12. DEMONSTRATION OF MULTIPLE BEAM INTERFERENCE IN MICA.

READING: Strong, Ch. 11.8-11.10, 12.4. For more detailed treatment, see M. Born and E. Wolf, Principles of Optics, (1959, Pergamon), Ch. 7.6.7, and also S. Tolansky, Multiple-Beam Interferometry of Surfaces and Films, (1948, Oxford, Clarendon Press).

In this demonstration, small changes in the thickness of mica (resulting from the natural crystalline layers of which it is built) are shown. Changes smaller than the wavelength of light are made apparent by observing multiple beam interference between partly silvered mica and partly silvered glass. Although here the experiment is purely qualitative, it can be made quantitative as the advanced references explain.

APPARATUS

1) Mercury arc and green filter.

2) Condenser lens.

3) Mirror about 2 inch square.

4) Low power microscope (20-40x).

5) 70-90% reflecting glass (transmission at least 10%) about two inch square.

6) Mica sheets silvered to reflect 70-90% and transmit at least 10%.

PROCEDURE

The demonstration is conveniently set up as shown in Fig. B12-1. Filtered green light from the mercury arc is condensed onto the semi-transparent mirror sandwich from below as shown. The transmitted light is viewed through the low power microscope. The mica sample can be pressed lightly against the glass with the two silvered sides facing each other. Rather fine Fizeau type multiple beam interference bands should become apparent when the microscope is correctly focused. The general appearance is shown in Fig. B12-2. Various areas of the mica are inspected and the

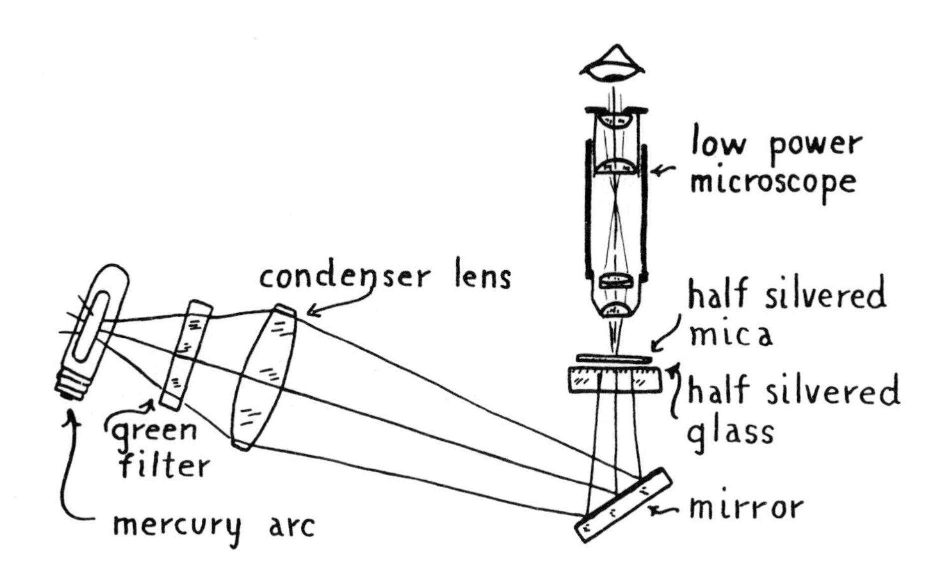

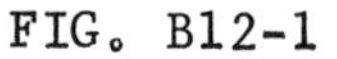
FIG. B12-1

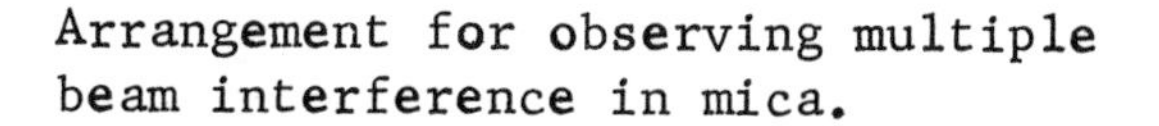
Arrangement for observing multiple beam interference in mica.

fringe pattern followed until a change in thickness is noticed. Sketch typical patterns. Some samples of mica cleave so well that there are large uniform areas which show no terraces. For this reason, it is well to silver several samples of mica to find one which clearly shows the desired terraces.

After inspection with green light, remove the filter and observe the fringes with unfiltered light. The fringes are more difficult to see now because the patterns for the different wavelengths may overlap.

The fringes may also be observed in reflection using a different arrangement of the optical parts.

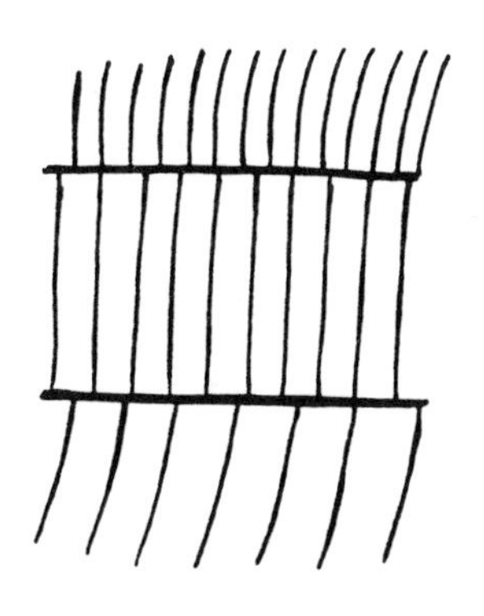

FIG. B12-2

Fringes in the region of mica terraces.

QUESTION

How could this demonstration be made quantitative?

EXPERIMENT B13. DEMONSTRATION OF FRESNEL DIFFRACTION BY CIRCULAR APERTURES AND OBSTRUCTIONS.

READING: Strong, Ch. 9-9.3; or Jenkins and White, Ch. 18-18.6.

This experiment illustrates the resolution of a spherical wave front into Fresnel zones by means of (1) Poisson's spot, (2) circular apertures, and (3) the zone plate.

THEORY

Figure B13-1 shows how a spherical wave front coming from point S is divided into half period zones. The spherical wave front of radius a intersects the line connecting S and the point of observation P at the point T as shown. The zones are apportioned so that the ring bounding each zone is half a wavelength farther from P than the previous ring. The distance QP is thus $b + 2\lambda$. The diameters of the various zone limits (rather than the radii) are given by

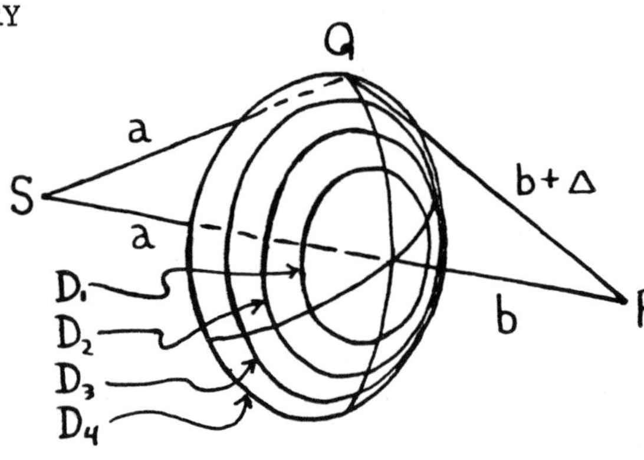

FIG. B13-1

Fresnel half period zones.

$$d_m = \sqrt{4m\lambda ab/(a+b)}$$

where m is the zone number.

If some central part of the spherical wave front is blocked by a circular obstruction, then all the large number of remaining zones contribute to the illumination at point P in the center of the shadow to form a spot very nearly as bright as the illumination by the unobstructed beam. This spot, first found by Maraldi and subsequently forgotten, was predicted by Poisson on the basis of Fresnel's theory (by which he sought to discredit the theory). The spot was promptly rediscovered by Arago after Poisson's prediction.

The contributions from successive half period zones tend to cancel, and so if the light from all zones save the first is obstructed, the illumination at P is four times greater than with the unobstructed wave! Adding a second zone to the aperture reduces the illumination (at P) to zero; a third zone restores it to that of one zone, and so forth.

A diaphragm constructed to block alternate zones is called a zone plate. Since alternate zones have opposite phases and tend to cancel each other's contributions, the elimination of either the even or odd zones allows all the contributions to add in phase so that the illumination at P is increased by the square of the number of zones. It makes no real difference whether the even zones or the odd zones are used in the plate (i.e. whether the central zone is opaque or transparent). If, however, a zone plate is made such that the alternate zones, instead of being blocked, are left clear and arranged to advance or retard the phase by just a half wavelength, their contributions are added in phase to that of the others, and the illumination at P is increased another fourfold. The construction of such a zone plate is discussed by R. W. Wood, Physical Optics, 3rd ed. (1934, Macmillan), p. 38. A zone plate has an effective focal length $f = d_1^2/4\lambda$.

APPARATUS

(1) Sodium or mercury arc (with green filter).

(2) Glass plate and thumbtack (or small bearing ball 1/8-1/4" dia.).

(3) Thin brass shim stock diaphragm:

The diaphragm is made for either of two distances, 25 cm or 1 meter with a equal to b. Holes are carefully drilled in the brass to transmit the light from 1, 2, 3, and 4 zones respectively. The dimensions and drill sizes are given in Table 1. The holes must be clean cut and round.

(4) Zone plates:

Satisfactory zone plates are made by photographing the zone plate pattern in Strong's book (p. 189) on a small scale. Use high contrast (therefore slow) film or plates. (If the pattern is 1/5 the original size, the focal length is about 2 meters).

(5) Magnifier lens about 1 in. dia. and 2 or 3 in. focus.

Table 1. Dimensions of holes for the Fresnel zone diaphragm.

	a = b = 25 cm			a = b = 1 meter		
Zone	dia. (mm)	dia. (mils)	drill#	dia. (mm)	dia. (mils)	drill#
1	0.50	20	76	1.00	40	60
2	0.71	28.3	70	1.41	56.6	54
3	0.86	34.4	65	1.73	69.2	50
4	1.00	40	60	2.00	80	46

PROCEDURE

(1) Poisson's spot

Cover the arc source with a pinhole in a card or aluminum foil -- the hole should be about half a millimeter in diameter. For observation of the Poisson spot, a concentrated arc (white) source may also be used. Mount the thumbtack on the glass plate with wax or cement to provide an obstructing disk. Keep the edge clean and round. The plate and disk are put about 3 meters away from the point source in a darkened room and the Poisson spot observed at about 3 meters on the other side of the thumbtack. Use the magnifier and observe its shadow carefully. Outside the shadow will be seen a series of concentric rings -- colored if a white light source is used -- and at the center of the shadow will be seen the Poisson spot also surrounded by a few rings of light.

(2) Fresnel zones

Mount the diaphragm with the series of holes at the design distance -- 25 cm or 1 meter from the pinhole arc source, and observe at the same distance on the other side. The smallest hole should be bright at the center, and the others should alternate dark and bright. If the patterns are not circular, but rather somewhat triangular, then the holes were not drilled circular -- look at them under a microscope.

(3) Zone plates

The zone plate may be either positive or negative, that is the central zone may be either clear or opaque -- the effect is the same. Set up the zone plate and illuminate it with the pinhole at a distance which is calculated to be 2f. Move gradually away from the zone plate and locate the image spot. Is the distance equal to that calculated? Is there more than one image spot?

Mount the zone plate at the end of a long tube of length f, blocking out all the light save that which passes through the zone plate. Use a long focus lens to make a telescope to examine a distant light source.

Does the zone plate obey the thin lens formula, $1/S + 1/S' = 1/f$?

EXPERIMENT B14. QUANTITATIVE STUDY OF FRESNEL DIFFRACTION BY A SLIT.

READING: Strong, Ch. 9.4-9.9; or Jenkins and White, Ch. 18.7-18.15.

REFERENCE: R. W. Wood, Physical Optics, 3rd ed. (1934, Macmillan), opposite p. 829 is an enlarged scale drawing of Cornu's spiral.

A photomultiplier unit is used in this experiment to obtain a quantitative measurement of the intensities in the Fresnel diffraction pattern of a single slit aperture. The measured pattern is to be compared with the theoretical curve. The necessary theory for the calculations is given in the reading. The numerical values are computed either from the table of values of the Fresnel integrals (given in Jenkins and White) or from the Cornu spiral (given by Wood).

APPARATUS

(1) Two meter optical bench. (A one meter bench could perhaps be used if necessary).

(2) Mercury arc and green line filter. (No special regulation of the power supply is ordinarily required).

(3) Condenser lens.

(4) Slit S_1 of good quality, 1 cm long.

(5) Low power eyepiece.

(6) Slit S_2: 2 mm x 5 cm.

(7) Traveling microscope carriage or precision, calibrated optical bench cross slide.

(8) 1P21 photomultiplier unit with power supply and galvanometer.

(9) Small high quality slit S_3: 1 cm long.

PROCEDURE

First, clean the slits with lens paper to make sure there is no oil or grease on the jaws, and then smoke the jaws with soot. Either a candle flame or a turpentine flame (obtained from a small wad of cotton soaked in turpentine and held with tweezers) may be used for smoking.

Mount the apparatus as shown in Fig. B14-1. First position the arc, slit S_1, and an observing eyepiece (in the position where the photomultiplier slit S_3 will go later) on the optical bench. Slit S_1 should be at least four times the focal length of the condenser lens from the arc. (Why?). It is important that the various parts be accurately in line. Add slit S_2 mounted on the traveling microscope frame or a precision cross slide half way between the slit S_1 and the observing eyepiece. Narrow the slit S_1, add the condenser lens to focus the arc on the slit correctly oriented for minimum spherical aberration (see Experiment A10), and the green filter. Observe the Fresnel diffraction pattern through the eyepiece and adjust both the slit width (S_1) and the orientation of either this slit or S_2 to obtain the best diffraction pattern. (Obviously S_2 must be mounted so that its length is perpendicular to the direction in which it is translated by the cross slide or traveling microscope unit).

Substitute the third slit S_3, 0.05 mm wide and the photomultiplier unit for the eyepiece. Close slit S_1 altogether, block out all possible extraneous light, and adjust the photomultiplier unit. First zero the galvanometer with the zeroing controls (with the photomultiplier shutter closed). Now open the shutter fully so that light may pass through the slit into the photocell. (It is important to make sure no other light reaches the cell). Now carefully open slit S_1 until the galvanometer reads about half scale. Translate S_2 to a point where an intensity maximum falls on the photomultiplier, and adjust the slit S_1 to give nearly full scale deflection. Carefully adjust the tilt of S_3 at the photomultiplier to give the maximum signal, that is to make the slit as nearly parallel to the other slits as possible. Now mask off all but 1 mm length of the slit and readjust slit S_1 to give nearly full scale deflection again.

Begin with S_2 at a position far enough off center so that the pattern does not fall on the photomultiplier, and move in small steps so as to record the galvanometer readings in terms of the slit position. At places where the intensities change fast, be sure to take enough readings for a good curve. Record data for the entire pattern, not just one side.

Measure and record also the distances a and b, the slit width S_2, and the slit width S_3.

RESULTS

Plot on a sheet of rectangular coordinate graph paper the galvanometer readings (proportional to intensity) vs. the slit position readings. It will probably be found that the curve is not entirely symmetrical, but it should be reasonably good.

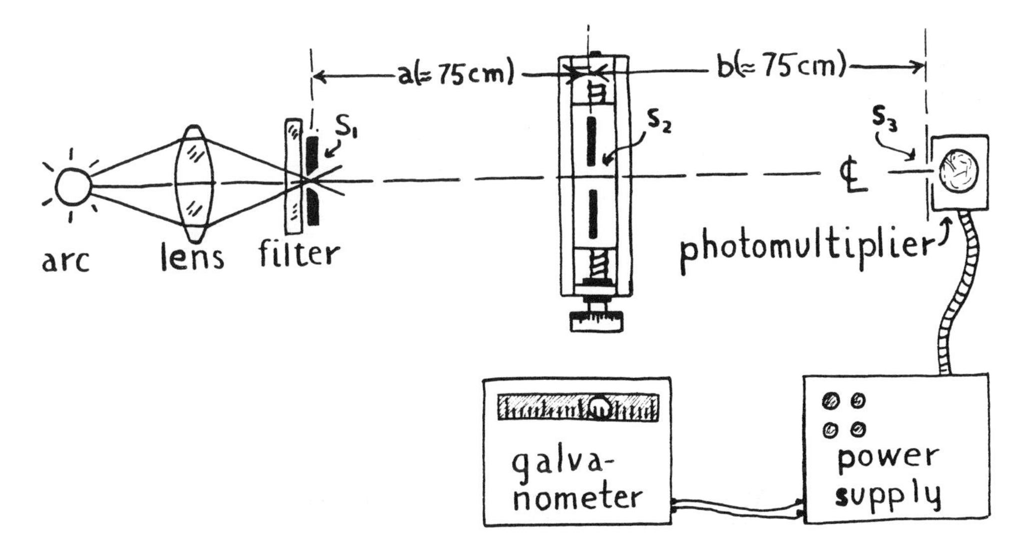

FIG. B14-1

Apparatus for photoelectric study of Fesnel diffraction.

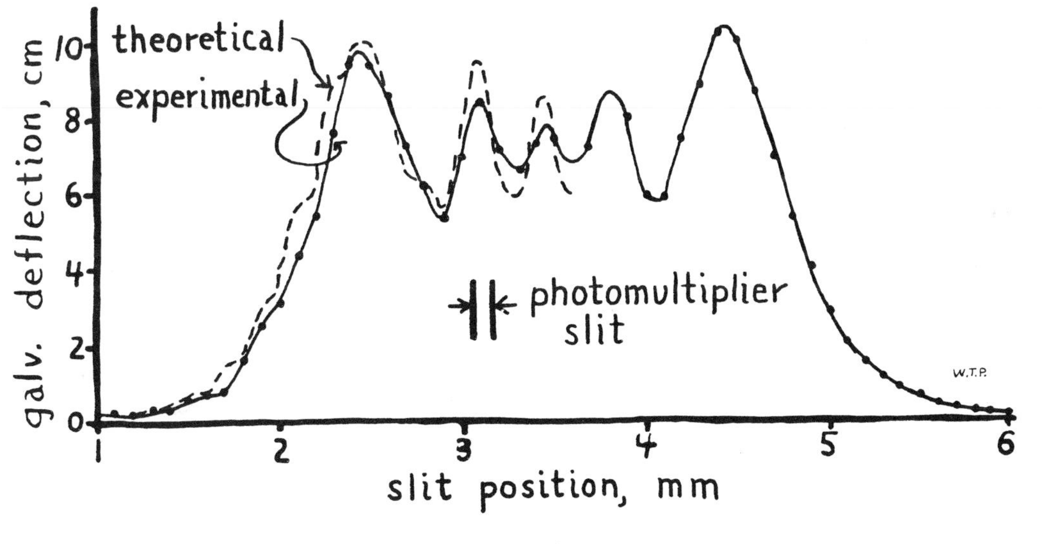

FIG. B14-2

Fresnel single slit diffraction.

Calculate by means of a table of Fresnel's integrals or from the Cornu spiral (see R. W. Wood, Physical Optics, 3rd ed. (1934, Macmillan) opposite p. 829, for a good spiral) the theoretical intensity curve for the pattern measured. Plot the theoretical curve on the same graph for comparison. It is necessary to adjust the theoretical intensity scale to fit the experimental data, so multiply the theoretical intensities by a constant factor to make the curves agree at the first significant maximum on one side of the pattern. Indicate also the slit width and the geometrical shadow edge. Figure B14-2 shows a sample curve.

Compare the theoretical and experimental curves. Why are the swings of the experimental curve less than those of the theoretical curve? If your curve is not symmetrical, explain the possible causes.

EXPERIMENT B15. FRAUNHOFER DIFFRACTION BY A SINGLE SLIT.

READING: Strong, Ch. 10-10.2; or Jenkins and White, Ch. 15-15.5.

In this experiment, a quantitative study of the Fraunhofer diffraction by a single slit will be made. Experiment B14 was concerned with the Fresnel diffraction by a single slit. The difference between the two is the shape of the wave front where the diffraction takes place. In the Fresnel case, a spherical or cylindrical wave front is obstructed in some way. In the Fraunhofer case, the obstructed wave front is plane; it comes from a collimated beam, and it is observed with a telescope.

The experiment may be done either with an optical bench (Procedure A) or with a spectrometer (Procedure B).

THEORY

The intensities for the Fraunhofer single slit diffraction pattern, shown in Fig. B15-1, are given by the relation

$$I = I_0 \sin^2\beta / \beta^2 \qquad (1)$$

where I_0 is the intensity of the principal maximum, and β a convenient general variable given in turn by

$$\beta \equiv \pi b \sin\theta / \lambda \qquad (2)$$

in which b is the slit width, θ the angle of diffraction, and λ the wavelength.

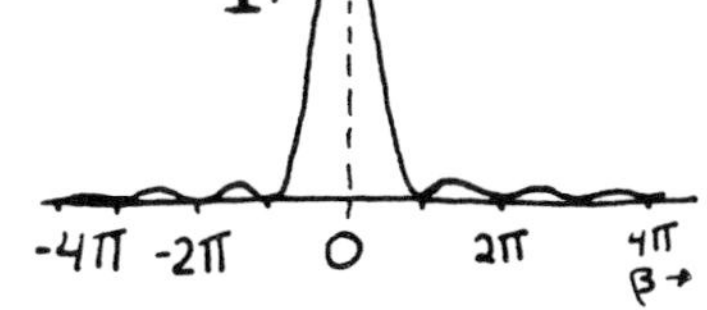

FIG. B15-1

Fraunhofer single slit diffraction.

The intensity minima occur at $\beta = \pm\pi, \pm 2\pi, \pm 3\pi$, etc. The intensity maxima occur approximately halfway between the minima, but not exactly (see reading).

APPARATUS

A. (1) One meter optical bench.

(2) Mercury arc and green line filter.

(3) Condenser lens.

(4) Adjustable slit.

(5) Two good achromatic lenses 1 or 2 in. dia. of about 10-15 in. focus.

(6) Traveling microscope.

(7) Precision adjustable slit.

Alternative list

B. (1) Precision student spectrometer equipped with micrometer telescope drive and with auxiliary high power eyepiece.

(2) Mercury arc and green line filter.

(3) Condenser lens.

(4) Precision adjustable slit.

(5) Traveling microscope.

PROCEDURE

A. Optical bench arrangement

Arrange the apparatus as in Fig. B15-2. First line up the arc, the slit S_1 and the traveling microscope on the optical bench. Mount lens L_1 (correctly oriented for minimum spherical aberration -- see Experiment A10) to focus the arc on the slit. Add the filter as shown, and mount lens L_2 correctly oriented. By autocollimation (see Experiment A1) make the slit S_1 coincide with the primary focal plane of the lens. Then mount L_3 a few inches from L_2 (orientation!) and slide the traveling microscope unit to the point where a sharp image of the slit S_1 is seen. Adjust the position of the crosshair so that there is no parallax between the slit image and the crosshair when the eye is moved slightly either way. Finally, mount the precision slit S_2 in the center of the collimated beam with its length parallel to that of S_1. Make slit S_1 just wide enough to let through an adequate amount of light, and adjust S_2 to give a good diffraction pattern. It should be possible to see at least twenty clearly defined fringes. Adjust the tilt of the slits and their width if necessary.

Begin with the traveling microscope on one side of the pattern and with the backlash removed. Record the positions of the minima and of the central maximum, identifying the number of the minimum (positive or negative) from the center. At least twenty minima should be measured.

Remove the green filter and look at the mercury "white" light fringes, or remove the arc and substitute an incandescent lamp. Describe the appearance of the white light fringes briefly.

Measure the focal length of lens L_3 to about 1% and then remove it from the optical bench. Slide the slit S_2 to the focal plane of the traveling microscope (or vice versa) and measure the slit width. The microscope should be carefully focused on the slit which is illuminated from behind with diffused white light. Measure the width along the center line of the field of view where the fringe position was measured; this is particularly important if the slit jaws are not strictly parallel. (Why?). Make at least five determinations of the slit width.

B. Spectrometer arrangement

If the spectrometer is used, the first step is to put it into adjustment (see Experiment A5). After it is adjusted -- only one prism face need be used -- remove the Gauss eyepiece, substitute the high power eyepiece, and focus on the crosshair which should be at the secondary focus of the objective. Check by turning the telescope to look directly into the collimator and making sure the entrance slit is in sharp focus and that there is no parallax between the crosshair and slit image.

Mount the slit S_2 as shown in Fig. B15-2b on the prism table and adjust it so that the slit is parallel to the collimator slit. To make this adjustment, it may be desirable to use a level on the collimator slit and on the prism table. Adjust the slit width to some convenient value so that 20 minima are distinctly seen when the spectrometer is illuminated with monochromatic green light as indicated in the figure.

After determining the calibration of the micrometer drive on the telescope, measure the positions of the twenty minima and the central maximum beginning on one side of the pattern to avoid backlash.

Substitute white light and describe the appearance.

Carefully remove the slit from the prism table of the spectrometer and measure its width with a traveling microscope, taking care that the slit width is not altered in transferring it to the microscope stage. Measure the slit width at the top, at the middle, and at the bottom, making two measurements at each place, and average the results.

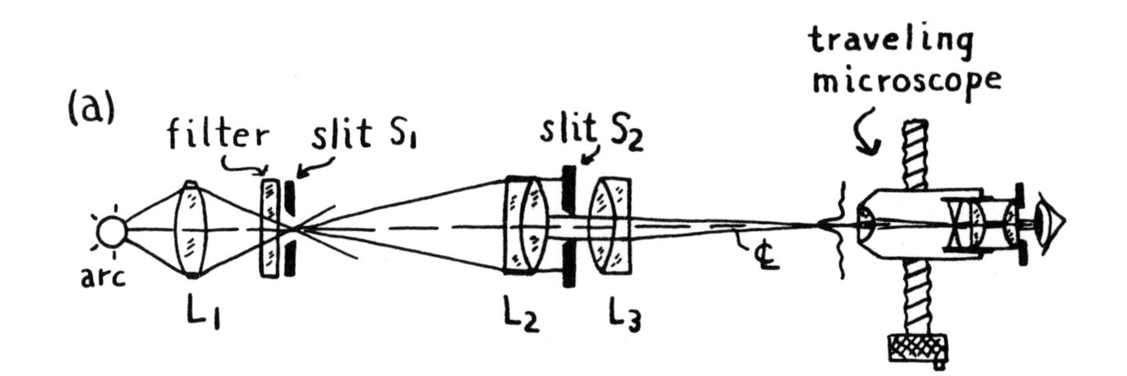

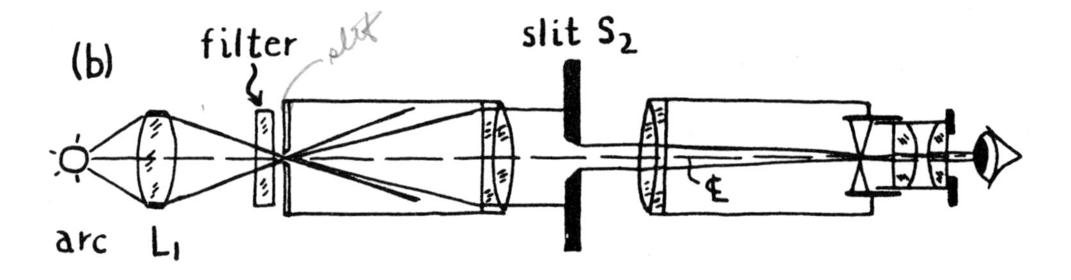

FIG. B15-2

Apparatus for Fraunhofer diffraction study. (a) optical bench arrangement, (b) spectrometer arrangement.

RESULTS

Compute angles θ for each minimum and evaluate the quantity $\pi \sin \theta / \beta$ (see equations (1) and (2) in each case). Average the results. Average also the slit width values. Determine for both these averages the mean deviation. Finally, compute the wavelength and the probable error. Compare the calculated wavelength with the accepted value. Does your measurement fall within the limits of experimental error? If not, why not? Discuss the sources of error in the experiment. Why were minima measured rather than maxima? Be sure to describe the apparatus used in the experiment.

EXPERIMENT B16. FRAUNHOFER DIFFRACTION BY A CIRCULAR APERTURE. THE RESOLVING POWER OF A TELESCOPE AND OF THE HUMAN EYE.

READING: Strong, Ch. 10.1-10.2, 10.4; Jenkins and White, Ch. 15.5-15.6, 15.8-15.9.

In this experiment, a comparison is made of the actual resolving power and the theoretical resolving power both of a high quality telescope and of the human eye. The resolving power of a good instrument with a circular aperture is limited by the diameter of the Airy disk at the center of the Fraunhofer diffraction pattern. According to the Rayleigh criterion, two points are said to be just resolved if the principal maximum of one diffraction pattern falls on the first minimum of the other (and vice versa). The angular separation of two points satisfying this criterion is, for a circular aperture, $\theta = 1.22\lambda/D$ where λ is the wavelength and D the diameter of the clear aperture of the objective lens or mirror.

APPARATUS

(1) Concentrated arc source.

(2) Mercury or cadmium arc.

(3) Filters to isolate three spectral lines: mercury yellow, green, and blue or violet or cadmium red, green and blue.

(4) Targets:

a. Several sets of artificial stars pricked with a needle in brass shim stock. The sets should include one rather large single star, one or more closely spaced pairs less than half a millimeter apart, and perhaps a triplet. Alternatively, a larger pattern of stars may be drilled in the shim stock and reduced with a microscope objective as in Experiment B10.

b. Ronchi rulings having about 50, 100, and 150 lines per inch.

(5) High quality telescope of 1 1/2 to 2 inch aperture together with a good eyepiece to give a total magnification of 100 to 200x.

(6) Calibrated iris diaphragm having a range 3/16 to 1 1/2 inch or more.

(7) Traveling microscope.

(8) Mounting box for target illumination as shown in Fig. B16-1 or the equivalent.

PROCEDURE

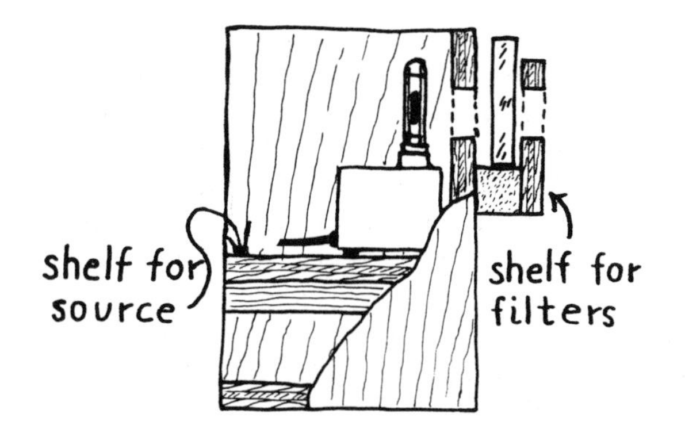

FIG. B16-1

Target illuminator and support.

(1) Diffraction pattern

Mount the telescope on blocks or clamp it to the table top at the other end. Focus the telescope carefully on the point source, and then mount the iris diaphragm over the objective lens, (either secure it to the lens mount or clamp it just in front of the mount). The iris should be symmetrical with respect to the lens, and no light should enter the lens save through the iris aperture. Close the iris to about half an inch aperture and carefully observe the Fraunhofer diffraction pattern from the point source. Note the Airy disk at the center. What color is the edge of this white disk? Describe the appearance of the rest of the pattern. How many rings can you see?

Defocus the telescope by moving the eyepiece both inside and outside focus and observe the appearance of the Fresnel diffraction patterns of the "star" image. Compare these patterns with those obtained in Experiment B13.

(2) Resolving power, qualitative

Set up the illuminating box (Fig. B16-1) at least 3 meters from the telescope and use either a white light source or an unfiltered arc source to illuminate the artificial star pattern. If the pattern was made on a large scale with drilled holes, use a microscope objective in reverse (as in Experiment B10) to obtain a small image of it. Focus the telescope on the edge of the box or any other target at the distance of the star pattern, then close the iris to its minimum size and point the telescope at the star pattern. Slowly open the iris as much as required to distinguish the components of the various pairs, singles, and triplets. Notice how the components of pairs gradually emerge from a single (perhaps misshapen) diffraction pattern as the aperture is increased from the smallest opening.

(3) Resolving power, quantitative

For the artificial star pattern, substitute suitable Ronchi rulings. This time, however, use monochromatic light, beginning with the green. The filter should be between the source and the Ronchi ruling target. Make five measurements of the minimum iris aperture required to resolve the pattern of lines for each of two rulings. Repeat for a blue or violet line and again for a yellow or red line. Measure the distance from the targets to the telescope objective.

Determine also at what maximum distance the lines of the two Ronchi rulings may be resolved by the eye alone -- the illumination may be yellow or white light. Record the results and identify the rulings used.

Put each of the Ronchi rulings on the traveling microscope stage and measure the distance between 25 or more rulings to determine the average spacing. Be sure that the rulings are perpendicular to the direction of travel of the microscope.

RESULTS

Describe your observations of the Fraunhofer diffraction pattern of a circular aperture and of the transition to Fresnel diffraction when the telescope is defocused.

Describe the qualitative results on resolving the star pairs.

Calculate the angle resolved by the telescope for each of the Ronchi rulings from the line spacing and the distance of the target. From these angles and three known wavelengths, calculate the theoretically required aperture diameter and plot two theoretical curves of theoretical aperture vs. wavelength. Average the measurements made with each of the targets at each of the three wavelengths and plot these six average values on the graph. These points should lie near the theoretical curve.

Calculate the resolving power of the eye from the measurements on the Ronchi rulings. Assume a reasonable diameter of the pupil of the eye and calculate its theoretical resolving power.

QUESTIONS

Is the Rayleigh criterion for resolving power realistic? Does it apply to the human eye? What limits the resolving power of the eye?

Strong mentions Sparrow's criterion for resolving power, (p. 669, problem 10-1). Is this criterion reasonable or perhaps more satisfactory than the Rayleigh criterion?

Can the actual resolving power ever exceed that given by the Rayleigh criterion?

EXPERIMENT B17. APODIZATION.

READING: Strong, Ch. 10.1-10.2, 10.4, Appendix E (P. Jacquinot).

This demonstration is concerned with apodization, that is, the modification of a diffraction pattern produced by special diaphragms. Apodization is used optically to improve the visibility of a faint light source near an overpowering bright one in cases where the resolving power is adequate. Strong's book mentions two specific examples, that of the star Sirius and its very faint companion and that of a helium spectral line and its faint companion. There are many such situations in astronomy and spectroscopy. Apodization of a diffraction pattern has already been used in Experiment B10 on the stellar interferometer; the double slit was made in the form of two cat eyes which apodizes the pattern and improves the visibility of the interference fringes to be observed. The process of apodization also has important applications in electrical engineering practice in the modification of the antenna patterns in, for example, radar or communications systems.

APPARATUS

(1) Mercury arc (or carbon arc).

(2) Pinhole, about 1/2 mm diameter in brass shim stock (#76 drill or use needle).

(3) Telescope of high quality and 1 1/2 to 3 in. aperture equipped with eyepiece, giving magnification of about 100x.

(4) Glass plate 1/4 to 1/2 in. thick, free of scratches or striae.

(5) Assorted diaphragms as illustrated in Fig. B17-1. They may be cut from cardboard with a razor blade.

PROCEDURE

Mount the mercury arc as far from the telescope as practical -- at least three meters. Cover the arc with the brass sheet having the 1/2 mm hole (no filter is needed). The arc is used in preference to a concentrated arc because the glass envelope of the latter scatters too much light.

Fix the telescope rigidly on a bench top so that the point source will be in the middle of the field of view. The room should be quite dark

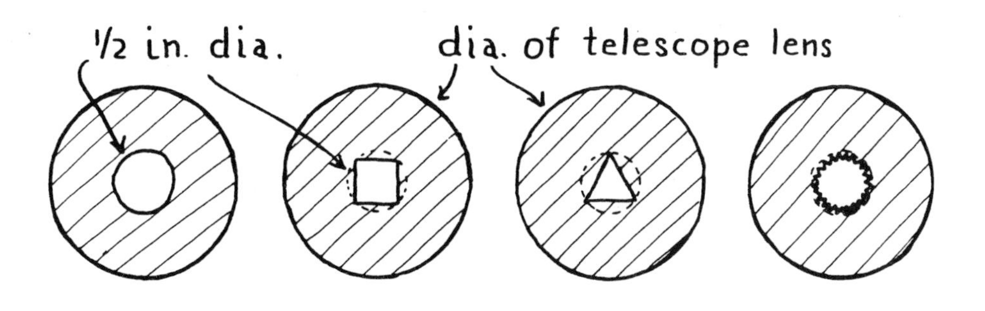

FIG. B17-1

Apodizing screen patterns.

to avoid difficulty with extraneous light. Use the full aperture of the telescope with a magnification of about 100x to focus carefully on the point source. Now put the 1/2 inch circular aperture over the objective and observe the same Fraunhofer pattern as in Experiment B16. Observe the ring pattern and note particularly the size of the Airy disk. Sketch the pattern roughly and substitute the square aperture. Sketch this pattern also, paying special attention to the central spot. Repeat for both the triangular and sawtooth apertures. Other shapes of possible interest are suggested in the reading, and if these shapes are used, sketch them also.

An artificial satellite star is now made using the glass plate as shown in Fig. B17-2. Since the glass reflects about 4% of the incident intensity at small angles of incidence, two reflections will give a satellite of about 0.0016 of the intensity of the parent. The glass plate is first cleaned with acetone and detergent solution if necessary and then rinsed with clear water. Dry carefully with paper towels and lens paper. Mount the clean plate just in front of the pinhole and tip it five or six degrees, if it is 1/2 inch thick or twice this much, if it is 1/4 inch thick.

Look at the star with the half inch circular aperture to see if the satellite can be detected. In any case, the satellite will be hard to see. The position of the satellite (if it can be seen) should be adjusted (by tilting the glass plate) so that it falls at about the second or third diffraction ring.

Now substitute the square aperture and look for the satellite, which, though very faint, should be clearly evident if the square is properly oriented. The primary diffraction pattern will not interfere significently.

Make a sketch of the location of the satellite with respect to the diffraction pattern of the primary. Vary the tilt of the glass plate to see how close the satellite may be to the primary star. Measure the approximate angle of the tilt of the plate and its thickness. Measure also the distance of the stars from the telescope.

RESULTS

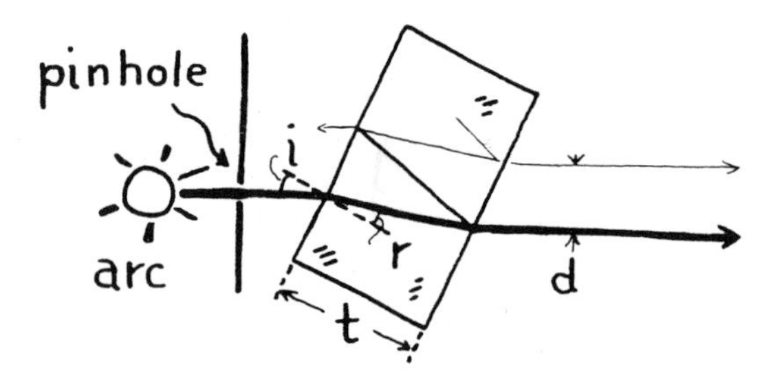

FIG. B17-2

Method of producing a weak satellite with a glass plate.

Calculate the theoretical position of the satellite with respect to the diffraction pattern (see Strong, Ch. 10.1). Calculate the maximum intensity of the primary diffraction pattern near the point where the satellite was observed. Compare this intensity with that of the satellite. Calculate also the intensity of the nearest ring of the circular aperture diffraction pattern, and compare this value to the others. Describe briefly how apodization improves the visibility of faint satellites. Is apodization practical if the primary and satellite are separated by an angle which can barely be resolved by the full aperture? Explain.

Derive the formula for d, the separation of the satellite and the primary star (see Fig. 17-2). Assume a refractive index of 1.5 and calculate the approximate value of d used in the experiment.

EXPERIMENT B18. DEMONSTRATION OF THE ILLUMINATION IN THE AIRY DISK.

READING: Strong, Ch. 10.2; (more advanced) M. Born and E. Wolf, Principles of Optics, (1959, Pergamon), Ch. 8.5.2.

In this experiment, the illumination at the center of the Airy disk is measured as a function of the objective aperture.

Figure B18-1 illustrates the appearance of the Fraunhofer diffraction pattern of a circular aperture. The angular dimensions of the pattern as seen from the telescope objective are suggested by the angular radius of the first dark ring, $\theta = 1.22\lambda/D$, where λ is the wavelength and D the diameter of the objective aperture. The bright disk at the center of the pattern, known as Airy's disk, contains 84% of the total illumination in the pattern, 60% of it within half the radius of the first dark ring. Put another way, the illumination within the Airy disk falls off rapidly: at a radius of 1/6th that of the first dark ring the illumination has fallen to 90% of the maximum, at 2/5 that radius to 50%. The linear dimensions of the pattern depend, of course, on the focal length of the telescope and are equal to the angular dimensions multiplied by the focal length of the objective (assuming there are no other lenses between the objective and image).

If the diameter of the objective is doubled, and the focal length kept fixed, the fourfold increase in the luminous flux collected by the telescope is concentrated within a diffraction pattern of one fourth the area, so that the illumination in the Airy disk should be 16 times greater. In other words, doubling the telescope aperture should increase the photographic speed 16 times. In practice, this is rarely the case, for one must contend with atmospheric "seeing." The "seeing" blurs the starlight theoretically concentrated in the diffraction pattern so that it is smeared out over a larger area. For this reason the 200 inch telescope does not photograph 16 times faster than the 100 inch, but perhaps even less than 4 times faster. But on a small scale, in the laboratory, the theoretical factor of 16 can be approached.

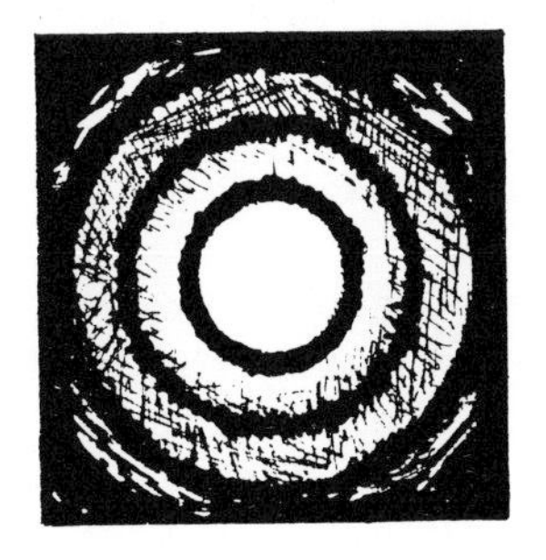

FIG. B18-1

Fraunhofer diffraction pattern.

APPARATUS

(1) Photomultiplier unit with 1P21 phototube and galvanometer.

(2) Good optical bench.

(3) Mercury arc with pinhole aperture (#80 drill used), filter to transmit green line -- the filters should be polished.

(4) Microscope objective of about 16 mm focus.

(5) Achromatic doublet of 1-2 in. dia. and 10-20 in focus.

(6) Pinhole to put over photomultiplier head (#80 drill).

(7) Stops: 1mm, 2mm, 4mm (use #60, #46, #20 drills).

(8) Low power eyepiece (10x).

PROCEDURE

Set up the apparatus as in Fig. B18-2. Line up the pinhole source, the telescope lens and the microscope objective. Examine the image of the pinhole (several meters away) with the low power eyepiece. The image should be clear and sharp with little trace of diffraction. In front of the objective, insert one of the diaphragms and the diffraction pattern should be evident. It is assumed that good optical components are used so that the diffraction pattern will be clean cut. Replace the eyepiece with the photomultiplier unit. Center the exit pinhole over the front of the photomultiplier so that the image with no stop is centered on the hole and sharply in focus. Cover the photomultiplier with the filter. Note that the alignment of the parts must be very accurate and must remain so -- do not lean on table! Build a light shield around the system to keep out as much stray light as possible.

With the photomultiplier shutter closed, increase the sensitivity slowly while adjusting the dark current balance. The galvanometer should read zero with the sensitivity maximum. Now turn the sensitivity control off and open the shutter. Center the largest diaphragm hole (4mm) in front of the lens. Adjust the sensitivity control to make the galvanometer read full scale deflection. With a piece of cardboard, interrupt the light beam to make sure that only light through the objective is measured. If all is well, record the galvanometer reading -- the average value if it fluctuates a few percent. (Why might it fluctuate?) Reduce the aperture to 2 mm and record the galvanometer reading. Reduce it further to 1 mm and record the reading. Again put the cardboard in to cut off the light and check the galvanometer zero.

Typical readings are 120, 13, 1mm (the latter is uncertain by about 10%).

Put the apertures on a traveling microscope and measure the hole sizes of the three apertures.

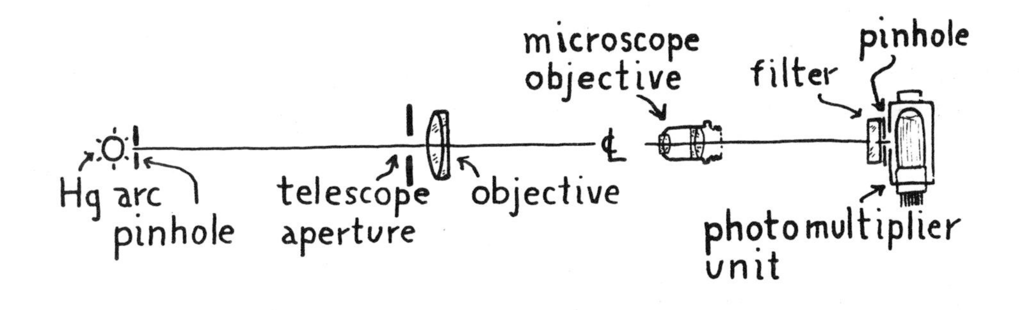

FIG. B18-2

Apparatus to measure the illumination in the Airy disk.

RESULTS

Calculate the ratios of the three galvanometer readings and determine the theoretical ratios for the holes.

Would you expect the theoretical ratios to obtain? Why? What effect does the finite size of the pinhole source and of the pinhole over the photomultiplier have on the results?

EXPERIMENT B19. DEMONSTRATION OF MISSING ORDERS.

READING: Jenkins and White, Ch. 16-16.6.

This demonstration illustrates the phenomenon of missing orders both for a double slit and for a multiple slit or grating. As explained in the reading, if the ratio of slit separation to slit width is a small integer, certain interference orders become so weak that they cannot be observed at all or are seen only faintly.

APPARATUS

(1) Student spectrometer.

(2) Monochromatic source, preferably yellow or green.

(3) Ronchi ruling having about 100 lines/inch.

(4) Set of special slits.

Fig. B19-1 shows a set of double slits having simple ratios of slit spacing to slit width. The ratios are indicated in the figure. In the first case, obviously, the slit widths are equal to the spacing so that the interference-diffraction pattern is the same as the diffraction pattern of the single slit. In the succeeding pairs, every 2nd, 3rd, 4th and 5th interference orders respectively, will be missing. The figure may be photographed on a scale of about 1/10 on very high contrast, fine grain film to produce a satisfactory series of slits. It is suggested that a 35 mm camera be used and the slit pairs photographed one at a time by masking out the undesired slits with white paper. Mount the processed film slits on pieces of cardboard and mark the ratios of slit spacing to slit width on them.

PROCEDURE

Adjust the collimator and telescope focus of a student spectrometer and illuminate the instrument with monochromatic light. Set the telescope so that the slit image falls on the cross hairs. Hold the various slit pairs in turn between the collimator and telescope. Begin with the single slit. Adjust the slits parallel to the collimator slit by tilting slightly in either direction about the vertical to obtain the sharpest pattern. Proceed from the single slit through the remainder of the set, noting which orders are unobservable or very weak.

The Ronchi ruling has equal width transparent and opaque spaces. What orders should be missing? Mask the grating so that 20 to 50 full slits are exposed, but no parts of slits. Hold the masked grating in the collimated beam and orient it for the best pattern. What can be said about the interference-diffraction pattern?

RESULTS

Record your observations in detail. Explain why the "missing orders" are weak but not totally absent in general. Are the observations in accord with theory? Explain.

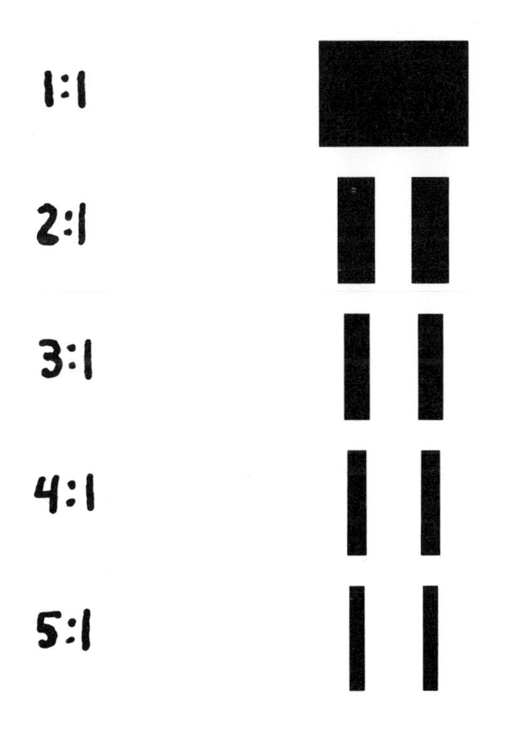

FIG. B19-1

Double slit pattern to study missing orders.

EXPERIMENT B20. SPECTROSCOPIC RESOLVING POWER.

READING: Strong, Ch. 10.4-10.7; or Jenkins and White, Ch. 15.6-15.7, 17.9, 23.1.

The spectroscopic resolving power of both a prism and a grating are to be measured and the results compared with the theoretical values.

THEORY

Spectroscopic or chromatic resolving power is the ability of an instrument to distinguish nearly equal wavelengths. The smaller the wavelength interval which can be separated, the greater the resolving power. We define resolving power as the ratio of the wavelength λ to the minimum wavelength interval $\Delta\lambda$ which can be distinguished:

$$r.p. = \lambda/\Delta\lambda \qquad (1)$$

As proved in the reading, the resolving power of a prism is given by:

$$r.p. = b\frac{dN}{d\lambda} \qquad (2)$$

where b is the effective base of the prism and $dN/d\lambda$ is the dispersion. For the purposes of this experiment, it is more convenient to write the formula in an alternative form involving the quantities directly measured. Figure B20-1 shows the various quantities concerned. By Rayleigh's criterion, two wavelengths will just be resolved if the principal maximum of the one line falls on the first minimum of the other. For a rectangular aperture of width a, the first minimum will fall at an angle $\Delta\theta$ from the principal maximum such that $\Delta\theta = \lambda/a$. Combining this expression with the definition of chromatic resolving power we find:

$$r.p. = \frac{\lambda}{\Delta\lambda} = a\frac{d\theta}{d\lambda} \qquad (3)$$

where we have substituted a derivative $d\theta/d\lambda$ for the ratio of two small quantities $\Delta\theta/\Delta\lambda$ since they are equal for all practical purposes. Equations (2) and (3) are equivalent, provided the prism is used at minimum deviation.

It should be noted in passing that b and a are effective values; if the full aperture of the prism is not used, the resolving power is less, for both b and a are less than maximum. Conversely, if two or more prisms are used in series, or if one prism is used twice or four times by passing the light through it several times, the resolving power is greater than that calculated from $a\, d\theta/d\lambda$ for one prism alone. The effective base is the sum of the bases of a train of prisms, or the base multiplied by the number of passes in the multiple pass system. The width of the beam in equation (3) is fixed, but in the prism train or multiple pass arrangement, the derivative $d\theta/d\lambda$ is increased.

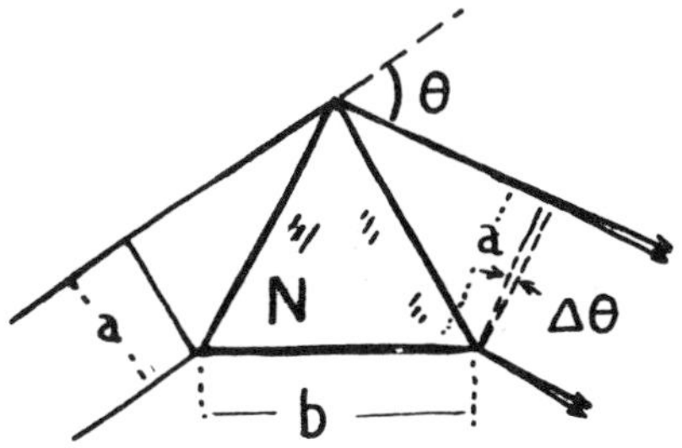

FIG. B20-1

Resolving power of a prism at minimum deviation.

APPARATUS

(1) Student spectrometer.

(2) Mercury arc.

(3) Sodium arc.

(4) 60° prism.

(5) Diffraction grating having 6000 to 15000 lines/inch.

(6) Calibrated slit which opens 0 to 1 cm (need not be very precise).

PROCEDURE

(1) Resolving power of a prism

Adjust the spectrometer as in Experiment A5. The prism must be adjusted so that two faces are properly leveled. Mount the calibrated slit between the collimator lens and the prism as shown in Fig. B20-2. Illuminate the spectrometer with a mercury arc taking care to fill both the entrance slit and the collimator lens (see appendix to Experiment A5).

Set the prism for minimum deviation for the yellow lines and measure the deviation by the prism for the yellow line of longer wavelength. Repeat for the green line. This step provides information for the calculation of $d\theta/d\lambda$.

With the prism again set for minimum deviation for the yellow, make ten measurements of the minimum slit width required to barely resolve the two yellow lines.

Close the slit almost all the way until the green and yellow lines merge. It is evident that the instrument resolving power is now less than that of the normal eye, since it is obviously possible for the observer to distinguish yellow and green despite the merging of the lines.

Substitute the sodium arc for the mercury arc, and use great care to illuminate the collimator correctly. Remove the auxiliary slit and use the full aperture of the instrument. With careful adjustment, it should be just possible to resolve the two yellow lines of sodium. If either the prism or the spectrometer are in bad condition, or if the collimator is not correctly illuminated or the focus incorrect, this test of resolving power will probably fail.

(2) Resolving power of a grating

Remove the prism from the spectrometer and substitute the grating (secured in the grating mount). The latter is fastened to the prism table. Remove the sodium arc and use the mercury arc. The spectrometer itself is in adjustment, but the grating must be adjusted so that the rulings are parallel to the mechanical instrument axis (i.e. vertical).

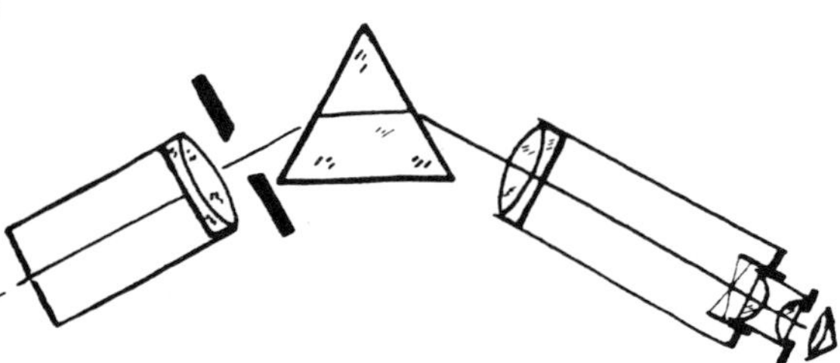

FIG. B20-2

Arrangement for measurement of resolving power.

Set the telescope approximately at a right angle to the collimator and turn the slit horizontal. Turn the prism table so that the protective glass covering of the grating reflects light from the collimator into the telescope. Adjust the tilt screws of the prism table to bring the image of the slit to the intersection of the cross hairs. The plane of the grating is now vertical, but the rulings themselves are probably not. Now set the telescope approximately in line with the collimator and rotate the prism table in both directions so that the spectral orders pass by the cross hair. The "spectra" are, of course, highly impure because the slit is turned the wrong way, but this is of no concern here. In general, it will be found that the colored lines on one side of the zero order will rise above the cross hairs whereas those on the other side will dip below. A few moments consideration should suggest which way the grating lines are tilted with respect to the vertical. To correct the error, adjust two leveling screws of the prism table simultaneously. They should be turned in opposite directions so as to tilt the rulings with respect to the vertical but to keep vertical the plane of the grating. After making approximately the right correction so that the colored lines pass the cross hairs at about the same level, check the plane of the grating as above to make sure it is still vertical. If it is not vertical, readjust it. Repeat the process until the colored lines all pass through the intersection of the cross hairs for any rotation of the prism table. At this point, the grating is correctly adjusted; rotate the slit to the vertical position, and set the grating normal to the collimator beam. If the telescope is set on the zero order, the plane of the grating may be rotated to be normal to the beam by autocollimation (meanwhile blocking the collimator beam). Finally, insert the calibrated slit as in Fig. B20-2.

The resolving power of the grating is so great that it is awkward to make resolution measurements with the mercury yellow lines. Accordingly, use the sodium arc and make five measurements of the minimum slit width required to resolve the D lines in the first order spectrum on each side of the normal. Average the results for each side and then make five more measurements for the second order. If there is any difference in the two sets of measurements for the first order, make the second order measurements on the better side -- otherwise, on either side of the normal. Also make five additional measurements for a higher order.

Measure the angle of diffraction for the D lines in the second order on each side of the zero order. This measurement determines the grating space.

RESULTS

Calculate the actual resolving power achieved for the mercury yellow lines and for the sodium D lines.

For the prism, average the slit width measurements and calculate the theoretical resolution of the instrument for this slit width. Compare this theoretical result with the actual resolution.

Use the grating equation and the known wavelength of sodium light (average of the D lines) to calculate the grating constant. Average the slit width measurements for the grating, and calculate the theoretical resolving power in each case. Compare these results with the actual resolution achieved.

Compute the theoretical resolution of both the prism and the grating (first order) when they are fully illuminated.

QUESTIONS

Suppose that a very large prism is used with the spectrometer so that the resolving power is determined by the circular beam emerging from the collimator. Would equation (3) still give the theoretical resolving power if the maximum illuminated prism width is used? If not, how would the theoretical value differ?

Frequently with gratings, the spectra in some orders are better defined and sharper than those in other orders. Can you explain why this might happen?

EXPERIMENT B21. DEMONSTRATION OF DIFFRACTION IN MICROSCOPE IMAGES.

READING: Strong, Appendix K, (F. Zernike) esp. pp. 525-530.

This demonstration of Professor Zernike shows the important role played by diffraction in the observation of very small microscopic objects. Although ray optics suffices in the description of images of large objects, diffraction (and therefore wave optics methods) become increasingly important as the object size becomes smaller, approaching the wavelength of light.

APPARATUS

(1) Standard microscope with low power (8-10X or 20-16mm focus) objective, and 10X eyepiece.

(2) 100 watt lamp source.

(3) Object plate with VERY SMALL opaque spots: Professor Zernike recommends finely pulvarized galena or india ink spots. It is rather difficult to get the ink spots small enough, and satisfactory spots may be made by spraying flat black paint (spray can) at a microscope slide from a distance of about two feet. Use just one shot of spray. The droplets of paint produced in this way are very small indeed.

(4) Diaphragm to fit the microscope objective as shown in Fig. K-2 in Strong's book.

(5) Glass disk with dark spot as shown in the same figure. The dark spot should be completely opaque. India ink is often not black enough, and it is a good idea to blacken both sides of the glass. A small disk of black paper is a good substitute.

PROCEDURE

The procedure is given in the reading in Strong's book. The smaller the objects to be viewed, the more impressive the demonstration. The 100 watt lamp must be several meters away, and it is best put behind a screen of some sort or a box so that there is very little light save from the direct rays of the lamp.

EXPERIMENT B22. THE IMAGE OF A PERIODIC STRUCTURE.

READING: Strong, appendix Q, (T. Williams) pp. 621-629; Ch. 10-10.3.

REFERENCES: A. B. Porter, On the Diffraction Theory of Microscopic Vision, Philosophical Magazine, Vol. XI, pp. 154-166 (1906); M. Born and E. Wolf, Principles of Optics, (1959, Pergamon), ch. 8.6.3.

This demonstration illustrates Abbe's theory of microscopic vision. It might be regarded as a continuation of the previous demonstration from a somewhat more sophistocated point of view. We consider in some detail the nature of the image formed by an optical system of a periodic grating structure. The following section on theory is a brief summary of Porter's paper.

THEORY

According to Abbe's theory of microscopic vision, the image formed by an optical system cannot be a truthful representation of the corresponding object unless all the light diffracted by small structures in the object is included in the image along with the direct rays. In fact, if the detail is so small or if the aperture of the optical system so restricted that no part of the diffracted rays are included in the image, then the detail is invisible, no matter what magnification is used. If only a part of the diffracted light is included in the image, then the image corresponds to a hypothetical object whose entire diffraction pattern is that which is transmitted to the image.

To understand how Abbe's theory works out in practice, we consider the image of a grating. To keep the mathematics relatively simple, we assume the grating to be of infinite extent; the error is negligible for our purposes. Figure B22-1(a) represents a crossection of the grating which, in our experiment, consists of a series of transparent stripes of width a separated by opaque stripes of width b. A parallel beam of monochromatic light having amplitude A is incident normally on the grating from below. Part (b) of the figure shows the distribution of light just as it emerges from the grating. The light has an amplitude f(x) equal to either 0 or A -- transmission or no transmission. Analytically we write:

$$f(x) = A \qquad 0 < x < a/2$$

$$f(x) = 0 \qquad a/2 < x < b + a/2$$

$$f(x) = A \qquad b/2 + a/2 < x < b + a$$

For values of $x > d$ or $x < 0$ we can obviously specify f(x) too, but we note simply that the function f(x) is periodic with the periodic distance d. (At the jumps, x = a/2, b + a/2, etc., we arbitrarily define f(x) to have the mean value A/2, to which the Fourier series converges).

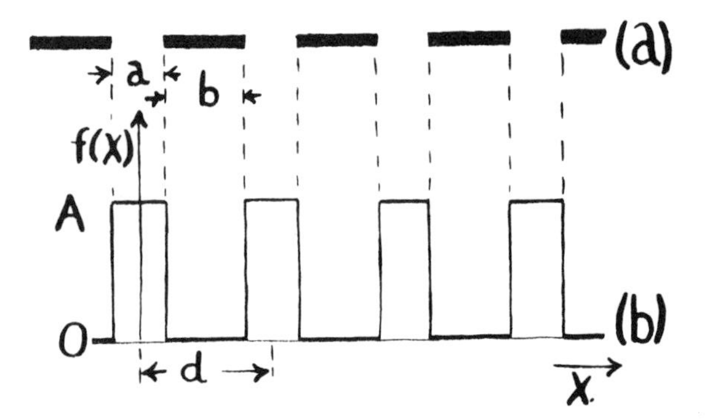

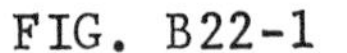
FIG. B22-1

Analysis of grating object. Part (a) shows the grating profile, part (b) the corresponding transmission function.

FIG. B22-2

Transparency of zeroth order simple "grating" and mth simple harmonic grating.

Following the methods described in Williams' appendix (pp. 621-629) we express the function f(x) in terms of an infinite series of sine and cosine waves. Added together, these waves give the original function specified above. We have chosen the origin for x in the middle of a transparent region so that the function will be symmetrical, an even function. Thus, in our case, only cosine terms will appear in the required series. (For a similar function, Williams places the origin at one of the jumps and thus obtains an odd function giving a sine series).

Thus we can write for the function:

$$f(x) = \frac{1}{2}a_0 + a_1 \cos\frac{\pi x}{\ell} + a_2 \cos\frac{2\pi x}{\ell} + \cdots + a_n \cos\frac{n\pi x}{\ell} + \cdots \tag{1}$$

where $2\ell = d = a + b$. We need to determine the coefficients, a_m, and they are given by:

$$a_m = \frac{2}{\ell}\int_0^{\ell} \cos\frac{m\pi x}{\ell}\, f(x)\, dx$$

$$= \frac{2}{\ell}\left[\int_0^{a/2} A\cos\frac{m\pi x}{\ell}\, dx + \int_{a/2}^{a/2+b/2} 0\cdot dx\right]$$

$$= \frac{2A}{\ell}\,\frac{\ell}{m\pi}\,\sin\frac{m\pi x}{\ell}\Bigg|_0^{a/2} = \frac{2A}{m\pi}\sin\frac{m\pi a}{2\ell}\,.$$

so that

$$a_m = \frac{2A}{m\pi} \sin \frac{m\pi a}{d} \qquad m \neq 0$$

$$a_0 = \frac{2}{\ell}\int_0^{a/2} A\,dx = 2A\frac{a}{d}. \tag{2}$$

(The value for a_0 also follows directly from a_m if we replace $\sin(m\pi a/d)$ by $m\pi a/d$ for small angles).
Thus we find f(x) can be written:

$$f(x) = A\frac{a}{d} + \frac{2A}{\pi}\sin\frac{\pi a}{d}\cos\frac{2\pi x}{d} + \frac{2A}{2\pi}\sin\frac{2\pi a}{d}\cos\frac{4\pi x}{d} + \cdots$$

$$+ \frac{2A}{m\pi}\sin\frac{m\pi a}{d}\cos\frac{2\pi m x}{d} + \cdots \tag{3}$$

The theory of the transmission grating (many identical slits) shows that the amplitudes of the spectral orders are proportional to $\sin\xi/\xi$ where $\xi = (\pi a \sin\theta)/\lambda$, θ being the angle of diffraction for the wavelength λ. The grating equation for normal incidence gives θ for the mth order spectrum as $m\lambda = d\sin\theta$. Thus $\xi = (\pi a/\lambda)(m\lambda/d) = m\pi a/d$, and the amplitudes are proportional to $(d/m\pi a)\sin(m\pi a/d)$. For the zero order, m = 0, the expression is indeterminate but since the sine of a small angle is equal to the angle (radian measure), we see that the expression approaches unity as m approaches zero. The average amplitude of the light transmitted is evidently just Aa/d, the amplitude of the zero order. Notice that this quantity is just equal to the first term of the expression for f(x), equation (3). The first order spectrum will have an amplitude $Aa/d \times d/\pi a \times \sin\pi a/d = (A/\pi)\sin(\pi a/d)$. Twice this amplitude is the total amplitude for the two first order spectra, and it is seen to be equal to the coefficient of the first cosine term in (3). Similarly, the second order spectra have a combined amplitude $(2Aa/d)(d/2\pi a)\sin(2\pi a/d) = (2A/2\pi)\sin(2\pi a/d)$, or the coefficient of the second cosine term in (3). Finally, the mth order pair of spectra have a combined amplitude of $(2A/m\pi)\sin(m\pi a/d)$, the coefficient of the corresponding cosine term in (3).

It appears, then, that we may regard the grating as a set of simple harmonic gratings equivalent to the original. Each of these simple harmonic gratings has a transparency varying as $\cos m\pi x/d$ and it gives rise to only one pair of spectral orders, + m and -m. Only one grating of the set contributes to the zero order spectrum, and it has a uniform partial transparency of a/d, independent of x. Each of the other simple harmonic gratings has a transparency, T(x), like that shown in Fig. B22-2. The amplitude of the curve, a_m, has the value given by equation (2), the sum of the amplitudes in the two mth order spectra. Negative transparency means the same thing as positive transparency except for the addition of a phase reversal as the light passes through the grating. Thus one could make the grating by preparing a sheet of the specified positive transparency and adding a phase reversal dielectric sheet over alternate stripes.

We note that if any of the simple harmonic gratings are missing in the grating description given by equation (3), then the corresponding spectral order is also absent. To put it another way, if we arrange an optical system (microscope) to look at the grating and we artificially remove some of the spectral orders, then the image we observe is equivalent to the combined set of gratings whose spectra are represented in the image. If, for instance, we remove all but the zero order spectrum, we see only the zero order simple grating without any stripes -- only uniform partial transparency, no grating lines at all. If we include the zero and first orders in the image, we see a simple harmonic grating having the proper number of lines per inch, but a false contour. If we remove the first order, and include only the zero order and the second orders, we see a simple harmonic grating having twice as many lines per inch as the original grating object. As more of the higher order spectra are included (as well as the zeroth and first) the faithfulness of the image increases -- the edges of the lines begin to be well defined and the grating profile begins to resemble that of the original.

The theory of image formation given here may be extended to include both transmission and reflection phase retardation gratings, but we confine our attention here to the simple case above.

APPARATUS

(1) Optical bench.

(2) Mercury arc.

(3) Green filter.

(4) Adjustable slit.

(5) Condenser lens.

(6) Collimator lens, 1-2 in. dia., 3-4 in. focal length.

(7) Ronchi ruling, 100-150 lines/in.

(8) Telescope lens, 1-2 in. dia., 7-9 in. focus.

(9) Eyepiece lens, 2-3 in. focus.

(10) Reticle and eyepiece; reticle scale calibrated in tenths of a millimeter, for ex., and telescope Ramsden eyepiece of 1 to 1/2 in. focus to be used with reticle.

(11) Cardboard, string, etc. for masks to block spectral orders as indicated in the Procedure.

PROCEDURE

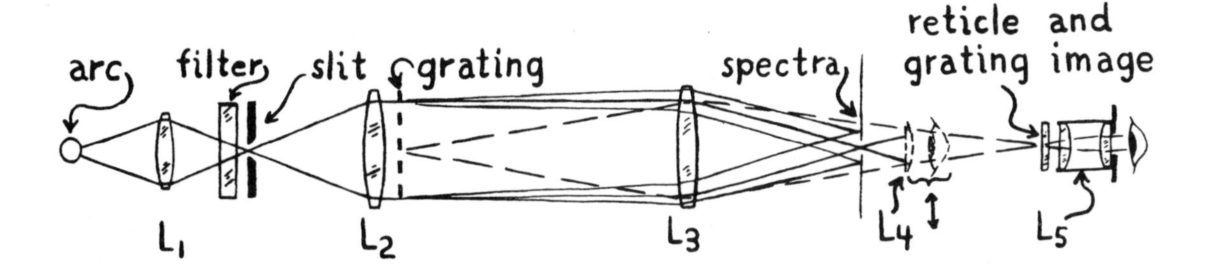

FIG. B22-3

Arrangement to view grating image in selected spectral orders.

The apparatus is set up as shown in Fig. B22-3. First mount the slit at one end of the optical bench and the reticle and telescope ocular at the other. The arc is mounted separately beyond the end of the bench. Line up the arc, slit, and eyepiece unit. Insert the condenser lens L_1 and then the collimator lens L_2. Mount the grating as close to the collimator lens as possible. Add lens L_3 and slide it along the bench to the position where a sharp image of the grating is seen in the eyepiece.

Lens L_3 serves a dual purpose: it images the grating on the eyepiece L_5, and, in addition, it forms spectral images of the slit in its secondary focal plane. It is in this focal plane that masks are to be placed to remove various spectral orders. Since the spectral orders are close together (because the grating is coarse), it is necessary to adjust the masks with care; for this reason, use a substantial mount for the masks -- an adjustable cross slide is helpful. It is convenient to use a rectangular aperture perhaps 1/2 in. high by 1 in. wide made of heavy cardboard or sheet metal as a support for the masks which may be taped or waxed across the aperture.

Finally, a low power eyepiece, L_4, is mounted on a tiltable bench slide so that one may look at the spectra or, alternatively, move this lens out of the way and view the grating through L_5.

Without the green filter in place, locate the zero order spectrum and the first few orders on either side. Record an estimated intensity for the orders from -5 to +5 as strong, moderate, or weak. Are any orders missing? (See Experiment B19).

Insert the green filter. With masks, remove all spectral orders except the zeroth. Use lens L_4 to observe the spectra in order to position the masks properly, and then move it out of the way. Look through L_5 at the grating and note that no lines can be seen. Proceed in this manner to view the grating in the orders indicated below. Describe the results in each case, indicating the image pattern by a small sketch which shows the intensity vs. distance along the image (x in theory section). Use the reticle scale and record in each case the number of completely black lines per millimeter. Observe the grating image in the following orders:

(1) 0th only.

(2) All orders except zeroth (dark field method). Block the zero order with thread or wire.

(3) + 1st only.

(4) + 2nd only.

(5) All + orders (oblique dark field).

(6) All + orders and 0th (half shadow).

(7) + 1, -1.

(8) + 1, 0, -1.

(9) +2, -2.

(10) +2, 0, -2.

(11) +3, -3.

(12) +3, 0, -3.

Finally, remove the filter and make a couple observations -- say (1) and (10) above.

RESULTS

Describe carefully all your observations and explain them. What is the significance of your observations? Does Abbe's theory hold for large objects too? Explain.

Discuss briefly the optimum focal length for the telescope lens, L_3.

APPENDIX

This experiment can also be used to demonstrate the principles of phase contrast. A phase retardation grating can be prepared from an ordinary photographed Ronchi ruling by bleaching the exposed silver grains; properly bleached, the black reduced silver grains can be washed out leaving the desired grating. (Consult a book on photography). The bleaching must not be continued too long, however, or the results are unsatisfactory. The phase retardation strip may be prepared either by etching a strip on a microscope slide using hydrofluoric acid (see Experiment B24) or by evaporation of MgF_2 or ZnS on the slide.

EXPERIMENT B23. DEMONSTRATION OF THE CHRISTIANSEN FILTER.

READING: Strong, Appendix O, (R. G. Greenler), pp. 583-585.

The characteristics and performance of a Christiansen scattering type filter are to be observed. For convenience, this filter is designed to operate in the visible part of the spectrum although its chief application is in the infrared where other suitable filters are often difficult to obtain.

THEORY

Suppose that a cell is filled with two substances, a finely ground solid powder and a liquid which does not dissolve the powder. If the refractive indices of the two substances are different, an incident parallel beam of light will be scattered in all directions as it passes through the cell. On the other hand, if the refractive indices are identical, the medium is optically homogeneous, and the cell will be quite transparent. As an illustration, an irregular piece of glass put in a container of liquid of the same refractive index is invisible. The principle is often used to determine the refractive index of irregular chunks of transparent substances, and regular sets of liquids with gradually increasing refractive indices from liquid to liquid are available for the purpose.

In the scattering filter, the two substances, solid and liquid, are chosen to have different dispersion curves which, however, cross at some wavelength. Thus, for all but the one wavelength where the curves cross, the two indices differ and the medium is optically inhomogeneous and scatters light. For the wavelength where the indices are the same, the medium is homogeneous and transparent. If a parallel beam of white light is incident on the cell, all but the one selected wavelength is scattered, the cell acting as a fairly narrow band pass filter.

APPARATUS

(1) Glass absorption cell 1/8th inch thick.
The cell is made by cementing together glass plates as shown in Fig. B23-1. One must use a cement which is not dissolved or softened by either benzene or carbon disulfide.

Commercially available epoxy cement is satisfactory. The glass pieces are laid out as in the figure so that there is little area of contact between the cement and the solutions. The glass parts should be thoroughly cleaned after they are cut so that the cement will adhere well. After the cell is assembled and the cement dry, it should be tested for leaks.

(2) Several cc of both carbon disulfide and benzene.

(3) Chunk of borosilicate crown glass -- half a cubic inch is more than enough.

(4) Metal mortar and pestle.

(5) Hydrochloric acid for cleaning the glass powder.

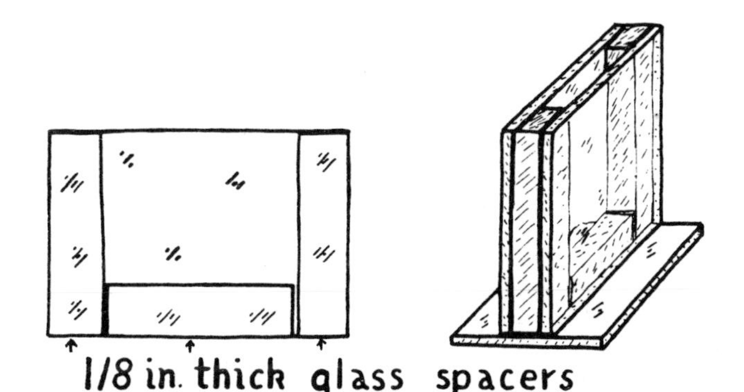

FIG. B23-1

Construction of cell for Christiansen filter.

PROCEDURE

Grind the borosilicate glass to a very fine powder in the mortar -- the finer the better. Wash the glass powder in dilute hydrochloric acid, rinse in water, and dry the powder by spreading it out on a clean paper towel or filter paper.

Fill the cell about two-thirds full of powdered glass, pour in carbon disulfide until the liquid reaches about half the depth of the cell. Stir carefully to eliminate the bubbles and look through the cell at a tungsten lamp placed several meters behind the cell.

Add benzene, a few drops at a time, mixing it thoroughly with the carbon disulfide. After each addition, observe the tungsten lamp. Continue adding benzene until red light can be seen through the cell. Now add still more benzene, noting the gradual change in transmitted color as benzene is added. Observe also the color of the scattered light. Look at it with a direct vision prism if possible.

QUESTIONS

Explain why the transmitted color changes as the benzene is added. Can the filter be made to transmit any color of the spectrum if the proper amount of benzene is added? Why?

Suppose the filter were put just in front of a camera lens and a snapshot of a distant landscape taken. What effect(s) would you anticipate in the developed photograph? How should the filter be used for best results?

EXPERIMENT B24. DEMONSTRATION WITH AN ETCHED TAPERED PHASE SHIFT PLATE.

READING: Strong, Ch. 9.5-9.7; Jenkins and White, Ch. 18.9-18.11; reference - R. W. Wood, Physical Optics, Second edition, (1911, Macmillan), pp. 250-252.

In this demonstration a phase-shift plate is used as a transparent straight edge diffraction screen to produce an unusual set of Fresnel fringes.

THEORY

In the reading, the theory of the intensity distribution at an opaque straight edge is discussed with the aid of the Cornu spiral. Figure B24-1(a) shows the Cornu spiral as applied to this case. The positive eye of the spiral, at Z, represents the last of the infinity of zones. At the geometric shadow edge, the amplitude vector has its tail at A, the origin of the spiral, and its tip at Z. The amplitude is clearly one half that of the unobstructed wave, and the intensity therefore 1/4 that of the unobstructed intensity. It is to be noted that the amplitudes obtained from the spiral must all be divided by $\sqrt{2}$ as explained in the reading.

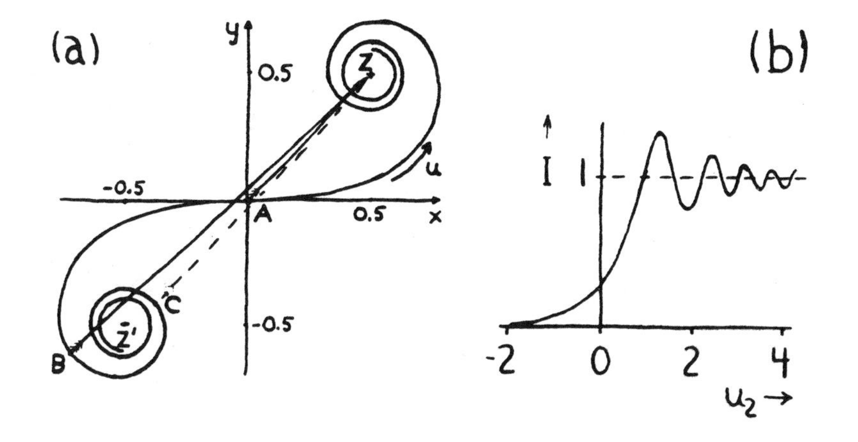

FIG. B24-1

(a) Cornu spiral applied to opaque straight edge, (b) resulting intensity distribution.

As the point of observation moves into the illuminated region, the tail of the amplitude vector includes more and more of the negative part of the spiral. Thus the tail of the vector slides along the curve to point B where the amplitude and intensity reach a maximum, greater than the unobstructed intensity. Still further in the illuminated region the tail slides to point C where the amplitude decreases to a minimum. Moving still more into the illuminated region, it is clear that the amplitude and intensity pass through a series of maxima and minima finally settling down to the unobstructed value represented by the amplitude vector from Z', the negative eye, to Z.

In the shadow region, as is evident from the spiral, the amplitude vector steadily decreases as the tail traces out the positive part of the spiral. The phase changes more and more rapidly in monotonic fashion. The intensity distribution is shown in Fig. B24-1(b).

Suppose now that the obstructing screen is replaced by a dielectric plate bounded by a straight edge. The plate divides the incident light into two parts -- one transmitted through the plate and one bypassing it. These two parts of the incident beam are of equal intensity and their amplitudes interfere constructively or destructively with each other. The symmetry is perhaps more evident if the "bypass" rays are regarded as traversing an air dielectric plate of the same thickness as the solid plate. Evidently, then, it is to be anticipated that the diffraction pattern (if any) will be symmetrical about the geometric boundary.

Let the plate have a thickness t and a refractive index N, and it will introduce an added phase shift in the transmitted wave of magnitude $\Delta = (N-1)t2\pi/\lambda$. If this phase shift is zero or any multiple of 2π, the two waves, the one bypassing the plate and the other transmitted through the plate, will interfere constructively, and the obstacle will not be detected. But suppose the phase shift is an odd multiple of π, then diffraction results.

Figure B24-2 shows a dielectric plate which introduces a phase shift of π in the transmitted wave. At the geometric boundary P, the bypass and transmitted waves interfere destructively to produce a black band. The nature of the diffraction pattern can be understood by reference to the Cornu spiral in Fig. B24-3(a). At the geometric edge P (Fig. B24-2) the bypass amplitude vector has its tail at A (the origin) and its tip at Z. The transmitted vector lies in the third quadrant with its tip at Z' and its tail at A. The vector is reversed because of the phase reversal introduced by the plate. The vector sum is

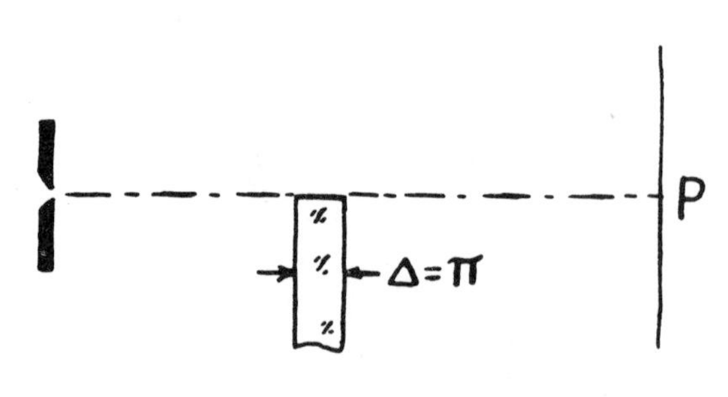

FIG. B24-2

Half wave phase shift plate

obviously zero for the geometric boundary. As the point of observation moves up into the bypass region, the tail of the bypass vector moves along the negative part of the spiral as for example at B. The tail of the transmitted vector is also at B, and the vector sum of the two vectors gives the resultant amplitude. The vector sum at point B gives an amplitude greater than the amplitude in the absence of the plate.

As the point of observation moves still farther into the bypass region, the amplitude vector tails move to point C where their contributions oppose. It can be seen that if the diffraction patterns of the opaque screen and the dielectric plate are compared near the geometric boundary, the pattern of the latter has brighter maxima and darker minima than the former. Unlike the opaque screen, the pattern for the dielectric plate is symmetrical about the geometric edge. The intensity curve is shown in Fig. B24-3(b).

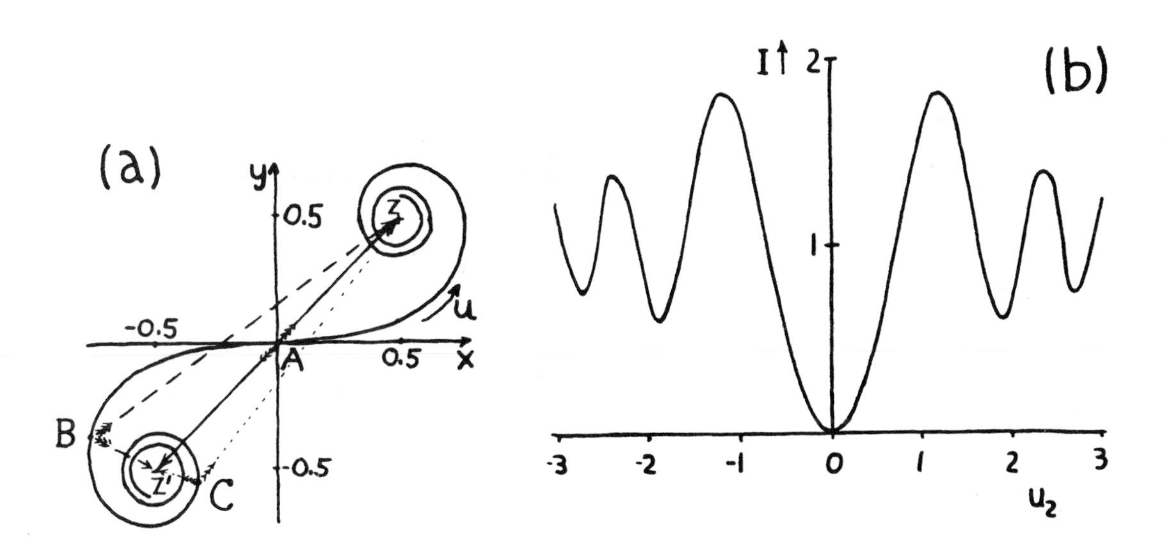

FIG. B24-3

(a) Cornu spiral applied to phase shift plate,
(b) resulting intensity distribution.

APPARATUS

(1) Etched phase shift plate

(a) Microscope slide: 2 in. x 3 in. x 1/16 in.

(b) Beaker: about 3-5 in. dia. and 4 in. deep.

(c) Hydrofluoric acid.

(d) Paraffin wax.

(e) Acetone.

(2) Hg. arc, green filter, and concentrated arc.

(3) Slit.

(4) Eyepiece (1 in. focal length).

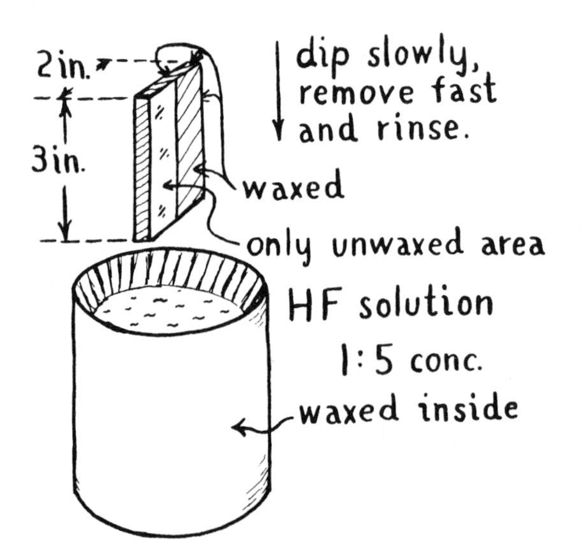

FIG. B24-4

Arrangement to make phase shift plate.

Coat the inside and rim of of the beaker with a thin layer of paraffin wax making sure that there are no uncovered areas. Next, wax the microscope slide except for one half of one side as illustrated in Fig. B24-4. The dividing line between the waxed and unwaxed portions should be a straight line. In the beaker, put diluted hydrofluoric acid -- one part acid to five parts water. The slide is to be dipped into the acid in such a way that the straight dividing line remains vertical. A holder of some sort (clothespin) is used for the dipping, for the acid must not come in contact with one's skin. The slide should be dipped slowly and evenly into the solution taking about ten seconds from start to finish. Thus the bottom of the plate is etched deeper than the top. Withdraw the plate quickly and rinse with clear water. Repeat the dipping once more and rinse again. Now, remove the wax by heating the slide and wiping off all possible wax with a paper towel. Finally clean the plate with acetone and the plate should be ready for use.

A still better phase shift plate could be made by evaporating a layer of MgO of the right thickness on a glass plate.

PROCEDURE

Set up the phase shift plate as in Fig. B24-5. The fringes are observed with a low power eyepiece. The phase shift along the plate is not uniform, and the diffraction pattern will change accordingly. At the end where the phase shift is nearly zero, no fringes are seen. The point where the phase shift is about π will show very clear fringes, whereas beyond the point, the phase shift will increase further and the fringes will fade away again as the shift approaches 2π.

The diffraction patterns should also be observed in white light. The central dark fringe will then be evident although it will no longer be black. Why?

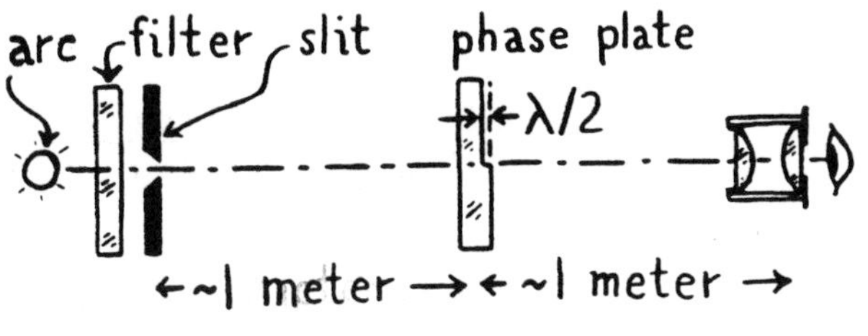

FIG. B24-5

Observation of Fresnel diffraction.

QUESTIONS

How could the phase shift plate be used in a precision alignment problem? How accurately could the three points be aligned? How would this method compare in accuracy with the one mentioned in Experiment B1?

EXPERIMENT B25. MEASUREMENT OF THE SPEED OF LIGHT BY THE MAXWELL COMMUTATOR BRIDGE.

READING: C. H. Palmer, Jr., Maxwell Commutator Bridge for Permittivity and Speed of Light, American Journal of Physics, Vol. 23, pp. 40-45, 1955.

The experiment described in the reading allows measurements of the speed of light to within about 1/2% by means of the Maxwell commutator bridge. This bridge, which has a definite resemblance to a Wheatstone bridge, is used to measure the capacitance of a special condenser in terms of resistance and time. The capacitance can also be calculated in terms of the permittivity of free space and the geometry and dimensions of the condenser. Thus by comparison of the measured capacitance and calculated capacitance, one can evaluate the permittivity of space. In the MKSQ system of units the value of the permeability of space is arbitrarily defined, and thus the speed of light, $c = 1/\sqrt{\mu_0 \varepsilon_0}$ is determined. (In other systems of units the definitions of permeability and permittivity differ, but in any case, the relation derived by Maxwell still holds).

To perform the experiment, one requires a special condenser which can be made from two brass disks which are machined on a lathe, and a special switch, two versions of which are described in the reading.

Additional apparatus includes a radio receiver to provide signals from station WWV, a cathode ray oscillograph, an audio oscillator, and components for a Wheatstone bridge using a long period galvanometer.

EXPERIMENT B26. THE PINHOLE CAMERA.

By W. T. Plummer*

READING: Jenkins and White, Ch. 18.

The pinhole camera is interesting because its operation must be understood jointly in terms of ray and wave principles. The intent of this experiment is to measure the angular resolution obtained with pinholes of various sizes, and with various colors of light, and thus determine optimum conditions of use. The camera will then be used to photograph an outdoor scene in ultraviolet, visible, and infrared light.

THEORY

The operation of the pinhole camera is based on the familiar principle that light travels approximately in straight lines. A small pinhole permits rays from each point of an object to fall only within a small circle on the photographic film, where an image is formed. As the pinhole is made smaller the image becomes more distinct, until the hole becomes so small that diffraction becomes important. The image then becomes worse as the hole is made smaller.

Figure B26-1 shows a cut away view of a pinhole camera. Two point objects at distance d from the pinhole subtend an angle θ at the hole. We wish to determine the minimum angle, θ , which can be resolved for various values of a, f, d, and λ . But if we multiplied each of these lengths by the same scale factor, the resolution, θ , would not be changed. We may therefore simplify the problem a little by studying θ in terms of the three "dimensionless" quantities λ/f, a/f, and d/f. For practical reasons, we may fix d/f to be some convenient value, such as 2.5, which is appropriate for the space and apparatus available. We will then study θ as a function of a/f, for several chosen values of λ/f.

For this purpose we will use a resolution test chart consisting of many sets of three parallel black lines having various spacings.

When the pinhole is large, so that diffraction may be neglected, we may expect two points on the object to be resolved when the centers of the "blur circles" they cast on the film are separated by about five sixths of their diameter. This condition may be written as:

*Laboratory of Astrophysics and Physical Meteorology, Johns Hopkins University.

$$\theta = \frac{5}{6}\,\frac{d+f}{d}\left(\frac{a}{f}\right), \quad \text{or} \quad \theta = \frac{5}{6}\,\frac{(d/f)+1}{(d/f)}\left(\frac{a}{f}\right). \tag{1}$$

(Derive this expression from the statement above).

On the other hand, we may use Rayleigh's criterion to define resolution when the pinhole is very small. A point on the object will then cast the familiar diffraction pattern of a circular aperture upon the film. When two points of the object are separated by an angle θ such that the central bright spot of the pattern cast by one falls upon the first minimum of the pattern cast by the other, then the two points will be considered to be at the limit of resolution. This condition for physical resolution may be written as:

$$\theta = 1.22\lambda/a \quad \text{or} \quad \theta = 1.22\,(\lambda/f)/(a/f). \tag{2}$$

For the arbitrary value $d/f = 2.5$, the criterion for geometric resolution may be simplified to

$$\theta = 1.12\,(a/f). \tag{3}$$

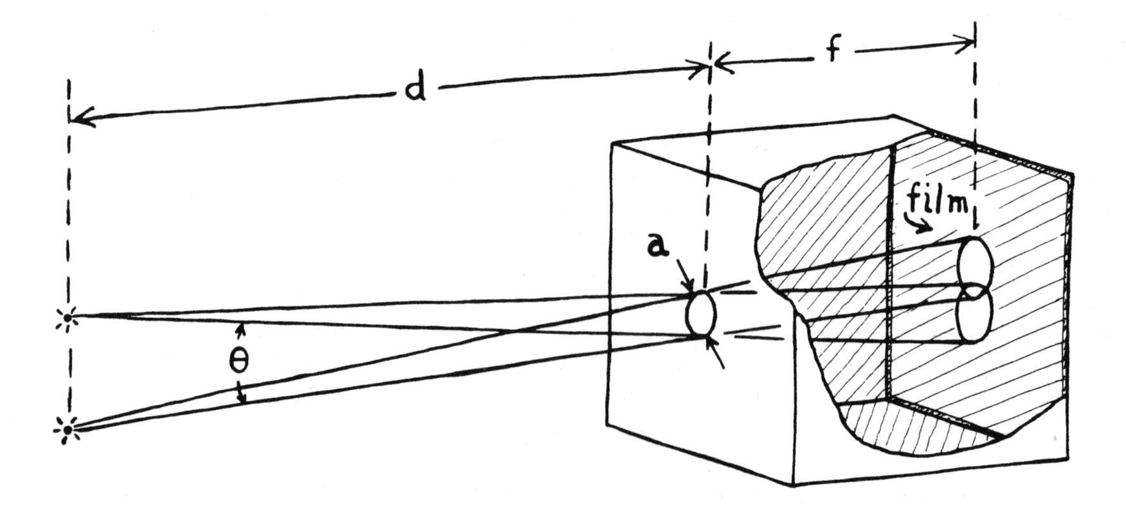

FIG. B26-1

Pinhole camera limited by ray-optical considerations.

APPARATUS

(1) Pinhole camera. The camera may be a simple light-tight plywood box to hold 3 1/4 x 4 1/4 inch cut film packs.

Provision is made to tape pinhole stops on the front of the camera and a simple shutter is provided. A pinhole to film distance f equal to 100 mm is convenient. The inside of the camera is painted flat black.

(2) Pinholes: A set of pinholes can be made by puncturing a piece of 0.004 in. thick brass shim stock with a sharp needle. A set of holes 0.1, 0.2, 0.4, 0.8, and 1.6 mm diameter is prepared. In addition, to fill in the interesting part of the curve, another set having nominal sizes 0.28, 0.35, 0.45, 0.50, 0.55, 0.62, and 0.72 mm dia. are used. Each of these pinholes should be measured accurately with a traveling microscope or with a fine scale etched on a microscope slide.

(3) Resolution test chart consisting of sets of three parallel black lines. This chart is on page 9 of the National Bureau of Standards Circular 533, "Method for Determining the Resolving Power of Photographic Lenses," obtained from the U. S. Government Printing Office, Washington 25, D. C. at $1.25.

(4) Mercury H3 lamp.
General Electric 250 watt infrared heat lamp.

(5) Kodak Infrared Film.
Kodak Royal Pan
Kodak Commercial

(6) Corning glass filters (polished).
3389
3480
3384
4305
5840

PROCEDURE

Pictures are to be taken in the infrared, visible and ultraviolet regions of the spectrum. Table I lists suitable conditions for the three spectral regions together with the approximate average wavelength. The green light has a wavelength of approximately 0.00055 mm. The ultraviolet light used is also monochromatic with a wavelength of 0.00037 mm. The infrared radiation is a broad band, but the response of the film is limited to a fairly narrow region. Its wavelength may be taken as 0.00084 mm, a representative average, and treated as if monochromatic. The filter for the ultraviolet pictures is placed over the pinhole rather than over the mercury source because the resolution chart may fluoresce and produce blue light to which the film is sensitive.

TABLE I. Arrangements for infrared, visible, and ultraviolet pictures.

Spectral region	Infrared	Visible	Ultraviolet
Wavelength	0.00084 mm	0.00055 mm	0.00037 mm
Source	Heat lamp	Mercury arc	Mercury arc
Film	Infrared	Royal pan	Commercial
Filters	3389 + 3480	3389 + 3384 + 4305	5840

The camera and resolution chart are placed on the optical bench so that d, the distance from pinhole to chart, is 2.5 f, or 250 mm. The proper filters for one region of the spectrum are placed between the pinhole and chart, and a loaded film pack put in the camera. A pinhole aperture is put in position, and the shutter closed. (Use the 0.4 mm pinhole for the first picture). The appropriate lamp is placed about 30 cm from the chart, and slightly off to one side. The film pack has two slides at one end. The slide toward the pinhole is pulled out and then the shutter is removed to make a test exposure. Try exposures of 1 min. for infrared, 1.5 min. for green, and 12 min. for ultraviolet. If these times are too long for convenience, use additional lamps.

When the test exposure has been made, replace the shutter, replace the slide (with its shiny side toward the pinhole to indicate "Exposed Film"), remove the film pack, and develop the negative in total darkness according to the printed instructions. For your next exposure, double or halve the time if necessary to get a distinct image.

Exposure times for other pinhole sizes are inversely proportional to the square of the hole diameter (with f constant), so one test exposure for each kind of film is enough. Pictures made with the smallest holes may tend to be a little underexposed, particularly the ones taken with infrared film, because photographic materials are less efficient at low light levels.

The test chart described is calibrated with the numbers 2, 3.5, 4.2, etc. These numbers, which we may call R, are the numbers of lines of the various widths which would fill 13 mm on the chart. On the negative find the narrowest lines which are resolved, and record R for them. The separation between adjacent lines is (13mm/R), and so

$$\theta = \frac{13\ \text{mm}}{R\, d} \qquad \text{or approximately } \theta = \frac{0.05}{R}$$

In this way obtain θ for each value of (a/f); a separate series for each value of (λ/f).

A deceptive result may occur when you are measuring the resolution from the large pinholes. The largest lines on the chart will probably be resolved. Lines of some smaller size will be just slightly resolved, or may blur together. But lines a lot smaller may again appear resolved! This false resolution can be explained geometrically, and is readily identified because the wrong number of lines will appear in the image. For our purposes R is to be read for the set of lines which are just resolved correctly, so that a set of three parallel lines appears as a set of three parallel lines. (This false resolution may be observed dramatically if you photograph a special chart on which three black lines start with a wide spacing at one side and taper away to a point at the other side. To see why the false resolution occurs, make a circular window of diameter $(d + f) \cdot (a/f)$ in a sheet of paper, and slide it across the pattern on the chart. The pinhole camera averages all of the light within such a projected circle on the plane of the object).

RESULTS

On a sheet of log log paper (two cycles each way), plot theoretical curves for the geometrical and physical resolution. That is, plot θ vertically in the range from 0.001 to 0.100 and the ratio a/f horizontally from 0.00025 to 0.0250. (The horizontal scale will be read in units of 0.00025). Equations (2) and (3) give straight lines, the latter giving one line for each wavelength. The intersections of the lines of the two limits of resolution will form large "V"s.

Plot your experimental values of θ vs. (a/f) on the same graphs. (If necessary the range of (λ/f) may be extended by using additional cameras with different values of f, always keeping d/f the same. The result is effectively the same as using a broader range of wavelengths).

Discuss the transition from physical blurring to geometrical blurring. What occurs between (a/f) = 0.002 and (a/f) = 0.008? Why does the curve have this general shape? How might the transition be approached from a theoretical point of view? Explain the various sources of error in your work, and explain how each would affect your graph.

Using all of your data, and considering the nature of the curves you have obtained for different values of (λ/f), choose a new pair of dimensionless variables so that <u>all</u> of your data points will lie along a single curve. (Hint: the curves will be properly superimposed if they are translated appropriate distances along a line bisecting the angle of the axes). In this way, it is possible to obtain a single curve for each value of (d/f). A similar procedure may be used to superimpose the various curves corresponding to different values of (d/f). What would be the change in your graphs now, if you changed (d/f)? Two new dimensionless variables would yield a <u>single</u> experimental curve for all possible values of a, f, d, and λ. Show that the variables might be:

$$\left[\theta\sqrt{\frac{df}{\lambda(d+f)}}\right] \quad \text{vs.} \quad \left[a\sqrt{\frac{d+f}{\lambda df}}\right].$$

The accompanying graph is a summary of all possible pinhole camera situations in terms of these variables. (See how your data fits this curve). In this way data obtained from a complicated experiment may frequently be reduced to a simple, general form for presentation. Figure B26-2 shows a sample curve.

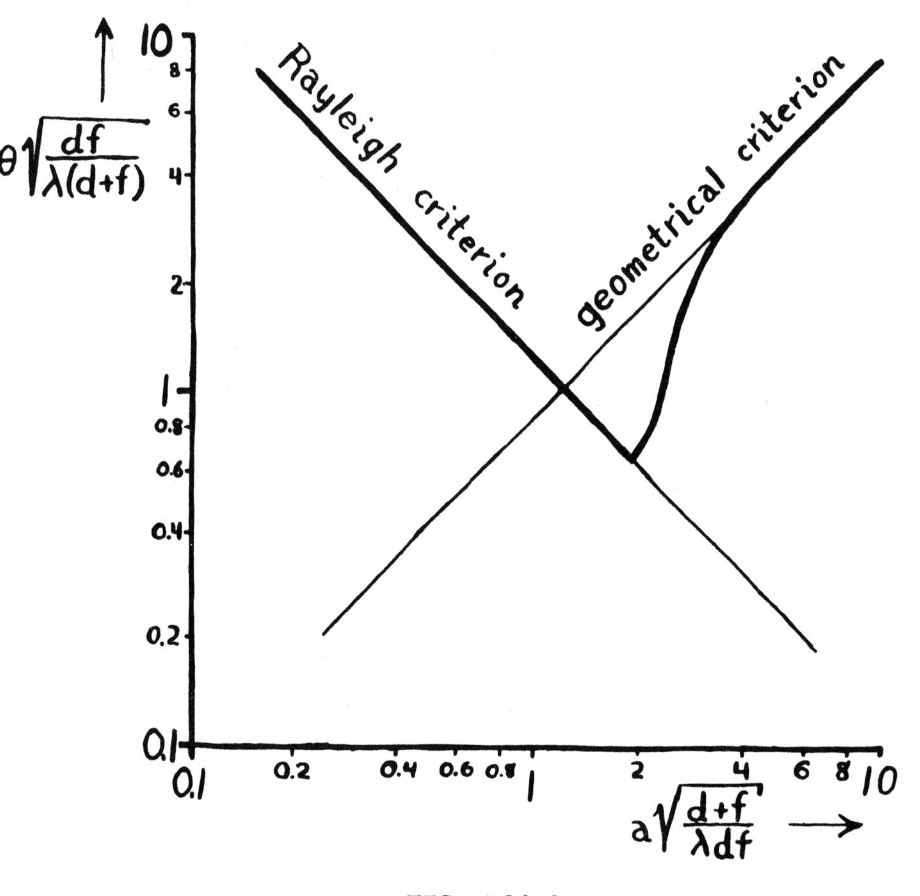

FIG. B26-2

Resolution obtainable with a pinhole camera.

Comment on the form of the region of transition. Is there ever a case where red light might give better resolution than blue?

From the final graph it can be seen that the optimum resolution occurs when $a\sqrt{(d+f)/\lambda df} \approx 2$. We may rewrite this as $a^2(d+f)/\lambda df = 4$, or $(d+f)/df = 4\lambda/a^2$, so that $1/d + 1/f = 4\lambda/a^2$. Remember that d and f are respectively the distances of the object and image from the pinhole. This equation has the same form as the simple lens equation, $1/S + 1/S' = 1/f$ so the pinhole may be treated approximately as a simple lens with a focal length of $a^2/4\lambda$. Thus for a particular wavelength and pinhole size, a pinhole camera may be focused as a lens camera would be. Ordinarily the distance from pinhole to film is kept fixed, and the aperture is changed.

- - - - - - - - - - - -

APPENDIX (optional experiment)

Select the optimum pinhole size for each of the three spectral regions. We may assume that these sizes do not change radically as d is increased, but a more extensive analysis will show that the optimum diameters will increase by a factor of 1.18 when d/f increases from 2.5 to a very large number. Use this correction to select a set of pinholes which will give the best photographs of a distant scene. If the day is clear, use these pinholes to take pictures of a suitable outdoor scene in ultraviolet, visible, and infrared light. The picture should include clouds, trees, grass, and concrete. Also photograph a special setup prepared as follows: find a place where the shadow of a building falls on grass, and is well defined. Lay a sheet of thin white paper along the shadow so that part is in the sunlight and part in the shade. Include the paper and some grass around it in the photographs. Make prints from the negatives when they are dry. Discuss suitable outdoor scenes for the pinhole camera. Compare the brightness of leaves or grass with that of white paper, and discuss the relative contrasts between sun and shade in the infrared and ultraviolet prints.

EXPERIMENT B27. STEFAN-BOLTZMANN LAW.

By W. T. Plummer*

READING: Strong, pp. 78; 476 (H. W. Yates).

This experiment provides a test for the validity of the Stefan-Boltzmann black body radiation law.

THEORY

The energy density radiated by a black body at absolute temperature T in the spectral range $d\nu$, centered on the frequency ν, is

$$\rho\, d\nu = \frac{8\pi h}{c^3} \frac{\nu^3 d\nu}{e^{h\nu/kT} - 1} \quad \text{ergs/cm}^3$$

To obtain the total energy u, we integrate this expression over all frequencies, and obtain

$$u = aT^4 \text{ ergs/cm}^3, \text{ where } a = \frac{8\pi^5 k^4}{15 c^3 h^3}.$$

Therefore the energy radiated in one particular direction, or upon a particular detector, by a black body depends on the fourth power of the temperature. If we use a thermocouple to detect the radiation the voltage produced will be, over a small range, proportional to the temperature rise of the thermocouple. A sensitive galvanometer will show a deflection proportional to that voltage.

But as the thermocouple is warmed by the radiation falling upon it, it loses energy more rapidly to its surroundings, which are at room temperature, and equilibrium will be reached when the amount lost by the thermocouple to its surroundings equals the amount radiated to it by the black body. If the final thermocouple temperature T_t is θ °C above room temperature T_r, the rate of energy loss is

$$W_1 = K_1\left[T_t^4 - T_r^4\right] = K_1\left[(T_r + \theta)^4 - T_r^4\right]$$

or

$$W_1 \approx 4K_1 T_r^3 \theta \approx K_2 \theta \quad \text{ergs/sec} \qquad \text{(Derive)}$$

*Laboratory of Astrophysics and Physical Meteorology, Johns Hopkins University.

The rate at which energy is being received from a source at temperature T_s is

$$W_2 = K_3(T_s^4 - T_t^4) \approx K_3(T_s^4 - T_r^4).$$

But at equilibrium $W_2 = W_1 = K_2\theta = K_4 d$, where d is the final galvanometer deflection. Then by substitution we obtain a simple relation:

$$(T_s^4 - T_r^4)/d \approx \text{constant}$$

which should apply for a fairly large range of T_s. This is the form of the Stefan-Boltzmann law which we will test experimentally. We can test this relation simply by observing the response of a thermopile detector to radiation from a black body whose temperature can be varied.

Because no real surface is actually close to being "black," we will point our detector at a cavity whose opaque walls are held at some known temperature. Such a cavity is a good approximation to a black body because any radiation which is not absorbed upon striking a wall inside will most likely be reflected onto another part of the wall, and so on. If a large fraction of the radiation is absorbed each time, there will be little unabsorbed radiation remaining after a few reflections. Consequently, the aperture leading into the cavity may be thought of as a "black" area, an area which absorbs all of the radiation falling upon it. Such a "black" area will also radiate perfectly, the intensity being appropriate to its temperature.

APPARATUS

1) Black body

a) For sensitive detector

Cut a hole in the middle of the side of a tin can, and solder a small cone inside, with its open end toward the hole, to serve as a cavity as shown in Fig. B27-1. Blacken the interior of the cavity. The cavity should be an inch or more in diameter. (A concave mirror will be used to focus the emitted radiation onto the detector surface).

b) For less sensitive detector

For larger signals needed for a less sensitive detector, use a cone shaped heating element and measure the radiation emitted from the open end of the cone. It is necessary to attach a strip of metal to mask the single hot wire which runs down the inside of the cone. The interior of the cone should be blackened with a paint made of sodium silicate solution (water glass) and lampblack. The temperature within the cone may be measured with a simple thermocouple thermometer, of the type used with small ceramic kilns. This thermocouple could in principle be calibrated by immersion in various molten metals, so there need be no confusion of physical laws arising from the use of thermocouples in two places. A suitable thermocouple is mounted in the cone, touching a wall, so that it can be read without disturbing the optical setup. The temperature inside the heating coil may be 700° C or more. It may be reduced by inserting various arrangements of similar coils in series and in parallel with it electrically. Or it may be more convenient to use an autotransformer (of sufficient power capacity) to lower the voltage.

2) Mirror, about 10 cm dia., 25 cm focal length.

3) Detector

The detector is a thermopile, a number (about 20) of thermocouples connected in series, with all junctions of one kind blackened and exposed to the radiation, and all balancing junctions shielded from the radiation. This detector should produce at least 20 microvolts when exposed to a black body at 100°C.

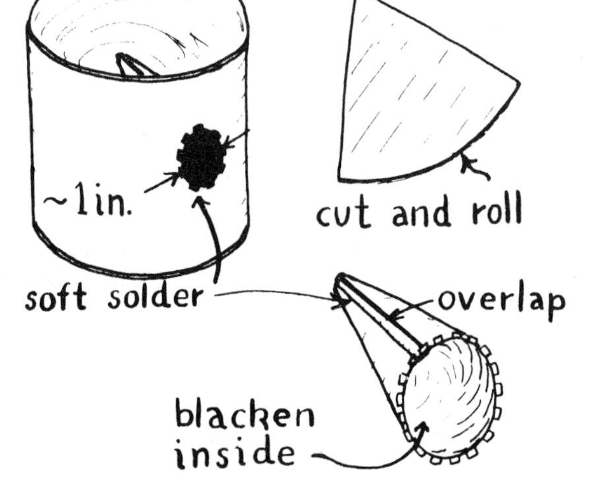

FIG. B27-1

Moderate temperature black body made from tin cans. Parts are soft soldered together to make watertight joint.

4) Galvanometer

A mirror galvanometer which will indicate with a sensitivity of about 0.2 microvolt/millimeter, perhaps with the help of a D. C. amplifier, is suitable for use with the first kind of black body, the can with a cavity. If a less sensitive galvanometer is available, the experiment may be performed over the greater temperature range, with correspondingly higher output voltages. With the heater-coil black body a galvanometer which will indicate 10 or 20 microvolts per millimeter is adequate.

5) Cardboard or wooden box to eliminate external heat and light (See Fig. B27-2).

6) Dry ice and alcohol or ice and salt.

PROCEDURE

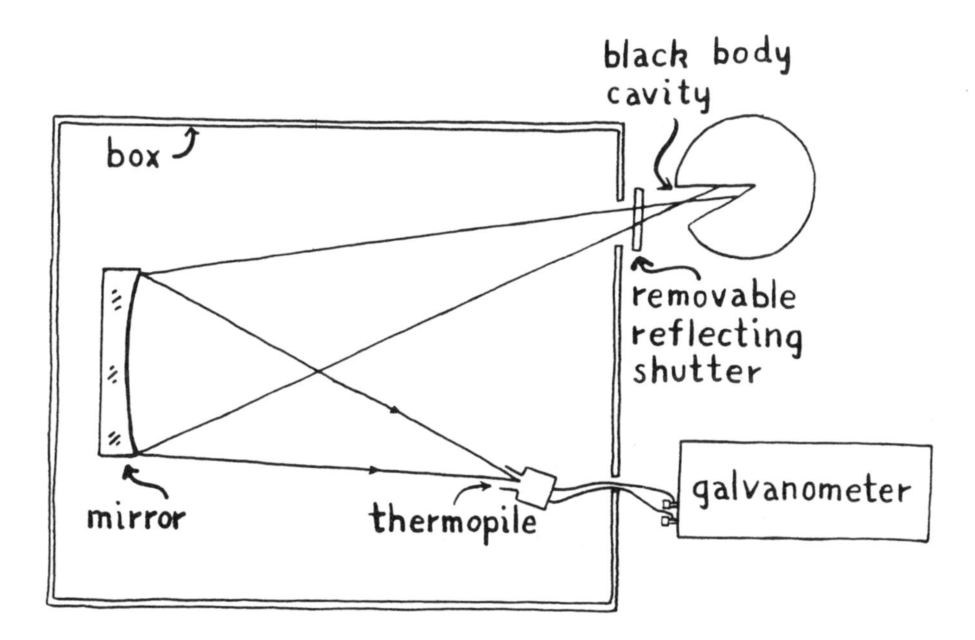

FIG. B27-2

Set-up to verify the Stefan-Boltzmann law.

Set up the black body source, mirror, and detector as shown in Fig. B27-2. The source should be tipped slightly so that a uniformly heated part of the source will be imaged on the detector. Use a small light source such as a flashlight bulb held at the center of the black body aperture to adjust the mirror so that the radiation will fall on the sensitive area of the detector. Remove the light and block off stray heat and light from the room. The enclosing box itself will, of course, radiate, but that radiation is constant. Leave a small opening in the box where it is against the cavity aperture, and place a removable shutter over this opening. Both sides of the shutter are covered with shiny aluminum foil to block the radiation without raising the temperature of the shutter.

If your detection system is sensitive enough to operate in the range from room temperature to 100°C, set the can-type black body in position in front of the mirror, fill it with boiling water, and suspend a thermometer in the liquid next to the cavity. Keeping the shutter closed, set the galvanometer to zero. Remove the shutter, wait a half minute for the thermopile to reach its equilibrium temperature, and record the maximum galvanometer deflection. Close the shutter. Record the water temperature and the room temperature (near the thermopile). Repeat the measurements of water temperature at intervals of 10°C until the water has cooled to room temperature. Repeat with a mixture of dry ice and alcohol in place of the water, or with a mixture of ice and salt if dry ice is not available.

If you are using the heating coil black body, connect the coil to the power line and wait until it has reached its highest temperature, as indicated on the thermocouple. Read its temperature, and room temperature (near the thermopile), then open the shutter and record the maximum galvanometer deflection. Close the shutter. Reduce the voltage on the coil, wait for it to reach its new equilibrium temperature, and repeat the measurements. Readings should be made for temperature intervals of about 50°C.

For each galvanometer reading calculate $(T_s^4 - T_r^4)/d$, and plot these values against T_s.

Discussion: Is the curve approximately that of a constant? Discuss any variations in the value of $(T_s^4 - T_r^4)/d$. What are the effects of the various approximations made in the derivation? Discuss the effects which would be observed if the mirror surface were less reflective, or if the relative distances from the mirror to the cavity and detector were changed. If you can justify calling $(T_s^4 - T_r^4)/d$ a constant, give its value for the equipment you have used, and indicate the range of variation.

Section C

POLARIZATION OPTICS

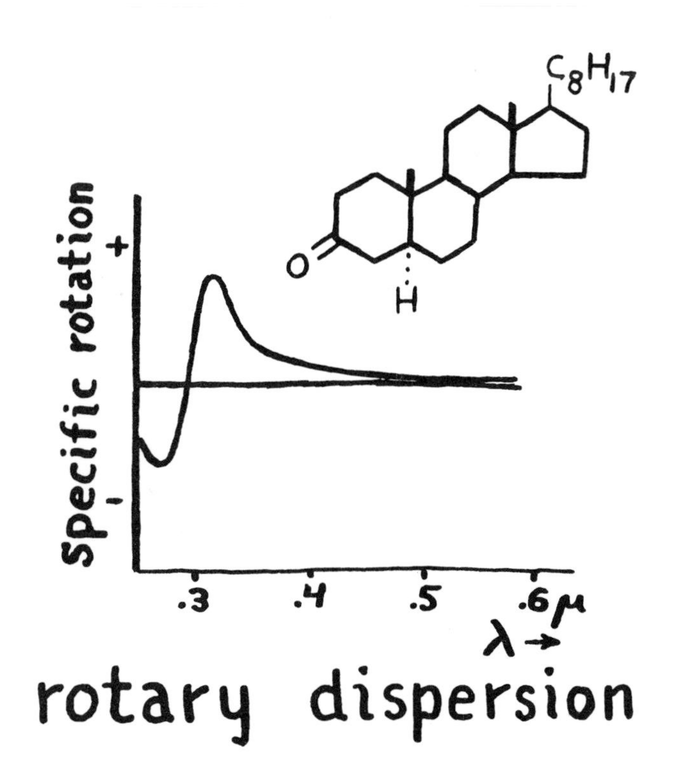

rotary dispersion

EXPERIMENT C1. REFLECTION OF POLARIZED LIGHT FROM DIELECTRICS.

READING: Strong, Ch. 3.7-3.9; or Jenkins and White, Ch. 25-25.5, (25.13).

The purpose of this experiment is to study the reflection of polarized light from a dielectric and to compare the results with those predicted by Fresnel's equations.

THEORY

The vibrations for which the electric vector lies parallel to the plane of incidence are called p vibrations; those for which the electric vector is perpendicular (senkrecht) to this plane are called s vibrations.

Fresnel's equations give the ratio of the reflected amplitude to the incident amplitude for the two polarizations s and p. These equations are:

$$\frac{R_s}{E_s} = -\frac{\sin(i-r)}{\sin(i+r)} \quad (1) \qquad \frac{R_p}{E_p} = +\frac{\tan(i-r)}{\tan(i+r)} \quad (2)$$

where E_s and E_p are the incident electric vector amplitudes for the s and p components respectively, and R_s and R_p are the corresponding reflected amplitudes. Angles i and r are the angles of incidence and refraction respectively.

If the incident light is plane polarized with the electric vector at 45° with respect to the plane of incidence, i.e. $E_s = E_p$, it follows that

$$\frac{R_p/E_p}{R_s/E_s} = \frac{R_p}{R_s} = -\frac{\tan(i-r)}{\tan(i+r)} \times \frac{\sin(i+r)}{\sin(i-r)}$$

or

$$\frac{R_p}{R_s} = -\frac{\cos(i+r)}{\cos(i-r)}$$

It is the object of this experiment to compare the measured value of R_p/R_s as a function of i with the theoretical value given by the equation.

Since the theoretical value of this ratio involves r as well as i , it is necessary to determine r in each case. The best procedure is to calculate r using Snell's law. The refractive index of the dielectric must be given or measured.

APPARATUS

(1) Student spectrometer equipped with:
- 60° prism.
- Uncalibrated polarizer (Polaroid or Nicol prism) mounted on collimator.
- Analyzer calibrated in degrees rotation mounted either on the telescope objective or eyepiece.

(2) Mercury arc with green line filter.

(3) (Optional) Auxiliary collimator -- useful for measurements at small angles of incidence.

PROCEDURE

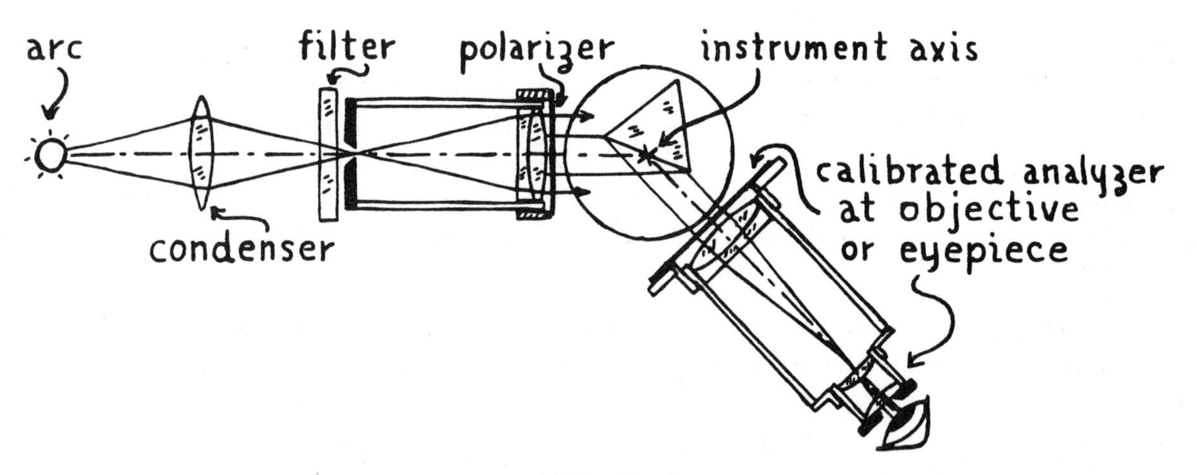

FIG. C1-1

Arrangement for reflection measurements.

The spectrometer is put into adjustment, i.e., the collimator and telescope are focused and aligned perpendicular to the mechanical axis of the instrument (See Experiment A5). If an auxiliary collimator is used as in Fig. C1-2, the leveling screws of the spectrometer are next adjusted so that the telescope can 'look' directly into the auxiliary collimator. One face of the carefully cleaned prism is made perpendicular to the telescope axis. (If the refractive index of the prism is to be measured, two prism faces must be adjusted). One face of the prism should be placed so that the mechanical instrument axis passes through it, approximately, as shown in Fig. C1-1, since the measurements involve reflections from the prism.

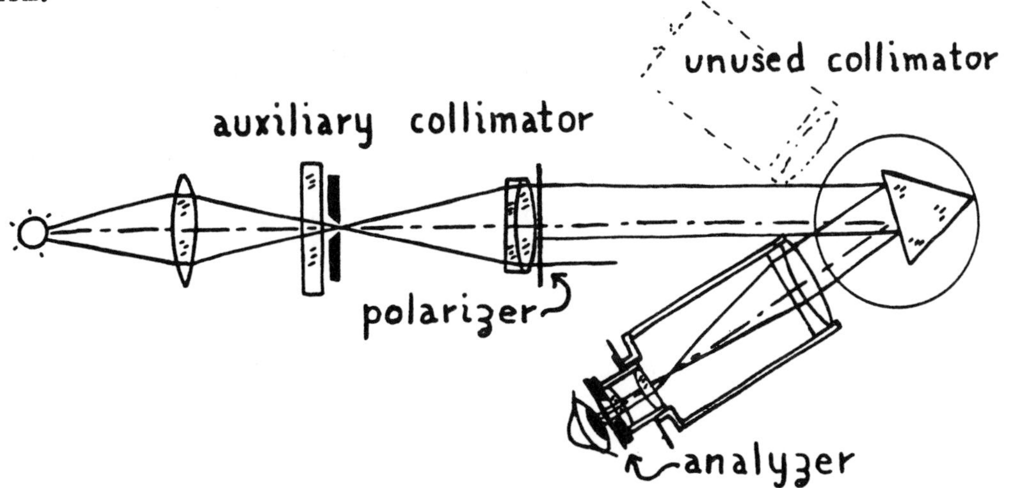

FIG. C1-2

Use of auxiliary collimator to allow measurements at small angles of incidence.

Set the telescope to 'look' into the collimator -- either the one belonging to the spectrometer or the auxiliary one. Adjust the telescope so that the collimator slit image (rather wide slit) is centered on the telescope cross hairs, and set the prism table scale to read 180°. The scale should now be clamped and left for the remainder of the experiment. The angle between the telescope and collimator is then directly read at any time.

With the polarizer removed from the collimator, set the telescope at an angle ($\sim 115°$) corresponding to roughly twice Brewster's angle. Turn the prism by rotating the prism table, but leaving the scale clamped, so that light is reflected from the prism face into the telescope. Next, set the analyzer scale (angle θ) to read $\theta = 0°$ and clamp or hold while the analyzer is turned to extinguish the light as nearly as possible. Small adjustments in $\angle i$ and in $\angle \theta$ should produce a rather sharp minimum of intensity. The light reflected off the prism face ($\angle i$ = Brewster's angle) is plane polarized with the electric vector perpendicular to the plane of incidence (S component). The analyzer angle θ should now read 0°, and this means that when no light is transmitted to the eye, θ is the angle of the plane of polarization of the electric vector with respect to the vertical or s component.

Return the telescope to $2i = 180°$ and set the slit image on the telescope cross hair. Set the analyzer angle θ to read 45°, insert the polarizer for the collimator, and rotate this polarizer to achieve extinction. At this point, the s and p components of the incident light are equal in magnitude, $E_s = E_p$.

Precautions should be taken to keep out all unwanted light from the telescope and to assure a very bright monochromatic parallel beam incident on the prism face.

Set the telescope at various angles of incidence and in each case, turn the prism table (<u>not scale</u>) to reflect the beam into the telescope. Record both i and θ needed for extinction. Use values of i from the minimum possible (about 15° unless an auxiliary collimator is used) to nearly 90°, using 10° steps except near Brewster's angle, where additional readings are desirable.

If the refractive index of the prism is to be measured, do so after the other measurements are completed and with the prism set for minimum deviation. The index need be determined for only the one wavelength concerned.

RESULTS

Plot values of $|R_p/R_s| = \tan\theta$ as a function of i. On the same graph, plot a theoretical curve of $|R_p/R_s|$ vs. i. The values of r must be calculated for each theoretical point from the refractive index and i. Figure C1-3 shows the general shape of the curve obtained.

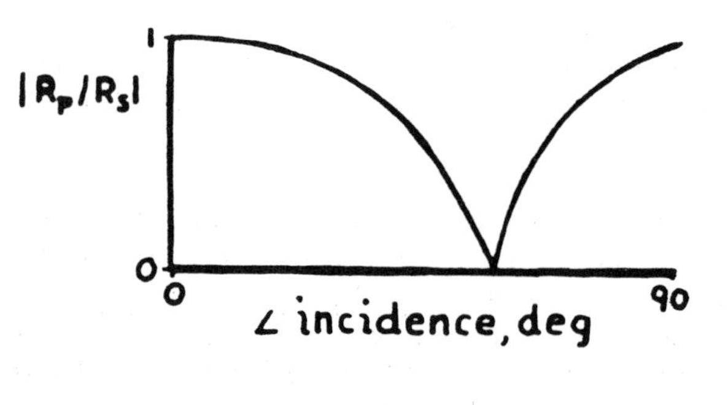

FIG. C1-3

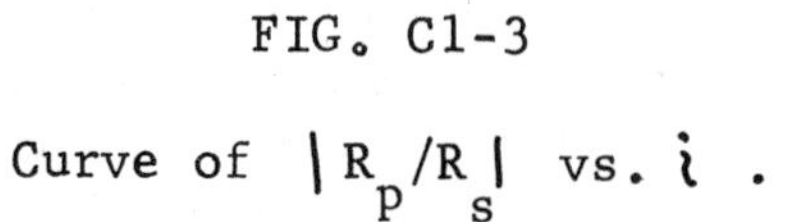

Curve of $|R_p/R_s|$ vs. i.

EXPERIMENT C2. HAIDINGER'S BRUSH AND THE OBSERVATION OF POLARIZATION IN NATURE.

READING: Strong, Ch. 6-6.2; and M. Minnaert, The Nature of Light and Color in the Open Air, (1954, Dover), pp. 254-257.

This experiment is a demonstration of the Haidinger brush phenomenon, which makes it possible to discern the polarization of light without any polarizing device save the eye itself. The Haidinger's brush is quite faint, and in order to see it at all, one should proceed in several steps.

APPARATUS

(1) Polaroid sheet.

(2) Sheet of white paper or white card.

(3) Small piece of glass plate.

PROCEDURE

The polarizer is first marked to indicate the axis of easy passage. Reflect light from the glass plate at Brewster's angle (about 57°) so as to obtain plane polarized light with the electric vector parallel to the plane of the glass surface (s component). Observe this light through the Polaroid and rotate it for extinction (or minimum light). The axis of easy transmission (the direction of polarization of the transmitted electric vector) should be marked with ink on the clean Polaroid surface.

Use the Polaroid analyzer to observe with one eye the sheet of paper or card which must be brightly illuminated with white light. After careful observation for a minute or so it will be noticed that there exists a pattern in the center of the field of the eye which is more or less in the form of a four-leafed clover. Two opposite blades are generally bluish and the other pair yellowish or brownish -- the pattern is caused by the dichroism of the yellow spot of the eye and forms the Haidinger brush. To make sure of the observation, turn the analyzer about the line of sight axis and note that the pattern rotates in synchronism. Determine the direction of the electric vector with respect to the direction of the bluish tint of the pattern.

Now look through the Polaroid at an area of blue sky in a direction at about right angles to the direction of the sun. The Haidinger brush should be found in a minute or so and can again be checked by rotating the analyzer.

After the pattern is clearly discernible in this way, remove the analyzer abruptly while looking at the pattern and it will be discovered that the pattern can still be seen, though it is now very faint. The orientation of the pattern will, in general, change, for the polarization transmitted by the analyzer will not be that of the sky alone. Note the direction of the bluish tint of the pattern and determine the direction of polarization of the sky light. Is this direction in accord with that described in the reference in Strong's book?

Finally, observe the sky without the analyzer and look for the Haidinger brush pattern. Considerable patience may be required. After you believe you have found the pattern, make sure by interposing the analyzer to verify the location of the pattern. The experiment should be tried repeatedly until it becomes easy to find the pattern.

If a body of water is available, try to find the brush in the light reflected by the water.

Make a sketch of the pattern as it appears to you and indicate the direction of polarization and the approximate angular dimensions of the pattern.

EXPERIMENT C3. SIMPLE POLARIZATION DEMONSTRATIONS.

READING: Strong, Ch. 6-6.8, 7-7.3, 7.7-7.9; or Jenkins and White, Ch. 24, 26-26.5, 27-27.3.

In this demonstration, we will show some of the methods for producing and analyzing polarized light.

APPARATUS

(1) White light source

(2) Polariscope

(3) Polarizing elements:

a) black glass mirror
b) glass plate pile
c) Nicol prism
d) Polaroid sheet (mounted on cardboard)
e) Calcite crystal
f) cellophane (cigarette package wrapper)
g) mica sheet (0.1 mm or more thick)
h) two quarter wave plates
i) plastic U to show strain double refraction
j) yellow filter

PROCEDURE

1. PLANE POLARIZED LIGHT

A. Polarization by reflection

Arrange the polariscope as indicated in Fig. C3-1 to produce plane polarized white light. The glass plate M1 should be clean in order to obtain good results. Make the angle of incidence approximately 57° (Brewster's angle for glass). For the analyzer, mount the black glass mirror M2 in the graduated upper support (disregard stage and mirror at bottom). Set the angle of incidence for this mirror also at about 57°. Orient top mirror so that plane of incidence is at right angles to plane of incidence for polarizer M1. Now adjust both mirrors carefully to obtain minimum light when looking into the beam reflected from the analyzer.

Note that the positions of the mirrors are quite critical. Rotate the analyzer stage and observe the positions of maxima and minima. Describe the results. Is the polarization complete? Explain the presence of any residual light. Describe the orientation of the electric vector (plane of vibration).

B. Polarization by transmission

Replace the analyzer with the glass plate pile. Set at the proper angle for maximum polarization (minimum light be reflection). Look through the pile and study the transmitted light as the stage is rotated. Describe the results. Why is the polarization incomplete?

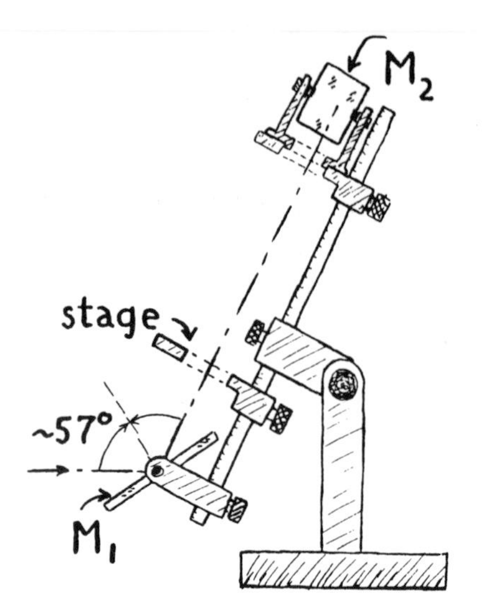

FIG. C3-1

Polariscope.

C. Nicol prism

Insert the Nicol prism in place of the pile. Study the light as the analyzer is rotated. Notice that the minima can be determined much more accurately than the maxima. Can you explain why?

D. Polaroid

Finally, insert a Polaroid analyzer. Mark on the cardboard ring the direction of the electric vector for transmission. Use a straight edge for the purpose.

2. DOUBLE REFRACTION

Place the calcite crystal over a small pencil dot on a piece of paper. Observe the motion of the two dots as the crystal is rotated. Use the analyzer to study the polarization of the two dots. Sketch the crystal face, show the position of the two dots and draw a short line through each dot to indicate the direction of vibration of the electric vector. Notice which dot appears to stand still as the crystal is rotated. This is the O ray, the other is the E ray. Which dot appears nearer the surface? Is the index greater for the O or the E ray? Is calcite a positive or negative crystal?

Place the Polaroid back on the polariscope and set for extinction. Put the mounted calcite crystal on the stage below and observe the effect of rotating the calcite on the stage. Notice that there are two positions for which extinction occurs. Compare these directions with respect to the directions of vibration for the two dots previously studied. For other orientations of the calcite no extinction is possible. Explain.

Set the Polaroid analyzer for extinction. On the stage below place a small sheet of cellophane. Rotate the stage and describe the results. How many maxima and minima are there? Rotate the cellophane to get a minimum and then 45° more in either direction. Now rotate the analyzer. The cellophane is obviously doubly refracting.

Place a thick sample of mica on the stage. Note that there are two orientations for the mica at which the analyzer can extinguish the light. These two positions determine the slow and fast axes for mica. If the mica is in any other position, the light changes color as either the mica or the analyzer is rotated, but in no case is the light extinguished.

Hold the plastic U in the polarized beam. Orient for minimum transmitted light (analyzer set for extinction with no sample). Squeeze the ends together and observe the strain pattern. When glass or plastic is strained, it becomes doubly refracting.

3. CIRCULARLY POLARIZED LIGHT

A. Place a quarter wave plate on the stage. Set the stage scale to read zero and orient the quarter wave plate so that no light is transmitted (two possible directions). Use a yellow filter in the beam. Rotate the stage 45° in either direction. In this position, rotation of the analyzer will produce little if any change in intensity. The light emerging from the quarter wave plate is circularly polarized.

B. By inserting a second quarter wave plate in the beam with its axes correctly oriented the beam may again be extinguished. If it is extinguished when the analyzer is in the normal extinction position, the two quarter wave plates are oriented so that the fast axis of one corresponds to the slow axis of the other and the net effect is the same as with no plate at all. On the other hand, if the beam can be extinguished at some other angle, the slow axes of both plates are parallel and the result is a half wave plate.

C. Use either the pair of quarter wave plates oriented as a half wave plate (or a half wave plate) and verify the fact that the plane of vibration is rotated by an angle 2θ where θ is the angle between the incident vibrations and the principal section. Set analyzer in normal extinction position, and half wave plate for extinction of beam. Rotate the wave plate through angle θ and note the angle through which the analyzer must be rotated to produce extinction.

EXPERIMENT C4. DOUBLE REFRACTION IN CALCITE.

READING: Strong, Ch. 7-7.6; or Jenkins and White, Ch. 26-26.6.

In this experiment, the double refracting properties of calcite are to be studied. Calcite is representative of the class of uniaxial crystals, and it is convenient to study because of its simple crystal habit, its insolubility in water, and especially because of the large difference in the ordinary and extraordinary indices of refraction. These same properties make calcite a useful optical material.

APPARATUS

1) Two calcite crystals of about the same thickness, at least 1 in. x 1 in. x 1/3 in. thick. Two faces at least should be clear and transparent, free from pits, crystal plane steps, or too many scratches. These faces may be either natural faces, or they may be polished, but if so, they must coincide with the natural faces.

2) A cut crystal, specially prepared to allow observation along the optic axis (desirable, but not essential).

3) Small piece of Polaroid.

4) Millimeter scale (15 cm. long) and protractor.

PROCEDURE

A. Shape of the calcite crystal

Examine the shape of the natural crystals provided. The crystal can be cleaved in directions parallel to the natural faces. How many kinds if corners are there? Describe each variety of corner in terms of the number of acute angles and obtuse angles made by the three edges.

On a sheet of paper, lay both a crystal and the millimeter scale. Hold the crystal to the paper so that it does not slide and place the millimeter scale so that one edge makes contact with one face of the crystal. Draw a line about 6 inches long using the other edge of the scale, (i.e. parallel to the crystal edge). Repeat for the other three edges so as to obtain a parallelogram whose dimensions are those of the crystal plus two widths of the millimeter scale. Measure the angles of the resulting parallelogram.

Finally, make a three dimensional sketch of the crystal.

B. Double refraction

Put a dot on a piece of paper and place a crystal over it. Sketch the top face of the crystal and put one dot in the center of the face. Carefully indicate the position of the second dot, observing both the angle with respect to the first and the separation. (The observer is assumed to be looking vertically down, normal to the crystal face). Identify which dot corresponds to the O ray and which to the E ray. To identify the two, rotate the crystal and observe which dot moves. Use a piece of Polaroid on which the axis of transmission has been marked. If not already marked, do so (with ink) by using the known polarizing properties of glass at Brewster's angle. Indicate on your sketch the direction of vibration for each dot by means of a short dash through each dot. Identify each of the two dots as O or E.

Which dot appears to be deeper? What do your observations indicate about the relative indices of refraction for O and E rays? (See Experiment A2 regarding the theory, if needed). A positive uniaxial crystal has $N_o < N_e$, and negative one, vice versa. Is calcite a positive or a negative crystal?

Verify that the separation of the dots both in apparent depth and in horizontal separation increases linearly with the thickness of the crystal.

Measure the angle of deviation of the E ray in traveling through the crystal. A satisfactory method for this measurement is to draw two almost parallel lines about 6 inches long, which meet at one end and are about 1/4 in. apart at the other. Lay the crystal on the lines and orient it to give maximum separation of the lines. Slide the crystal along the lines and ascertain at what point two of the four lines merge so as to give three in all. Measure accurately the separation of the lines at this point, and measure the thickness of the crystal (not slant height). Calculate the E ray deviation at normal incidence.

C. Double refraction in two calcite crystals of approximately the same thickness

Place one crystal over a very black but small dot on a piece of paper and orient it so that the blunt corner of the top side is toward the upper left. Place the second crystal on top so that its surfaces are parallel to those of the lower crystal. Make a sketch representing the crystals by parallelograms, the lower one dashed, the upper solid. Indicate the positions of the dots and their polarization (use the Polaroid).

Rotate the upper crystal only by 45° clockwise and again indicate the dots and their polarizations. Make similar observations for each 45° rotation of the upper crystal until it has been rotated a full 360°. To help keep directions straight, it is suggested that the student crosshatch the solid line which, in the first sketch, was at the top of the parallelogram.

If three equally thick crystals were stacked on top of each other over a dot, and if the middle crystal were rotated 45° with respect to the bottom one, and the top crystal at 45° with respect to the middle, how many dots would be observed? What would their polarizations be?

If a specially cut crystal is available to permit observation along the optic axis, identify this axis. Along the optic axis, there is no distinction between O ray and E ray. Is this the same thing as saying that the optic axis is the direction in which there is no double refraction? Explain. Does the optic axis join two opposite corners of the crystal? Explain.

E. Measurement of refractive index (Optional)

Use the method of Experiment A2 to measure two refractive indices corresponding to the O and E rays. The crystal should be about 1/4 in. thick -- if much thicker, it will not be possible to focus the microscope on the images of the dots, and if much thinner, the change in focus will be too small. A Polaroid sheet over the eyepiece will improve the contrast and make the focusing on the two dots easier. Distinguish O and E rays in the measurements. Obviously these measurements determine the O index. Do they, in fact, determine the E index? Can the E index be calculated from this and perhaps other simple measurements on the crystal? Explain.

RESULTS

Include all sketches, diagrams, and measurements with probable accuracy. Answer the questions in the sections of the experiment you have performed.

EXPERIMENT C5. THICK WAVE PLATES AND CHANNELED SPECTRUM IN QUARTZ.

READING: Jenkins and White, Ch. 27-27.7; and Strong, pp. 141-144; pp. 585-586 (R. G. Greenler).

In this experiment a thick quartz plate is used to produce a channeled spectrum with closely spaced bands.

A thick wave plate placed between crossed or parallel polarizers can produce a channeled spectrum by interference like the channeled spectrum of the thin dielectric plate in Experiment B7. The mechanism, however, is somewhat different. With the thin dielectric plate it is the two beams reflected from the front and back of the glass or cellophane plate which interfere with each other to produce the channeling. The incident light may be unpolarized. The thick crystal plate, on the other hand, produces interference in the transmitted beam. An incident plane polarized beam is resolved by the crystal into two components vibrating parallel to and perpendicular to the optic axis, and they travel at different speeds through the non-isotropic crystal. The two emerging rays, vibrating at right angles to each other, cannot interfere, but they are combined to give plane, circular, or elliptically polarized light. A second polarizer selects the components parallel to its direction of transmission, and these components interfere constructively or destructively.

An important application of the channeling produced by thick wave plates is found in the Lyot-Ohman narrow band filter. The construction of this device is explained in the reading. The filter, often built of ammonium dihydrogen phosphate crystals, is used to study the sun in hydrogen, calcium, or helium light. The following experiment shows the properties of one stage of such a filter.

THEORY

The number of bands in the channeled spectrum produced by a thick wave plate depends upon its thickness, the wavelength interval, and the E and O refractive indices. Suppose a parallel beam of plane polarized light falls normally on the crystal plate, as in Fig. C5-1, so that the plane of vibration is at 45° with respect to the optic axis of the crystal. The crystal resolves the incident light into E and O components parallel and perpendicular, respectively, to its optic axis. These components, traveling through the crystal at different speeds, c/N_e and c/N_o, recombine upon emerging from the plate. If the retardation of the E ray relative to the O ray corresponds to an integral number of wavelengths $m\lambda$, the two rays recombine to give the same polarization as the incident light. If, on the other hand, the relative retardation is $(m+1/2)\lambda$, the recombined light is plane polarized at right angles to the incident light (the plane of polarization is rotated through twice the angle between the polarizer and the crystal optic axis). Thus, if a second polarizer follows the crystal, it will transmit wavelengths for which the relative retardation is $m\lambda$ and block those for which it is $(m+1/2)\lambda$ provided both polarizers are parallel. If they are crossed, clearly the situation is reversed; the previously transmitted rays are blocked and vice versa. If the relative retardation is neither integral nor half integral the emerging light is elliptically polarized, and one component of it is transmitted by the second polarizer.

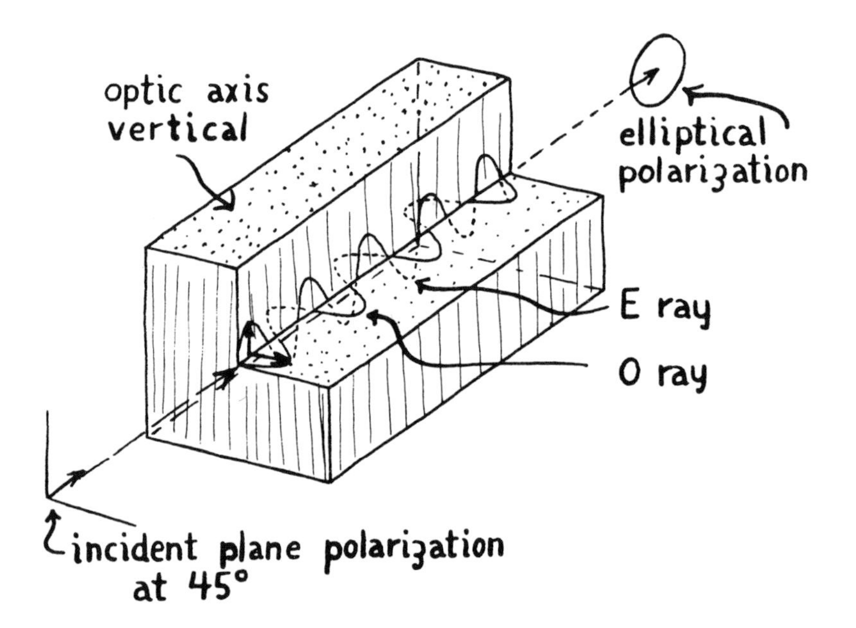

FIG. C5-1

Plane polarized light incident on crystal quartz plate.

Let $\varepsilon_1 = N_{e1} - N_{o1}$ be the difference in refractive indices for E and O rays at wavelength λ_1 and similarly $\varepsilon_2 = N_{e2} - N_{o2}$ at wavelength λ_2. Assume that the polarizers are crossed. Then if the thickness of the wave plate is t, dark bands occur for two wavelengths if

$$m_1 \lambda_1 = t\varepsilon_1$$

$$m_2 \lambda_2 = t\varepsilon_2$$

The number of dark bands occurring between the two wavelengths is then

$$\Delta m = m_1 - m_2 = t\left(\frac{\varepsilon_1}{\lambda_1} - \frac{\varepsilon_2}{\lambda_2}\right).$$

If two arbitrary wavelengths are chosen, the corresponding values of m for a given crystal thickness will not, in general, be integers. It is practical, however, to estimate the fractional parts of bands so that known wavelengths (corresponding to known mercury spectral lines, for instance) may be used.

APPARATUS

1) Student spectrometer equipped with a diffraction grating of 10 to 20 thousand lines per inch.

2) Quartz crystal to make a wave plate 2 or 3 cms. thick. The optic axis must, of course, lie in the face. Natural crystals with reasonably good faces are satisfactory. Alternatively, a cut, polished quartz wave plate of about this thickness can be used.

3) Two lenses, preferably achromatic, of 2-5 in. focus and about an inch diameter.

4) Condenser lens (L_3), and plane piece of glass roughly an inch square (M).

5) Two sheets of Polaroid about an inch or two square.

6) Auto lamp source.

7) Mercury arc source.

PROCEDURE

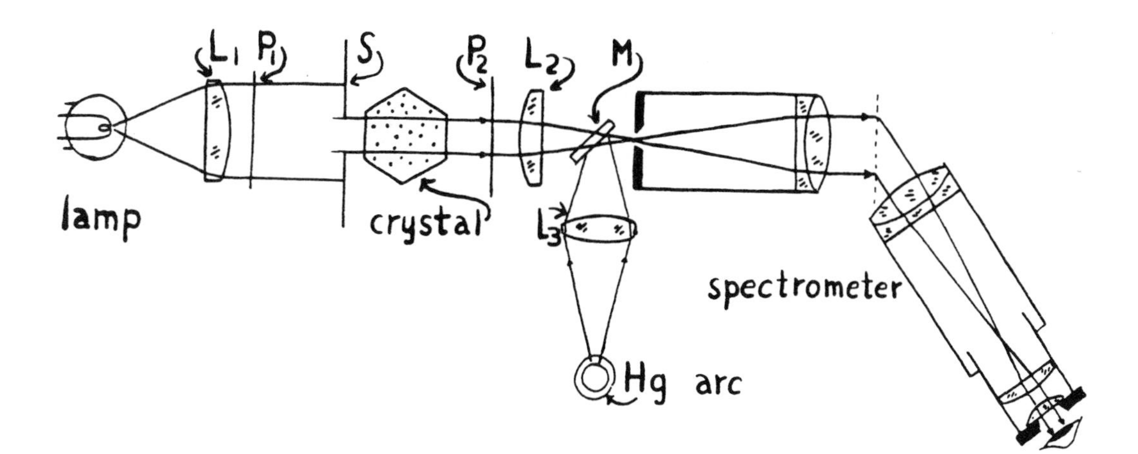

FIG. C5-2

Apparatus to study channeled spectrum in quartz.

The apparatus is set up as shown in Fig. C5-2. Put the spectrometer into adjustment (see Experiment A5) and use a grating to obtain adequate dispersion for the estimation of fractional parts of the bands. Put the lamp in line with the collimator at a distance several inches greater than the sum of the focal lengths of the two lenses. Position lens L_1 (orient for minimum spherical aberration) so that the lamp filament is at its primary focal point. Insert L_2 to focus the parallel beam on the collimator. The spectrometer should now show a smooth, bright continuous spectrum. If horizontal lines attributable to the windings of the lamp filament are seen, try changing the focus of L_2 slightly or using a piece of finely ground glass at the slit.

Mount the two Polaroid sheets so as to leave room for the crystal between them. The transmission direction of the first Polaroid is set at 45° to the vertical (either of the two possible angles). Cross the second Polaroid with respect to the first so that no light enters the spectrometer. Now mount the crystal so that the optic axis is vertical and its faces normal to the incident beam. The spectrometer should now reveal a channeled spectrum, though the bands may not be very clear and straight. If the bands have bends in them, examine the crystal for uniformity of thickness. Quartz frequently has rough striations running perpendicular to the optic axis. Use a cardboard or paper stop to restrict the incident light to the best part of the crystal face so that the interference bands are clear, sharp, and straight.

Set the telescope cross hair on an arbitrary black band. Rotate the second Polaroid and observe the effect. Explain. Return the Polaroid to its original crossed position.

Set up the comparison spectrum using an unsilvered piece of glass for the mirror M. The plate may be placed directly in the beam if desired so that it will superpose the mercury spectrum on the channeled spectrum, or it may be arranged so that the comparison spectrum is above or below the channeled spectrum.

With the aid of the known wavelengths provided by the comparison spectrum, determine accurately the number of black bands between the two wavelengths. (Estimate the fractional band at each end). Choose wavelengths for which the number of bands will be between about ten and thirty. It is prudent to make a diagram showing the black bands in relation to the comparison spectrum.

Use a micrometer to measure accurately the thickness of the quartz crystal -- if there is any variation in thickness, measure that part of the crystal which was used.

RESULTS

From a table of refractive indices for quartz (see Jenkins and White, p. 542) determine the indices for the two wavelengths used in the experiment. Calculate the interference orders m_1 and m_2 (non-integral) and find the theoretical number of bands between these wavelengths. Compare the result with the experimental determination. Are the results satisfactory? Explain. Discuss the errors in the experiment, and estimate their magnitude.

Is it practical to perform this experiment with a calcite crystal wave plate instead of quartz? Why? How would the results differ if a crystal of ammonium dihydrogen phosphate were used instead of quartz?

In making a Lyot filter, what tolerances would have to be placed on the thickness of the various wave plates?

EXPERIMENT C6. OPTICAL ACTIVITY AND BIREFRINGENCE IN CRYSTAL QUARTZ.

READING: Jenkins and White, Ch. 28; and Strong, Ch. 7.10, 7.11.

In the first part of this experiment measurements are made of the rotary dispersion of crystal quartz along the optic axis. From these measurements, the difference in left and right hand indices is evaluated as a function of wavelength. The remainder of the experiment shows, within the limitations of the apparatus available, the change in polarization of rays transmitted along or near the direction of the optic axis.

APPARATUS

1) Student spectrometer equipped with calibrated analyzer.

2) Two crystal quartz plates cut perpendicular to the optic axis. One plate should have a thickness of 1-2 mm and the other 5-10 mm.

3) Mercury or mercury cadmium arc.

4) White light source.

5) 60° glass prism.

6) 60° crystal quartz prism cut to allow transmission along the optic axis (that is the optic axis lies in the plane of the triangular ends). Alternatively, a Cornu prism which can be separated into its two component prisms.

7) Polaroid sheets.

PROCEDURE

I. Rotary dispersion in quartz

The apparatus is arranged as in Fig. C6-1. First put the spectrometer into adjustment using the methods of Experiment A5. The dispersing element is the 60° glass prism set at minimum deviation for the green. Clamp the prism table. The source is either the mercury arc or the mercury-cadmium arc. Open the collimator slit as wide as possible while keeping the

spectral lines clearly separated.

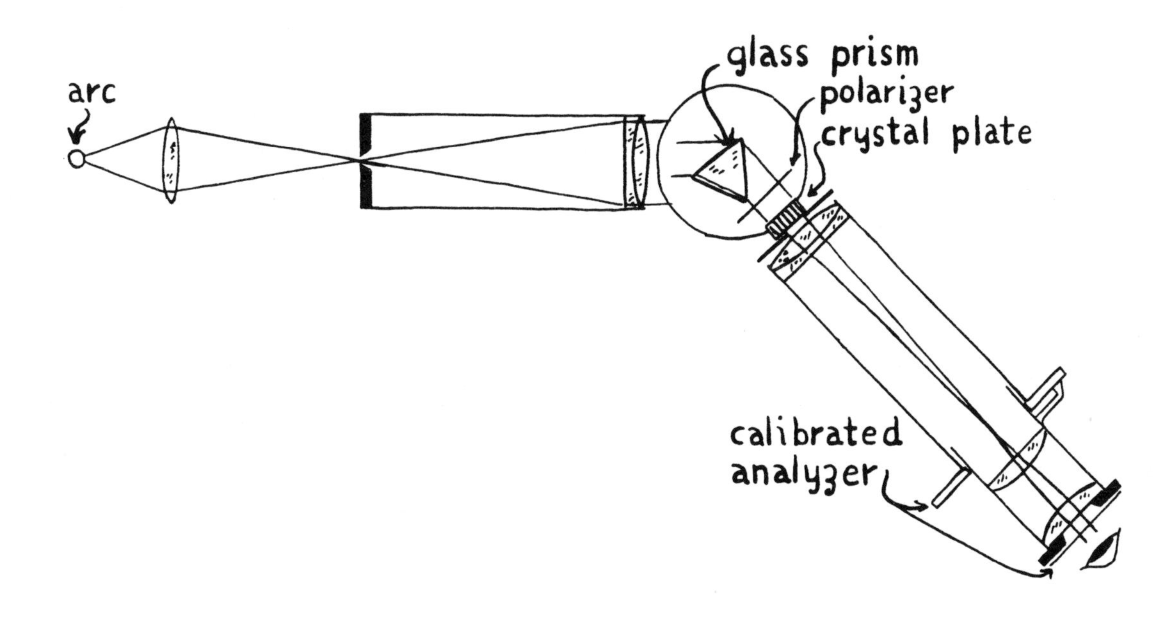

FIG. C6-1

Measurement of rotary dispersion.

Substitute the calibrated analyzer for the regular eyepiece. (If the analyzer is to be mounted over the telescope objective, move the prism closer to the collimator to leave room for the polarizer and crystal plates). Mount the Polaroid polarizer close to the prism. (Any azimuth is all right, but the polarizer should be approximately normal to the beam). Set the analyzer for extinction and record the angular setting. Use a micrometer to measure the thickness of the two plates(being careful not to scratch them) and record the results. Mount the thinner crystal plate between the polarizer and the telescope, either on the prism table or at the end of the telescope.

Put the line of longest (known) wavelength at the cross hair and set the analyzer for extinction or minimum transmission (crossed Polaroids have some residual transmission, especially in the red and violet). Record the angle and determine the rotation. Repeat for a green line, a blue line, and a violet line. Plot a rough curve of specific rotation vs. wavelength.

Remove the thin crystal plate, check that the analyzer calibration has not been disturbed, and substitute the thicker crystal plate. Make accurate measurements of rotation for all the prominent known spectral lines from red to violet. The results for the thin plate will resolve any uncertainty in the number of half revolutions involved. Remove the thick crystal plate and check the analyzer calibration once more.

Hold the thin crystal plate between two pieces of Polaroid and observe a white light source at a moderate distance. Rotate one of the Polaroid sheets with respect to the other. Describe your observations. Can you predict the result? Explain. Try the experiment with the thick plate. Is there any difference? What and why? What would you expect to observe if the plate were 5 to 10 cm thick? Would any colors be seen? How would the spectrum look if the light transmitted through this crystal was directed into a spectrometer? Could a narrow band filter be constructed by making use of rotary dispersion? How would it compare with the Lyot-Ohman filter?

II. Polarization at and near the optic axis

Remove the polarizing attachments for the spectrometer and substitute a crystal quartz prism for the glass one. If a 60° prism is used, set it for minimum deviation in the green. If half of a Cornu prism is used, set it so that the light is transmitted along the optic axis. (In a Cornu prism, the base is parallel to the optic axis). Narrow the collimator slit so that the spectral lines have minimum width. Adjust the focus carefully for the green. No doubling of the green line should be observed. (The reason will become apparent later). Check that the light is neither plane polarized nor elliptically polarized. Have you shown that the light is not circular polarized? Would you expect it to be circular polarized? Explain. How would you test for circular polarization?

Measure the angle of deviation of the green rays and calculate the refractive index approximately. Which index has been determined?

Rotate the prism slightly off the optic axis until doubling of the spectral lines is observed. Use a piece of Polaroid to determine the nature of the polarization of the lines. Describe the direction of the major axis of the ellipse if these lines are elliptically polarized. Continue to turn the prism until the doubled lines become very nearly plane polarized. Determine roughly the angle between the ray direction and the optic axis. Describe the plane of vibration of the two components and identify the rays as ordinary or extraordinary. Can you determine whether the crystal is right hand or left hand quartz? How? If not, how could one distinguish right and left handedness in the prism? Devise a simple practical test and determine which it is.

RESULTS

Plot an accurate curve of the specific rotation of quartz as a function of wavelength. Specify whether the wave plates are right or left hand quartz. Evaluate the constants of the Cauchy dispersion formula for the specific rotation, $\rho = A + B/\lambda^2$, and plot this curve on the same sheet. Is the fit good or poor? Explain.

From the experimental data, calculate the difference in refractive indices $N_l - N_r$ along the optic axis as a function of wavelength. Calculate the angular separation of the two circularly polarized green components in passing through the prism along the optic axis. Calculate the theoretical chromatic resolving power of the same prism at the same wavelength. Show why no doubling of the green line is observed along the optic axis. Specify whether the quartz prism is right or left hand.

Record the calculated refractive index along the optic axis and the probable error in the result. What index is this?

Describe the transition from optical activity to birefringence in quartz.

Answer the other questions in the section on procedure.

EXPERIMENT C7. OPTICAL ACTIVITY IN SODIUM CHLORATE.

REFERENCE: A. Holden and P. Singer, Crystals and Crystal Growing, (1960, Doubleday), Ch. 4.

Sodium chlorate which crystallizes in cubic form, is nevertheless optically active. Furthermore, the specific rotation is the same for all directions in the crystal. It is the object of this experiment to grow several crystals big enough to study, and to measure their specific rotation.

APPARATUS

1) Crystal growing apparatus:

    a) 1 lb. sodium chlorate powder

    b) 1 liter pyrex container

    c) filter and filter paper

    d) large watch glass or glass baking dish

2) Polarimeter.

3) Mercury arc and filters to isolate one or more lines.

4) Immersion liquid having a refractive index

$$N_D = 1.50 - 1.52$$

PROCEDURE

I. Crystal growing

Several crystals 5 to 10 mm thick are needed for the experiment. They are grown from aqueous solution. One method is to make about 50 cc of saturated sodium chlorate solution and let it evaporate in a watch glass over night. The dish should be put where the temperature is reasonably uniform and where it is protected from dust. With luck, half a dozen or more acceptable crystals can be grown over night. More specific directions and improved methods are described in the reference.

Select the best crystals from the remaining solution, remove them carefully, preferably with tweezers, and dry off the solution adhering to them. The crystals should have rather flat clear faces.

II. Measurements of optical activity

Select a good crystal and measure its approximate thickness with a micrometer. Set up the polariscope and set the analyzer for extinction. Record the angular reading. Put the measured crystal on the stage and determine the specific rotation for one or more wavelengths. Is the crystal dextrorotatory or levorotatory? Turn the crystal so that the rotation is measured in all three directions normal to natural crystal faces. Is the optical activity the same in these three selected directions?

Try other crystals to ascertain whether or not all crystals have the same handedness, right or left. Record your observations. Measure the rotation of several thicknesses of crystal to evaluate the specific rotation with moderate accuracy.

Make a flat bottomed cell using a piece of metal or glass tubing about half an inch diameter and half an inch long waxed to a microscope slide. Grind or pound to a powder several crystals of the same variety and fill the cell to a depth of several millimeters. Cover the powder with immersion oil and determine the optical activity of the mixture. What do the results indicate about the optical activity in directions other than normal to one or the other crystal face?

RESULTS

Discuss the observations and measurements made. Is sodium chlorate birefringent? What conclusions can be drawn about the wave surfaces in sodium chlorate? What is implied about the structure of the crystal?

What kind of molecular structure would give rise to optical activity in liquids?

Section D
SPECTROSCOPY

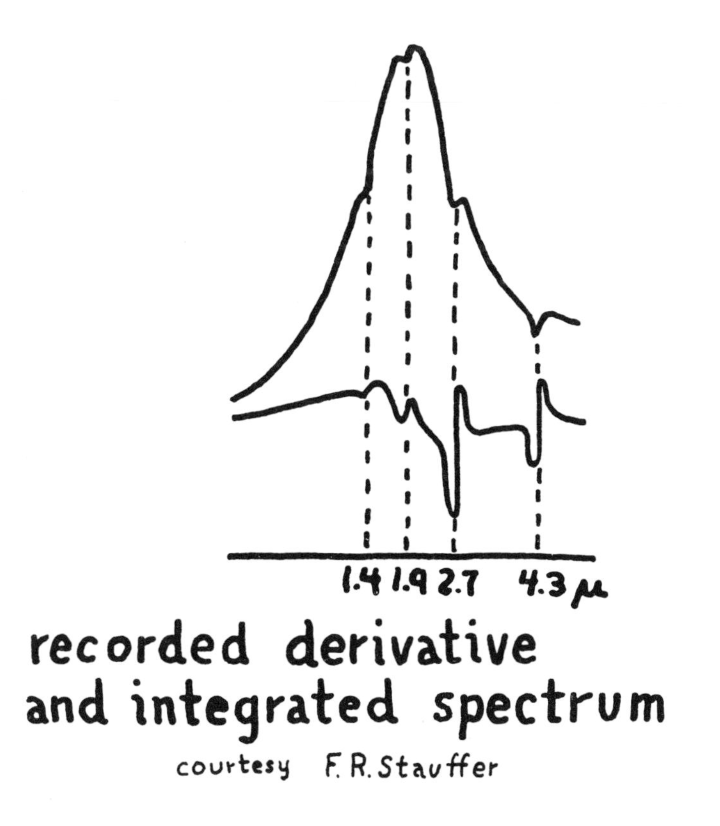

recorded derivative and integrated spectrum

courtesy F. R. Stauffer

EXPERIMENT D1. WAVELENGTH DETERMINATION BY THE PRISM SPECTROGRAPH.

READING: Experiment A5; R. A. Sawyer, Experimental Spectroscopy, (1946, Prentice-Hall), Ch. 3, Ch. 8, Ch. 9, sects. 88, 89.

In this experiment a prism spectrograph is used to determine the wavelengths of several unknown spectral lines in terms of three known wavelengths. In the first part of the experiment, a spectrum plate suitable for measurement is obtained, and in the following part, the spectrum is reduced and the wavelengths calculated with the aid of a desk calculator.

APPARATUS

1) Prism spectrograph. Any good quality prism spectrograph for the visible and near ultraviolet spectral region is satisfactory.

2) Mercury arc (AH3) source.

3) Helium or cadmium source.

4) Traveling microscope. A Gaertner traveling microscope with 5 cm travel is sufficient in most cases.

5) Desk calculator with six or eight figure capability.

6) Photographic plates. Suitable spectroscopic emulsions are described in the manufacturer's literature. Sawyer gives information on Eastman Kodak emulsions. A high contrast, fine grain, panchromatic emulsion is desired. Ordinary panchromatic plates or film may be used if necessary. Sealed plates or film may be stored for long periods in a refrigerator before they lose sensitivity. Before opening, they should be allowed to warm to room temperature.

7) Darkroom and processing chemicals and trays.

PROCEDURE

PART I. THE SPECTRUM PLATES

1) Adjustment of the spectrograph

The spectrograph must be put into accurate adjustment by using a procedure based on the principles studied in Experiment A5. Obviously the exact procedure to be followed will depend upon the construction of the instrument and the adjustments which can be made on it. The collimator, prism, and camera must all be in proper adjustment. The collimator must be in focus for the part of the spectrum to be studied, its slit parallel to the refracting edge of the prism (vertical), and the slit width and height proper (see below). The prism must be properly level. The camera must be in good focus over the whole spectrum to be photographed. It is assumed that the optical axes of the collimator and telescope are normal to the instrument axis. (In some instruments such as the Hilger constant deviation spectrograph, the collimator and camera tubes are permanently adjusted).

After the prism and collimator are adjusted, illuminate the instrument with mercury light. (See appendix to Experiment 4). The collimator should not be greatly overfilled as light might then be scattered from its walls and partially fog the plate. Adjust the instrument so that the mercury green line is about in the center of the ground glass screen of the camera -- for the Hilger constant deviation spectrograph, for instance, the red will fall at one end of the plate or ground glass and the near ultraviolet at the other. Three known wavelengths must be unmistakably identified -- yellow, green, and blue lines of mercury are convenient. (Alternatively, one might use three cadmium lines). The camera is first focused on the green line. Close the collimator slit completely and then open it just enough to see the line. Use a 5-10X magnifier and continue to open the slit very slowly. The line at first will simply grow brighter, but soon will also begin to broaden. The proper slit width is that for which the line is as bright as possible but not widened. Focus the camera as carefully as possible for the green line and then adjust the tilt of the plate holder so that both ends are also in the best possible focus.

Determine the proper settings for the plate holder so that with a slit height of perhaps 5 mm, eight or so spectra can be photographed on the plate without overlap.

2) Loading the plate holders

The plate holder(s) should be loaded in total darkness since the emulsion is panchromatic. The emulsion side may be distinguished from the anti-halation backing with a slightly moistened finger -- the stickier side is the emulsion. Alternatively, the corner of the plate may be put between ones lips. Perform this test at a corner of the plate!

3) Test plate

Insert the plate holder in the camera and position it so that the spectra may be photographed in sequence. Use the mercury arc and make exposures of 1, 4, 16, and 64 seconds. Repeat for the other source using a range of exposures which will bracket the correct exposure. Be careful not to disturb the instrument while opening or closing the slide on the plate holder. Remove the plate holder for processing.

Make up the developer, short stop, and fixer solutions according to directions. Small separately packaged amounts of the three required powders are convenient. Only a few ounces of solution are needed for small trays -- which should be filled to a depth of no more than 1/3 inch. Process the plate according to the directions. Rock the tray gently while developing the plate to agitate the solution so that the development will be entirely uniform. After the plate has been in the fixer about half the proper time, the room lights may be turned on briefly and the plate examined. Complete the fixing and then wash the plate in warm (but not hot) water for ten to fifteen minutes. Use a sponge to remove the excess water, being careful not to injure the emulsion. The plate may now be examined critically with a good magnifier (or it may be dried first if there is time) to make sure the focus is sharp. If necessary, make a second test plate after the focus is readjusted.

4) Analysis plate

The analysis plate may be planned as shown in Fig. D1-1. The spectra should partly overlap as shown so that there is no difficulty in identifying the source for each line and so that the combination spectrum can be measured in the region of overlap. Two sets of spectra should be made using different exposures. If the top part of the mercury spectrum in the overlap region is exposed 1/4 to 1/6 as long as the rest of the spectrum, the strong lines will be less broad and their centers more accurately determined under the microscope.

After the test plate is exposed, process it very carefully, using the full time in the fixer according to the processing directions. Wash the plate in lukewarm water for the requisite time, sponge off the excess water, and dry, preferably in a nearly vertical position so that the water may drain off without leaving water marks on the plate. The plate should be inspected to make sure that it will be satisfactory for measurement.

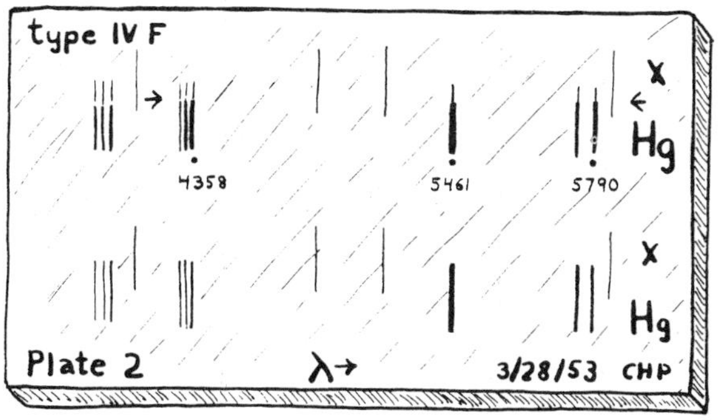

FIG. D1-1

Suggested arrangement and labeling for spectrum plate.

PART II. WAVELENGTH MEASUREMENT

1) Plate identification

It is assumed that the analysis plate is clean and dry. It should now be carefully inspected and the known wavelengths identified as

FIG. D1-2

Form for machine calculation of wavelengths by Hartmann formula.

$$\lambda = \lambda_0 + \frac{c}{d_0 - d} = \qquad + \frac{\qquad}{\qquad - d}$$

source	intens.	d mm	$d_0 - d =$ .... $-d$	$\frac{c}{d_0 - d}$ $c =$	$\lambda_{calc.}$ $\lambda_0 =$	λ_{table}	$\Delta\lambda$
equation check λ_1							equation check
λ_2							
λ_3							

FIG. D1-3

Form for machine calculation of Hartmann constants.

$$\lambda = \lambda_o + \frac{c}{d_o - d}$$

$\lambda_3 =$ $d_3 =$

$\lambda_2 =$ $d_2 =$

$\lambda_1 =$ $d_1 =$

$\lambda_2 - \lambda_1 =$ d $d_2 - d_1 =$

$\lambda_3 - \lambda_1 =$ $d_3 - d_1 =$

$d_3 - d_2 =$

$$a \equiv \frac{d_2 - d_1}{\lambda_2 - \lambda_1} = \qquad a - b =$$

$$b \equiv \frac{d_3 - d_1}{\lambda_3 - \lambda_1} =$$

$$\frac{c}{d_o - d_1} = \frac{d_3 - d_2}{a - b} =$$

$$\underline{\underline{\underline{\lambda_o}}} = \lambda_1 - \frac{c}{d_o - d_1} =$$

$$d_o - d_1 = (\lambda_2 - \lambda_o)\,a =$$

$$\underline{\underline{\underline{d_o}}} = (d_o - d_1) + d_1 =$$

$$\underline{\underline{\underline{c}}} = \frac{(d_o - d_1)(d_3 - d_2)}{a - b} =$$

$$\lambda = \qquad\qquad + \frac{\qquad\qquad}{\qquad - d}$$

in the figure, with black or india ink. Indicate with arrows the part of the spectrum to be measured -- probably limited to 5 cm by the measuring range of the traveling microscope.

2) Measurement of the plate

Put the plate on the traveling microscope stage, emulsion face up. Position the plate so that the part to be measured falls within the possible positions of the microscope and so that the focused microscope travels midway along the overlap region. If the ends of the mercury comparison spectrum have been given a reduced exposure, the microscope crosshair should pass over the dividing line between the more intense and less intense parts of the lines. Begin at one end of the spectrum and measure as accurately as possible the centers of all the lines -- of both sources. Use a form such as that of Fig. D1-2. Identify the source for each line and its estimated intensity ranging from 1 (nearly invisible) to 10 (extremely black). Record the positions of the three known lines in the upper section and the remainder in the lower. After making the measurements, repeat (at least roughly) to verify the readings and identifications.

3) Calculation of the Hartmann constants

The Hartmann formula for wavelength describes fairly accurately the dispersion of a prism in regions away from absorption bands:

$$\lambda = \lambda_o + \frac{c}{d_o - d}$$

The constants λ_o, c, and d_o are determined by substituting in the equation the three known reference wavelengths and their positions. Watch decimal points! Figure D1-3 shows a convenient form for making the required calculation by machine. (Prove that the operations indicated give the constants of the formula). Use the longest wavelength for λ_3. Results should be calculated to six figures where possible and the wavelengths to the nearest tenth of an angstrom unit. When the constants have been computed, use the final equation to calculate the three standard wavelengths as a check. These results should differ from the original wavelengths by at most one tenth of an angstrom; any larger error must be corrected.

Compute the remaining wavelengths by formula. From a handbook, find the accepted wavelengths, fill in the second to last column (Fig. D1-2) and evaluate the error.

RESULTS

Include all calculations and the resulting wavelengths. Discuss the accuracy of the results and the chief sources of error. Are there any errors which might arise from the processing of the plate? How could the accuracy of the experiment be improved? Explain.

EXPERIMENT D2. THE GRATING SPECTROMETER.

READING: Jenkins and White, Ch. 17; and Strong, Ch. 10.3, 10.6.

This experiment illustrates the use of diffraction gratings in wavelength measurement. The grating space of a coarse grating is accurately determined with a traveling microscope and the wavelength of a standard spectral line measured. This standard line is then used with a fine grating to measure the visible lines of the mercury spectrum. The phenomena of overlapping orders and the normal spectrum are also studied. (The resolving power of a grating is discussed in Experiment B20).

APPARATUS

1) Student spectrometer equipped with Gauss eyepiece and grating mount.

2) Coarse transmission grating having 500-1000 lines per inch. (If necessary, a Ronchi ruling of at least 100 lines per inch can be used). The grating should be mounted on glass or between two microscope slides.

3) Fine transmission grating having 10,000 to 15,000 lines per inch.

4) Traveling microscope.

5) Mercury arc source.

6) Filters to isolate several mercury lines. The filters should include both a combination to isolate the green line and a purple filter to transmit red and blue (e.g. Corning #9863).

PROCEDURE

I. DETERMINATION OF GRATING SPACE BY DIRECT MEASUREMENT

Mount the coarse grating on the traveling microscope stage with the rulings perpendicular to the direction of travel of the microscope and so positioned that the greatest possible width of the grating may be measured.

The microscope must be focused on the grating with more than usual care to avoid two sources of error. These difficulties in the apparently simple matter of focusing the microscope may be readily demonstrated by greatly exaggerating them.

A. Aperture limitation

Focus the microscope on the grating. Note that the depth of focus is rather small; the image is not sharp unless the focus is nearly correct. Now cut a narrow slit about a quarter millimeter wide in a file card or heavy paper, and hold the slit over the center of the microscope objective. The length of the slit is to be parallel to the rulings. Adjust the focus and notice the great increase in depth of focus. The apparent grating space can be changed considerably. Explain. If the grating space were determined by comparison with a reticle scale in the eyepiece, considerable error would result if the depth of focus were large. Since the grating space is to be determined by translating the entire microscope relative to the grating, a change in magnification or a slightly incorrect focus because of large depth of focus would not cause serious error in the measurement. Why?

B. Fourier error

Another type of focal error, which might be called a Fourier error because it results from the periodicity of the grating, is more important. Illuminate the grating and microscope with parallel illumination, say from an incandescent lamp about ten feet away, which is reflected by a plane mirror through the grating into the microscope. If the microscope focus is now changed, it will be seen that there are apparently several different focal positions for the microscope where the grating space changes by integral multiples. The appearance of the ruled surface changes qualitatively. The explanation of the phenomenon involves the theory of microscopic imaging discussed further in Experiments B21 and B22. In any case, it is apparent that if the illumination for the microscope is incorrect (as in the demonstration), gross errors can be made in the determination of groove space in the grating.

We now return to the problem of measurement of the grating space. The grating is well illuminated by a broad white light source and the microscope focused. Adjust the grating carefully so that the rulings are accurately perpendicular to the direction of travel of the microscope. Now focus the microscope accurately on the grating rulings at one end of the ruled area and then move the microscope to the other end of the ruled portion or to the end of its travel as the case may be, and check the focus. If the rulings are out of focus at this end of the grating, it will be necessary to block up one end of the grating or the other with thin pieces of paper until the whole grating is in focus.

Set the microscope at one end of its travel, and, after removing the backlash in the screw, set the crosshair on the edge of one of the rulings. Record accurately the micrometer screw reading (number 1). Translate the microscope while counting about 20 lines and set the crosshair on the last counted line. Record both the reading (number 2) and the number of lines to which it corresponds. Now without counting lines, translate the microscope a distance corresponding to about 150 lines, set on an arbitrary line, and record the screw reading (number 3). Now translate the microscope a distance of about a thousand lines (or the full width of the grating if it has fewer lines). Again set on a line and record the reading (number 4).

Calculate the accurate line spacing: Use the counted lines and screw readings 1 and 2 to establish the approximate spacing. With the approximate spacing, find the correct number of lines between readings 1 and 3. Divide the distance between these screw readings by the number of lines to obtain a better approximation to the line spacing. Use this second approximation to establish the total number of lines between readings 1 and 4. Finally, calculate the accurate line spacing. Record its value and the probable error.

II. MEASUREMENT OF A STANDARD WAVELENGTH

Put the spectrometer into good adjustment using the procedure of Experiment A5. Adjust the slit width for optimum resolution. Insert the coarse grating in the grating mount and adjust it so that the lines are accurately parallel to the mechanical axis (see Experiment B20). The grating is now set at a definite angle of incidence, normal to the beam unless the grating is very coarse ($\sim$100 lines per inch) in which case the angle of incidence should be as much as fifty degrees. First, set the telescope cross-hairs on the zero order spectrum (easily identified with the unfiltered mercury arc source). Block the collimator light and use the Gauss eyepiece to set the grating normal to the telescope. (For gratings of very high quality and well blazed, the reflection for the zero order may be extremely weak). The grating is now normal to the collimator. If another angle of incidence is desired, the grating is turned to the proper angle with the aid of the spectrometer scales.

The telescope is set on the zero order. Record both scale readings of the prism table. Insert a green filter at the entrance slit of the collimator to eliminate overlapping orders. Measure the angular position of 5 to 7 orders for the green line on either side of the zero order. Choose these orders so that an appreciable angle is covered -- for instance, with a 600 line/in. grating orders -10, -5, 0, +5, and +10. Calculate the wavelength of the mercury green line.

Note and record any unusual changes in the brightness of the green line in different spectral orders.

III. OVERLAPPING ORDERS

Illuminate the grating with unfiltered mercury light. Observe that as one progresses from the first order to higher orders the spectral lines begin to overlap, that is, they begin to get mixed up, not in proper spectral sequence. In orders above five or ten the spectral lines seem to be arranged in almost random order. Set the crosshair on the third order green line, and, with the aid of filters, show that the third order green and a violet line very nearly overlap. Assuming that the two lines were precisely superposed, determine the unknown wavelength from its order and the overlap with the third order green. Look at the same combination in the sixth order green and the ninth. Is the wavelength of the violet line greater or less than that calculated on the assumption of exact overlap? How could the small difference be evaluated?

Note also that the fourth order green nearly overlaps a fifth order blue line. Identify the presence of both lines with the aid of filters if necessary, and compute the approximate wavelength of the blue line.

The third order yellow line(s) nearly overlap a blue line. Identify the separate lines with filters if need be and determine the approximate wavelength of the yellow lines.

Use a purple transmitting filter (e.g. Corning #9863) which transmits the mercury red and blue-violet lines. Are both red and blue lines in sharp focus simultaneously? (Generally they will not be.) The residual chromatism of the optical system reveals a weakness in the method of overlapping orders for many instruments.

High spectral orders of a grating are of considerable use in the calibration of infrared grating spectrometers. If the bright green line (having known wavelength) is used as a reference, one can simply count orders to establish the scale for long wavelength spectrum. Orders as high as forty or fifty may be used in this way.

IV. NORMAL SPECTRUM

Remove the coarse grating from the spectrometer and substitute the fine one. Adjust the grating as before and make the slit width optimum for good resolution. Set the angle of incidence at a known value -- about 30° should be convenient. Record the angular position (note which side of grating normal) of all the visible mercury lines together with the spectral order. For each line estimate the intensity roughly on the basis of 1 to 10 where 1 is just visible and 10 very bright. The red and violet mercury lines can be more easily seen if filters are held between the eye and the eyepiece of the telescope. Use either a dark red or a purple filter to see the red mercury lines and a blue or purple filter for the violet lines. A few ultraviolet lines also may be seen if the spectrometer is equipped with a glass reticle which has a light coating of vaseline or oil to make it fluoresce. (The infrared spectrometer of the next Experiment can be used for the near ultraviolet also, q.v.).

Note carefully any unexpected changes in the brightness of spectral lines observed in different orders.

RESULTS

Record and evaluate the results of the grating space measurements of part I. Describe and if possible explain your observations for the two demonstrations in that part. Clearly if the rulings are not perpendicular to the direction of travel of the microscope, an error in the determination of the grating space will be made. How far from the correct orientation of the grating is tolerable if the error is to be less than 1%? Less than 1/10%?

Calculate the wavelength of the mercury green line; estimate the error.

Discuss the results of the overlapping order observations. (Under what conditions would high orders be useful for the calibration of an infrared spectrometer? How would you make the calibration? What limitations might there be in the method)?

From the measurements of part IV, determine the wavelengths of the mercury lines measured. Include an estimate of the uncertainty. The results should be put in tabular form -- (they may be of some future use).

Plot a curve on a large scale of the wavelength multiplied by the order as a function of angle (for part IV). The curve should be linear near zero angle of diffraction, and it will gradually deviate at large angles from the normal. Discuss the meaning of the term "normal spectrum."

Discuss briefly the relative merits of prisms and gratings.

If any unusual changes in intensities in the various orders of the two gratings were noted, discuss their significance very briefly.

EXPERIMENT D3. INFRARED PRISM SPECTROMETER.

REFERENCES: Strong, pp. 588-595 (R. G. Greenler), and Appendix I (H. W. Yates); Smith, Jones and Chasmar, The Detection and Measurement of Infra-Red Radiation, (1957, Oxford University Press); Ballard, McCarthy, and Wolfe, Optical Materials for Infrared Instrumentation, Report #2389-11-S, (1959, University of Michigan); W. Brugel, Physik und Technik der Ultrarotstrahlung, (1951, Curt. R. Vincentz, Hannover).

This experiment describes in general terms the construction, testing and calibration of a simple infrared prism spectrometer. After some preliminary remarks about infrared spectroscopy, two basic optical systems for such an instrument, Littrow and Ebert, are described. This spectrometer, although severely limited in resolving power, will nevertheless show several of the major atmospheric absorption bands; it should serve as a useful introduction to the study of infrared spectroscopy. Detailed instructions for the construction of a specific instrument are not given since the design will vary with the components that are available.

Before proceding further, it may be well to stop to consider briefly the tremendous importance of infrared spectroscopy. The infrared spectrum provides key information in the study of organic chemical bonds and hence, also in the identification of organic and biological compounds. The infrared absorption of carbon dioxide and water vapor largely determines the thermal structure of the stratosphere and the overall meteorology of the earth. Infrared methods are of the utmost significance in the study of flames, free radicals, rocket exhausts, and inorganic materials such as semiconductors.

DISCUSSION

A. Limitations on resolving power

Probably the most serious present limitation in infrared spectroscopy is the lack of adequate sources and detectors, especially in the far infrared region beyond 10 or 20 microns. (It may be that one day various solid state devices will remedy this difficulty). The only currently available continuous spectrum sources are thermal ones, and a careful study of black body radiation curves shows that even extremely hot sources emit relatively little energy at long wavelengths. In addition, there are no very sensitive detectors for long wavelengths; thermal detectors, thermocouples, bolometers, and pneumatic cells are the best available at present.

This combination of inadequate sources and detectors means that, in general, infrared spectrometers have a resolution far below their theoretical capabilities.

B. Filters

A second difficulty in infrared spectroscopy is the lack of sharp cut-off filters. A variety of optical properties are used in making infrared filters: transmission, absorption, selective reflection, and scattering, but in most cases, the cut-off is not sharp. The lack of sharp filters is particularly serious in grating instruments, especially when used in the second or third order. An important reason for the difficulty is that the undesired short wavelength radiation, which must be eliminated with a grating instrument, is many times stronger than the desired radiation. The proper selection of filters depends on the instrument, the tolerable level of unwanted radiation, the spectral region to be isolated, the spectral sensitivity of the detector, the type of source or its temperature, and the condition of the optical surfaces.

C. Choice of prism-detector combination

The prism material and detector determine the spectral region which may be investigated with the spectrometer. Clearly it is preferable to use whatever detector is most sensitive for the spectral region to be studied. Two combinations are suggested for this experiment. A quartz prism (either fused or crystal) can be used to wavelengths of about 3 microns, and a lead sulfide photoconductive detector is then best. Such detectors are obtainable for $15 to $20 (Eastman Kodak Co., Cetron, etc.). Alternatively, a sodium chloride prism (about the same price as a quartz prism) can be used to wavelengths of about 15 microns, but if used in this experiment, the practical limit will be about 7 to 10 microns. Sodium chloride is, of course, hygroscopic and must be protected from moisture at all times (keep in a dessicator or under a heat lamp except for brief periods). The detector with this prism should be a thermal one. (It will cost perhaps $75 to $100 or more; photoconductive detectors for this region are many times more expensive, and must be operated at liquid nitrogen temperatures).

D. Operation of infrared detectors

The infrared spectral region has a unique feature not shared by either the visible region or the ultraviolet, that is, the necessity to distinguish the radiation to be measured from the often equally strong or stronger unwanted radiation from the surroundings. The spectrometer walls, the slit jaws, imperfectly reflecting mirrors, partly transparent prisms, even the detector itself emit amounts of infrared radiation that are very appreciable when compared to the source radiation (especially at long wavelengths). To solve the difficulty one must periodically interrupt the radiation to be measured and record only the alternating component of the detector response.

E. Spectrometer optics

Two optical systems commonly used for infrared spectrometers are the Littrow system and the Ebert system. Figure D3-1(a) shows the Littrow arrangement. The radiation to be analyzed is converged on the slit at the left. The chopper which periodically interrupts the beam is placed at this point so that only light entering the system is modulated. Since the detector-amplifier unit responds only to changes in radiation, the system discriminates against all the steady radiation arising from within the instrument.

Radiation entering the slit is collimated by spherical mirror M_1 and reflected through the prism to mirror M_2. The latter mirror returns the light through the prism to M_1 which then converges the parallel beam to a focus at the exit slit or detector as the case may be. In the figure, an exit slit has been omitted because it is assumed that the detector has a sensitive area perhaps a millimeter wide and several millimeters long, and so the detector itself acts as the exit slit. If, however, the detector has a large sensitive area, an exit slit is placed at this focus and the light transmitted through the exit slit refocused on the detector. Mirror M_3 obviously serves the purpose of redirecting the beam in a convenient direction for the detector.

The light passes through the prism twice and is accordingly dispersed twice as much as if it went through only once. Thus the resolving power is nearly doubled. It is entirely practical to increase the number of passes through the prism to 4, 6, or 8. If this procedure is used, the resolving power is increased, but the chopper must be relocated to discriminate against the light which has passed only twice (or 4 or 6 times) through the prism. We confine our attention here to the simpler system shown.

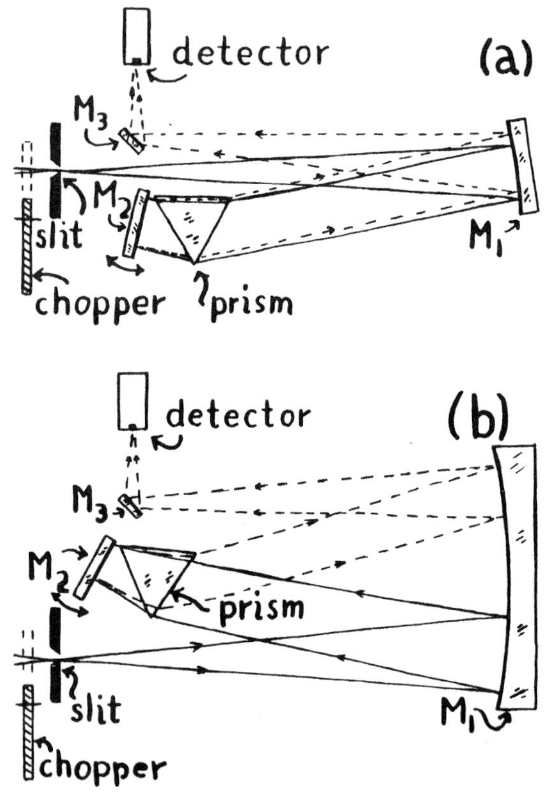

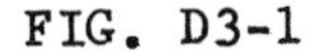
FIG. D3-1

Spectrometer optical systems, (a) Littrow, and (b) Ebert.

If a long entrance slit is used in this instrument, it is found that the images of spectral lines are curved (as in the case of the usual prism spectrometers used earlier in this book). Curved slits are often used in commercial instruments to compensate.

The Ebert optical system, once discarded as almost useless, has been recently rediscovered by Mr. William G. Fastie, now at Johns Hopkins. Far from being inferior, its performance is actually very superior, especially when used with curved slits (Fastie's invention). The system is shown in Fig. D3-1(b) with a prism instead of a grating. As in the Littrow system, light transmitted by the chopper and slit is collimated by the lower part of spherical mirror M_1. The collimated beam passes twice through the prism as indicated and is returned to M_1 which reconverges the light on the detector. The chief difference between the Littrow and the Ebert optical systems is that in the former the aberration introduced by the first off-axis reflection from the spherical mirror is doubled by the second reflection, whereas in the Ebert system, it is canceled. In the Littrow system, the two reflections from M_1 are off-axis the same way; in the Ebert system, the two reflections are off-axis in opposite directions.

Once again, the wavelength is varied by rotation of mirror M_2. The exit slit is also replaced by a detector having a narrow sensitive area as before.

F. Electronics

The alternating current output signal from the detector must be amplified and then measured. The kind of amplifier required depends on the detector used. Here we shall assume that the detector is either a lead sulfide (PbS) cell or a thermocouple or thermopile. In either case, it is desirable to restrict the frequency response of the amplifier to as narrow a frequency as possible, centered around the chopping frequency. The reason for limiting the bandwidth is that the electrical noise produced by the detector increases with the bandwidth. The thermal or Johnson noise produced across the terminals of any resistance, for example, is proportional to the square root of the bandwidth. Specifically, the mean squared Johnson noise voltage is given by $\overline{E^2} = 4KTR\,\Delta\nu$ where K is the Boltzmann gas constant, T the absolute temperature of the resistance and $\Delta\nu$ the bandwidth of the amplifier. Thus for any given detector, it is generally desirable to make $\Delta\nu$ as small as possible. The other factors we need not consider here.

A PbS detector is usually characterized by a high resistance, in the order of 100,000 ohms to a megohm. For this reason, it is convenient to connect its output directly to a vacuum tube amplifier. In fact, if a high gain cathode ray oscilloscope with adjustable bandwidth is available, the detector output (of the order of a millivolt) may be connected directly to the oscilloscope input. Figure D3-2 shows a convenient amplifier circuit for this detector. The amplifier frequency response is chosen to match the chopping frequency -- preferably 100 to 1000 cps. Details of the circuit are adapted to the situation. The circuit diagram has no particular claim to superiority; so-called synchronous amplifiers are much better, but also more complicated.

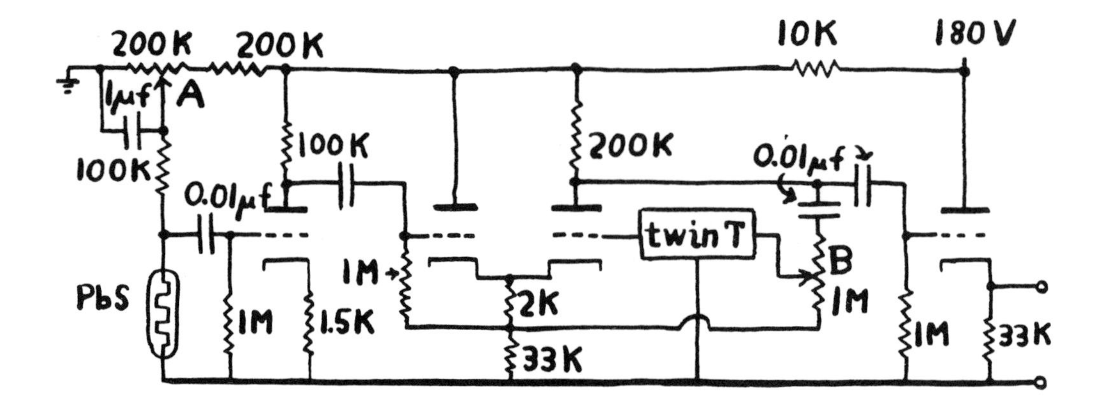

FIG. D3-2

Simple narrow band preamplifier for PbS cell. Tubes are type 12AY7. Rheostat A controls PbS cell bias. Rheostat B controls bandwidth. Overall gain approximately 200. The output cathode follower could, of course, be converted to an ordinary amplifier stage instead.

Thermocouples and thermopiles are characterized both by low resistance (1 to 50 ohms) and by very low frequency response. The proper chopping frequency for most thermal detectors is 5 to 20 cps. Because they have very low resistance, their output signals are in the order of microvolts, and considerable amplification is needed. Also, because vacuum tubes operate best with high input resistances, it is common practice to use a high quality (very well shielded) step-up transformer between the thermocouple and the vacuum tube input grid. The purpose is actually not so much to increase the voltage (though that is desirable too), but to match the input impedance of the amplifier to that of the detector for optimum power transfer.

The output of the amplifer may be rectified, filtered, and connected to a dc meter. Alternatively, if a resonant galvanometer is available, the light entering the spectrometer may be chopped at this frequèncy, the signal amplified, and the ac output connected directly to the galvanometer without being rectified. If such a galvanometer is used as the indicating device, the output impedance of the amplifier should be made low to match that of the galvanometer approximately. Unless the amplifier output stage is a cathode follower, it is desirable to use a step-down transformer to make this impedance match. A partial circuit diagram is shown in Fig. D3-3. As with the PbS detector, a high gain oscilloscope with good low frequency response may be used, but auxiliary amplification will probably be needed to obtain an adequate indication.

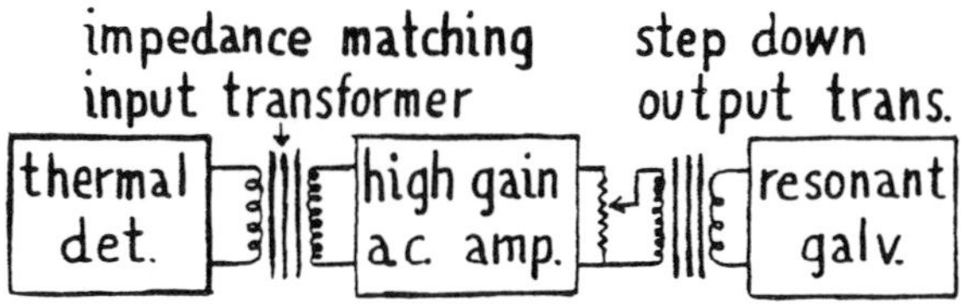

impedance matching input transformer
step down output trans.
thermal det.
high gain a.c. amp.
resonant galv.

FIG. D3-3

Thermocouple amplifier connected to low frequency resonant galvanometer.

APPARATUS

The following list of apparatus is intended to be suggestive rather than definitive.

1) Prism-detector

   A. 60° quartz prism about 3 cm on each edge.

   B. Lead sulfide cell detector. The sensitive area should be 1/2 x 2 mm preferably. If the width is greater it may be masked down to avoid the necessity of an exit slit.

   alternatively

   A. 60° NaCl prism about 3 cm on each edge.

   B. Dessicator for the prism when not in use.

   C. Infrared heat lamp.

   D. Thermocouple, thermopile, or bolometer detector (1 x 3 mm sensitive area).

   E. Input transformer to match detector to amplifier.

2) Other optical components. All mirrors must be first surface; glass becomes rapidly opaque beyond two microns wavelength in any appreciable thickness.

   A. Spherical mirror. For the Littrow system, the mirror may be about 2 inches in diameter with a focal length of 6 to 10 inches -- preferably the latter. For the Ebert system, the mirror should be about 4 inches diameter with a focal length of 10 to 16 inches.

   B. Plane mirror for M_2 in either optical system about 1 to 1 1/2 inch square.

   C. Auxiliary spherical and plane mirrors to illuminate the instrument and for M_3.

3) Mechanical parts

A. Plywood or metal baseplate for the instrument.

B. Mounting blocks of wood or metal for the various optical components.

C. Special mounting for M_2. This mounting may be a vertical post with some sort of sleeve bearings and a driving arm 6 or 8 inches long. The end of the driving arm may be driven with a 0 to 1 in. micrometer caliper or a depth micrometer. (A movement of about 1/2 to 3/4 in. is sufficient for a 6 inch driving arm).

D. Entrance slit, 5-10 mm long, 1/2 to 1 mm wide. (The optimum width is about equal to the detector width if the signal level is sufficient). A fixed slit may readily be made with a razor blade.

E. Chopper unit. The chopper is a motor driven sector disk. For a PbS cell detector, an 1800 rpm synchronous motor is convenient. The sector disk may be a metal plate about 5 in. dia. with about 30 holes uniformly spaced around the circumference. The hole size should be roughly the same as the separation between the holes so that the transmitting and blocking times are about equal. A toothed wheel may also be used, but it has the disadvantage of bearing an all too close resemblance to a buzz saw. For a thermal detector, use a slow speed motor or belt drive arrangement to obtain the desired chopping frequency.

F. Pot of beeswax-rosin mixture to mount parts.

4) Coelostat

An extremely useful auxiliary component is some sort of coelostat which is equipped with first surface mirrors. Additional mirrors may be used to reflect a beam of sunlight to the instrument. If sunlight is available, the problems of calibration are greatly reduced, for atmospheric transmission bands provide excellent landmarks as well as being of considerable interest in their own right.

5) Mercury arc

The arc is a useful source of visible and near infrared lines both for adjustment of the instrument and for its calibration.

6) Carbon arc or globar source

For the near infrared, an auto lamp source is satisfactory.

7) Useful infrared filters

A. Plexiglass sheet.

B. Corning 2540 filter.

C. Liquid water cell (glass) 1 cm thick.
Liquid water cell for 0.01 in. of water. A pair of glass plates with a drop of water between may be used.
1/4 in. thick plate glass
If the rocksalt prism is used, a sheet of polyethylene 3 mils. thick provides one or two useful known wavelengths for calibration.

PROCEDURE

A. Optical system

A base plate of duraluminum 1/4 to 1/2 inch thick or of plywood 1 inch thick is suitable. The dimensions should be chosen to fit the optical parts available. The length should be 3 or 4 inches longer than the focal length of mirror M_1 to allow for shielding and for mounting the chopper motor on rubber blocks of some sort. The width should be sufficient to leave room for both the detector and the wavelength drive. It is desirable to support the base plate itself on three small blocks. The optical components and mechanical parts can be mounted on vertical rods or blocks and secured with hot beeswax and rosin (see Experiment A20). Lay out the parts keeping in mind the mounting of M_2 which is most critical. This mirror can be waxed to a vertical shaft arranged with some sort of bearings at both ends; any kind of bearing is suitable provided only that there is little play possible. The shaft is rotated by a lever arm 6 to 10 inches long and driven by a micrometer as shown in Fig. D3-4. To allow greater wavelength range, slit the end of the drive lever and insert a thin metal or glass tongue to be driven by the micrometer spindle. Support the end of the lever on any smooth surface, and keep the tongue against the micrometer drive with a rubber band or weak spring.

Choose a convenient height (say 2 inches) above the base plate for the center of the optical beam. Fix the slit and adjust M_1 so that the slit is at its focus to within a millimeter or so. The slit itself may be cut from a thin piece of brass shim stock, or heavy aluminum foil, or it may be readily made from an old razor blade. Use a mercury arc to set up the optical system. Locate the prism (if NaCl is to be used, preliminary adjustments may be made with a glass prism) on a block at the right height with its base toward the entering beam. (Try the reverse orientation to see the increased curvature of the spectral lines). Adjust M_2 so that it reflects the visible spectrum back in the direction of the entrance slit. The focus can be checked easily at this point. Locate M_3 and put the detector sensitive area at the focal point by adjusting M_2 to place the

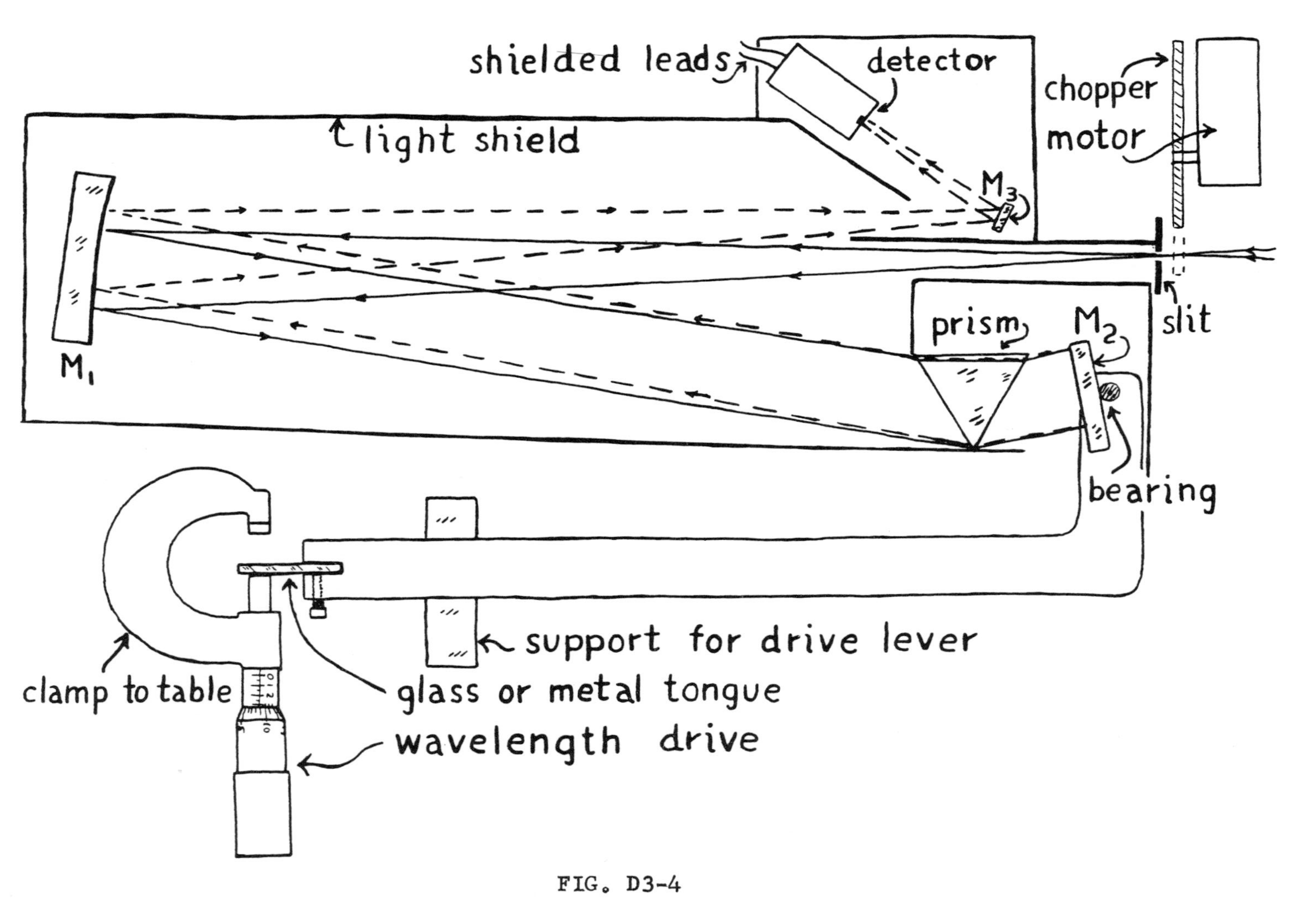

FIG. D3-4

Suggested parts layout for the Littrow optical system.

mercury green or yellow lines on the detector. The prism is set approximately at minimum deviation for the red end of the spectrum. (If a NaCl prism is used, use an infrared heat lamp about two feet away to keep it warm. For brief periods, the lamp may be turned off). Adjust M_2 so that with the micrometer near one end of its travel, the green line falls on the detector and the infrared will be scanned in the range of the micrometer. A travel of somewhat less than an inch should be sufficient to cover the whole usable spectrum.

Mount the chopper unit near the entrance slit and observe that it does indeed interrupt the light falling on the detector when slowly rotated. Finally mount cardboard shields (preferably blackened with paint) to keep out as much stray light as possible. The heavy line in the figure shows a possible placement of the light shields. (It is better to overdo the shielding than to have too little).

It should now be possible to scan at least part of the visible spectrum (as well as the infrared) and check the operation.

B. Electronics

Set the wavelength drive for the green or yellow mercury line (unless the detector happens to be insensitive to visible light as is the case with some specially prepared thermopiles) and connect the electronics. An oscilloscope of some sort is very useful in checking for signals. If the electronics are properly working and warmed up and the indicating unit is connected, a large signal should be noted when the chopper motor is turned on. If the detector is a lead sulfide cell and no signal is evident, check the bias on the cell. Check the amplifier by noting whether or not the meter gives a large irregular output with the input open circuited. Check for ground loops and improper connections.

If a signal is indicated, change the wavelength micrometer slightly in either direction to maximize the signal; reduce the amplifier gain if necessary. Momentarily, block the beam entering the slit to make sure that only the desired signal is seen; the signal should vanish and only random noise should be left. If the gain is low, the noise will not be evident.

Scan the wavelength slowly toward the red and infrared and note the presence of the several infrared mercury lines. These lines will assist in calibration.

If the spectrometer appears to be functioning properly, the instrument should be calibrated.

C. Calibration

The calibration of an infrared prism spectrometer depends upon the identification of at least a few known wavelengths, and suitable ones for the present instrument are scarce.

TABLE I: Principal emission lines of the mercury arc

Wavelength (microns)	Intensity
0.313	10
0.334	6
0.365	10
0.405	10
0.435	10
0.546	10
0.578	10
0.623	8
0.691	10
1.014	10
1.129	10
1.36	10
2.05 (Helium)	10

TABLE II: Useful Transmission Characteristics

Wavelength, μ	Material	Transmission vs λ
0.8 - 1.0	Corning type 2540 filter	T rises rapidly toward longer λ.
1.65	5 mm plexiglass	T decreases from 70% to 20%.
2.4	" "	Becomes opaque.
1.13	1 cm liquid water	T $\approx$ 20%, changing rapidly.
1.32	" " "	Becomes opaque.
1.75	1 mm germanium	T rises rapidly.
2.6	5 mm plate glass	Becomes opaque.
5 1/2 - 9	1 mm calcite	Opaque.
3 - 5	" "	O ray transmission shows interesting structure.
3.0	5.4 mm spinel	T drops from 80% on either side to 30%.
3.5 - 5	2 mm fused quartz	T drops from 90% at 3.5μ to 0% at 5μ.
3.5, 6.9, 13.8	0.1 mm polyethylene	T drops rather sharply to 0%.
3 1/4 - 3 1/2	0.02 mm plexiglass	T drops from 95% to 50%.
5.7	" "	T drops to zero.

Begin with the emission lines of the mercury arc. The green and yellow lines accurately establish one end of the calibration curve. If a purple filter or dark red filter is available, the λ6234 and 6907 lines may be set on the detector for calibration purposes. Some of the prominent mercury arc lines are listed in Table I. A good helium discharge tube has a strong emission line at 2.05 microns. The positions and rough intensities of all the infrared mercury lines should be recorded.

Further calibration may be made in terms of known transmission characteristics of various materials. (A list of some convenient wavelengths is given in Table II). A one cm. thick water cell, for example, shows a rather sharp decrease in transmission at about 1.13 microns, and it becomes opaque at about 1.3 microns. If the signal from the carbon arc, globar, or incandescent lamp is studied with and without the water cell, these two wavelengths should be identifiable. Other transmission characteristics are used in the same way. Most of the points thus established will not be very definite, however, for the exact point where the transmission either ceases or changes most rapidly is not very well marked. If an NaCl prism is used, a thin sheet of polyethylene should be used to establish two calibration points at 3.5 and 6.9 microns.

D. Atmospheric transmission

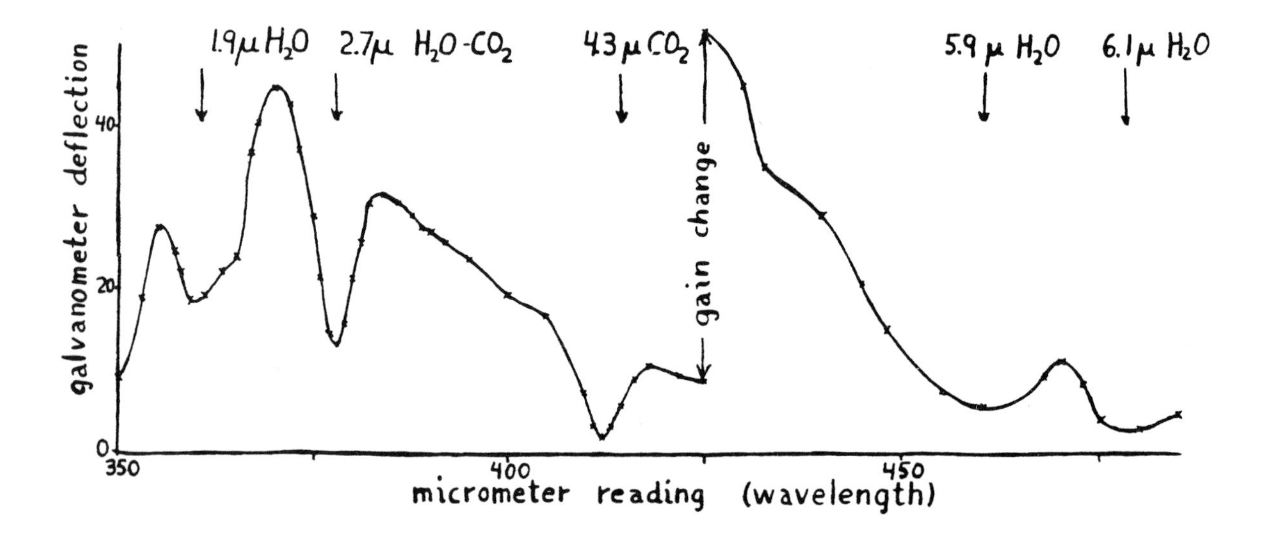

FIG. D3-5

Atmospheric absorption bands, 20 meter path, NaCl prism with thermopile detector.

If possible, a coelostat should be used to direct a beam of sunlight into the spectrometer. A quartz prism and lead sulfide cell detector will show a minimum of four definite absorption bands caused by atmospheric water vapor and carbon dioxide. More bands will be seen with a salt prism. If the sun cannot be used as a source, arrange the carbon arc so that a parallel beam from it traverses a path of 20 meters or more before entering the spectrometer. Figure D3-5 shows the atmospheric bands seen with a NaCl prism and a carbon arc source 20 meters away.

E. Residual rays

If the spectrometer is equipped with a salt prism, the transmission and reflection of both crystal quartz and calcite plates should be studied. Most crystals have their residual ray reflection farther out in the infrared -- calcite and quartz are exceptions.

APPENDIX

It is of interest to note that the infrared spectrometer of this experiment can, with slight modification, be used to observe ultraviolet lines in the mercury arc spectrum. For this purpose, the detector is removed and mirror M_2 readjusted so that the red end of the spectrum falls at a micrometer reading corresponding to the infrared limit. The wavelength range covered will now include some of the ultraviolet. A zinc sulfide fluorescent screen may be used to make visible the ultraviolet lines, and the spectrometer calibrated in terms of known ultraviolet mercury lines. Wavelengths longer than about 3000 Å may be observed.

APPENDIX A. THE EVALUATION OF EXPERIMENTAL ERRORS.

(In Collaboration with Trevor Williams*)

REFERENCES: H. Margenau and G. M. Murphy, The Mathematics of Physics and Chemistry, (1943, Van Nostrand), p. 487 ff; R. T. Birge, The Calculation of Errors by the Method of Least Squares, Physical Review, vol. 40, pp. 207-227 (1932); E. B. Wilson, Jr., An Introduction to Scientific Research, (1952, McGraw-Hill), Ch. 8.9; M. B. Stout, Basic Electrical Measurements, 2nd ed. (1960, Prentice-Hall), Ch. 2.

I. INTRODUCTION

In all quantitative experimental work, error is involved, and it is clearly an important matter that we be able to determine how much reliance can be placed on our measurements and calculated results. In this appendix, we discuss very briefly a few of the more salient points in this regard. The reader should consult the references for further details.

The proper treatment of error is exceedingly difficult, and it cannot be made on a purely mathematical basis. The trouble is that errors in experimental work are of several types. Suppose that we make a measurement of length by using a traveling microscope. One kind of error we make is a purely statistical one, and that one is amendable to mathematical description. A second kind of error is a "systematic error," an error resulting from the fact that the micrometer screw of the instrument is imperfect -- the threads are closer together in some places than others. This kind of error can only be estimated -- perhaps only "guesstimated." A third type of error is personal error in which the observer demonstrates a prejudice for certain readings or perhaps consistently overshoots the "true" mark in setting the traveling microscope. Finally, there is another kind of error which might be called a gross error or mistake -- the incorrect reading of a scale, etc. If the scale is calibrated in unusual units such an error is easy to make. With all these possible errors which are combined in various ways, it becomes a difficult matter to evaluate our results, though we must try.

II. STATISTICAL ERRORS

Let us first consider some of the basic principles of statistical errors, for these can be treated mathmatically in a rather satisfactory way. We may illustrate some of the basic principles in the analysis of error with

*School of Hygiene, Johns Hopkins University.

a very simple experiment in which the statistical errors are (one hopes) dominant. A ruler about a foot long graduated in inches or centimeters and a piece of heavy cardboard having a straight edge are needed. On the cardboard make two marks, short lines perpendicular to the edge, arbitrarily placed a few inches apart. The separation of these two marks, A and B, is to be determined in a special way. Place the ruler against the edge of the cardboard without regard to the scale markings on it. To improve randomness one may slide the ruler along the edge while looking away. The position of A on the scale should be random, and it is read to the nearest inch or centimeter -- no fractional part is to be noted. Read also the position of B to the nearest inch only and record the difference of the readings (obviously a whole number of inches or centimeters). A set of one hundred readings of this kind is needed. (The process becomes a bit tedious, but not unendurable). The differences of the readings are recorded on a tally sheet arranged in ten columns of ten numbers.

Take the grand average of all the hundred readings by adding each group of ten, adding the ten sub answers, and dividing the grand total by 100. Next read the distance AB with the ruler in the ordinary way as accurately as possible. You will probably find that the grand average agrees with the accurate measurement to within a tenth of an inch or centimeter. This, notwithstanding the fact that the individual measurements were made only to the nearest inch or centimeter. To understand this increase in accuracy, we must look more closely at the mechanism involved.

Let the true distance AB be $n + \theta$ where n is an integer or whole number and θ is a fraction between zero and one. The experiment can only lead to the values n or $n + 1$. In a few cases there may be some uncertainty as to the proper reading when either point A or B lies almost halfway between two centimeter or inch marks; in that case one may also get $n-1$ or $n+2$, but these readings should be few and they do not affect the argument.

From the way the experiment was set up, the "probability" that the reading obtained is n can be shown to be simply $1-\theta$, and the probability that it is $n+1$ is θ. We use the word "probability" since we have an intuitive feeling for it; but we need to make the term quantitative. It would seem reasonable that if we make a large number of measurements, say N, we would expect to obtain n about $N(1-\theta)$ times and $n+1$ about $N\theta$ times. Thus if we toss a coin 1000 times (fairly!) we expect about half heads and half tails, but surely not exactly 500 of each. In our experiment with A and B, we calculated the grand average by adding the N measurements and dividing by N. We ought to get about the same answer in terms of probability. We should have about $N(1-\theta)$ readings equal to n and about $N\theta$ readings equal to $n+1$. Thus we should have for the grand average

$$[N(1-\theta)n + N\theta(n+1)] \div N \qquad \text{or}$$

$$n(1-\theta) + (n+1)\theta = n + \theta \qquad (1)$$

That is, the expected value (also called the "mean value" or "expectation value") of the mean of N measurements is just equal to the true value. This jibes with the fact that the average of our hundred readings is very close to the true length AB -- closer than any of the individual readings.

But, although the expected value of the mean of N observations is the true value, we know that when we make N measurements we get various answers which cluster more or less about the expected value. To illustrate, take the original hundred readings in the experiment and average them four at a time successively, obtaining a total of 25 values. The only possible values (assuming $n-1$ or $n+2$ did not occur) are: n, $n+1/4$, $n+1/2$, $n+3/4$, and $n+1$. Thus the 25 values will be distributed among these five possible values; count how many of each there are. The number of each is called the observed "frequency." Plot the results by drawing a series of rectangles whose heights are equal to the different frequencies and whose bases are 1/4 cm (or inch) wide and centered at the values n, $n+1/4$, etc. Such a series of frequency rectangles is called a "histogram"; it shows at a glance how the various readings are distributed among the results. For example, if the true length AB was 7 1/3 cm, the histogram might look like Fig. AA-1. (Usually the frequencies are not labeled in the rectangles of the histogram; they are read from the scale). Alternatively, the frequencies may also be plotted as percentages of the total count as in Fig. AA-2.

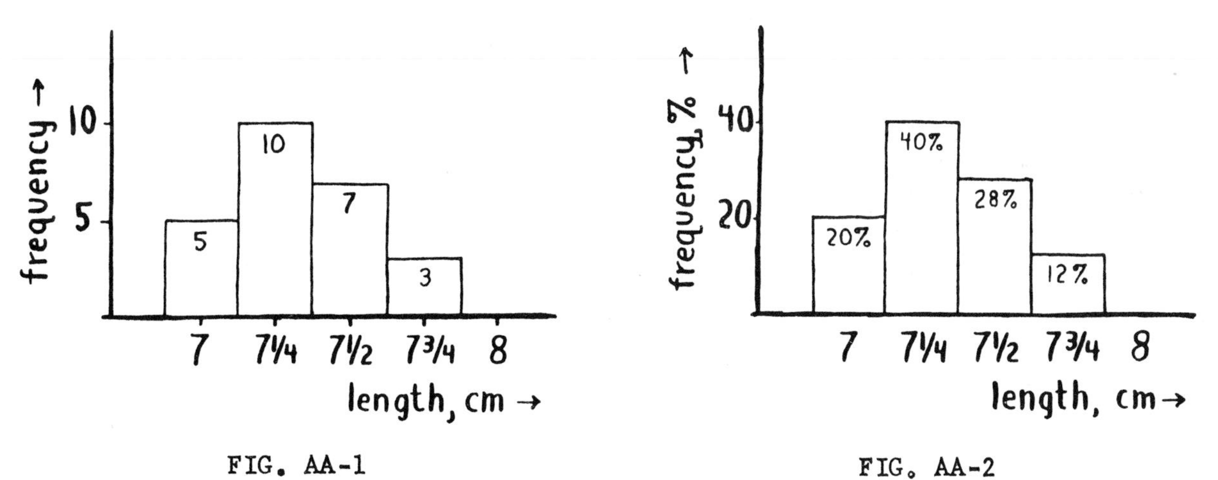

FIG. AA-1

Histogram, sample size 4.

FIG. AA-2

The same histogram expressed in percent.

In the two histograms of Figures AA-1 and AA-2, we are dealing with the mean of a sample of size 4 since each value was the average of four readings. We have 25 such values and we say we have a sample of the averages of size 25. The term sample has been used twice, but it has the same basic meaning in both cases -- a sample of size 25 of a variable defined as the mean of a sample of size 4. If we increased the number of observations indefinitely and plotted the histogram for the averages of sample size 4 -- the histogram would settle down to the limiting asymptotic values of Fig. AA-3. (This is called a binomial distribution of order 4). The reason

Fig. AA-2 differs from Fig. AA-3 is because the former has "sampling fluctuations." Fig. AA-2 is based on a sample of size 25, Fig. AA-3 presupposes an infinite sample size. We say Fig. AA-2 is a "statistical image" of Fig. AA-3.

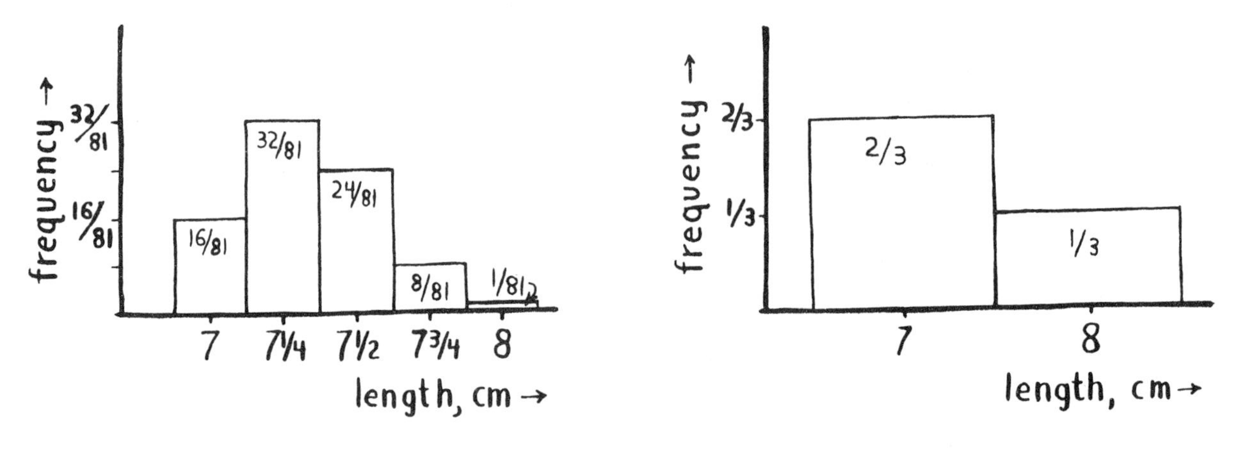

FIG. AA-3

Limiting value of histogram, sample size 4.

FIG. AA-4

Histogram, sample size 1.

Consider again the histogram or "frequency distribution" with which we began. We said there was a probability $1-\theta$ of reading the value n and a probability θ of reading $n+1$. In the example given, we have $n=7$ and $\theta = 1/3$. Hence we can draw the histogram of Fig. AA-4 to describe the outcome of a single reading. Compare Figs. AA-3 and AA-4. From Fig. AA-4 the strongest statement we can make is that there is a probability of 66 2/3% that a single reading will be 7. But from Fig. AA-3 we can make the much stronger statement that the probability is little better than 69% (32/81 + 24/81 in %) that the mean of four readings will be equal to either 7 1/4 or 7 1/2. Clearly we are much closer to the true value of 7 1/3 in the latter case than in the former.

It is evident that this behavior is going to become more and more pronounced as we consider the distribution of larger and larger sample sizes. To see this effect, go back to your original 100 readings and average them 10 at a time successively, obtaining ten values. The only values you can obtain are now: n, $n+0.1$, $n+0.2$, . . . , $n+0.9$, and $n+1$. Make a histogram of your results. In our example where AB = 7 1/3 cm, the histogram might be that of Fig. AA-5. Again, this histogram is a statistical image of the "parent distribution" in Fig. AA-6, (a binomial distribution of order 10). Fig. AA-6 shows that there is better than a 68% probability that the mean of a sample size of 10 will be 7.2, 7.3, or 7.4 (68% being the sum of the probabilities of the three readings). We are "converging in probability" on the true value 7 1/3 cm since it is becoming more and more unlikely, or improbable, of getting a value substantially different from the true value.

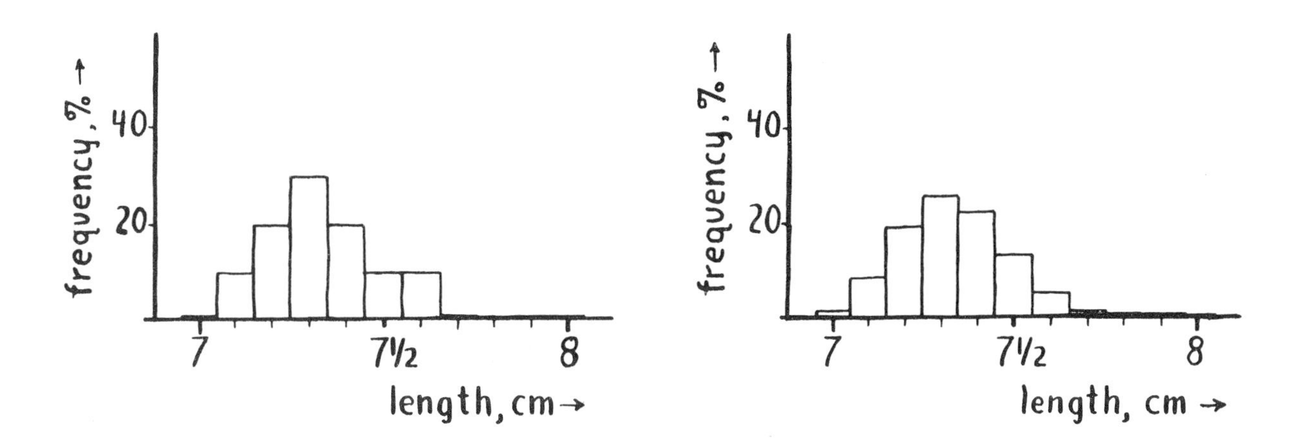

FIG. AA-5

Histogram, sample size 10.

FIG. AA-6

Parent distribution, sample size 10.

Another way to put this is that progressing from Fig. AA-4 to Fig. AA-3 to Fig. AA-6, that is as we progress from sample size 1 to 4 to 10, the distributions become successively narrower. A quantitative measure of the width of a distribution is given in terms of "variance." Suppose we have N numbers, x_1, x_2, ..., x_N. Their mean, $\overline{X}$, is given by

$$\overline{X} = \frac{1}{N}(X_1 + X_2 + \cdots + X_N) . \quad (2)$$

Consider, now, the N new numbers we get by subtracting the mean from each of the original numbers, called "deviations from the mean"

$$X_1 - \overline{X},\ X_2 - \overline{X},\ \ldots,\ X_N - \overline{X} .$$

Some of these differences are positive and some negative; they add up to zero (prove this). If the width of the distribution is small, then the deviations from the mean will be small; if the width is large, the deviations will be large. Since the sum of the deviations is zero, we square the deviations so that the sum will be positive and not zero. The variance, written as S_X^2, or if there is no ambiguity, simply S^2 is defined by:

$$S_X^2 = \frac{1}{N-1}\left[(X_1-\overline{X})^2 + (X_2-\overline{X})^2 + \cdots + (X_N-\overline{X})^2\right]. \quad (3)$$

Note that the mean in this case is found by dividing by N - 1 rather than N. The reason we do not consider there, but the reasonableness is seen by considering the case of N = 1. If the sample is of size 1, we have no information at all about the width of the distribution, and equation (3) warns us away from trying by giving the answer 0/0. (The actual reason for dividing by N - 1 involves what is called an "unbiased estimate" -- see a standard book on statistics).

From equations (2) and (3) we can write the variance as

$$S^2 = \frac{1}{N-1}\left[X_1^2 + X_2^2 + \cdots + X_N^2 - N\bar{X}^2\right] \qquad (4)$$

or, roughly, in words, the mean of the squares less the square of the mean. A more general result, which includes both (3) and (4) is

$$S^2 = \frac{1}{N-1}\left[(X_1-\mathbf{X})^2 + \cdots + (X_N-\mathbf{X})^2 - N(\bar{X}-\mathbf{X})^2\right] \qquad (5)$$

where X is any number at all. Taking $\mathbf{X} = \bar{x}$, equation (5) becomes identical with (3), while taking $X = 0$ gives (4). More important, (5) shows that the variance is independent of what point we choose as the zero-point of our scale, and is indeed, therefore, a measure of the spread of the distribution, quite independent of its position. If we slide the distribution back or forth along the line without changing its shape, we do not alter the variance. (Of course, if we shift the origin by X, we shift the mean by the same amount -- prove this). If the distribution is all concentrated at a point, then, and only then, the variance is zero. A small variance is thus indicative of a narrow distribution and a large variance of a broad one.

The equations for variance show that is is a homogeneous quadratic quantity, that is, if the x's are measured in cm, the variance is in cm^2. We want the measure of spread to be linear, so we take the positive square root of the variance, and the quantity S ($=\sqrt{S^2}$) is called the "standard deviation." To illustrate, consider the data of Fig. AA-5. For convenience let us choose the value 7 as the origin. First, the mean is

$$\bar{x} = 7 + \frac{1}{10}(0.1 + 2\times0.2 + 3\times0.3 + 2\times0.4 + 0.5 + 0.6)$$

$$= 7 + \frac{1}{10}(0.1 + 0.4 + 0.9 + 0.8 + 0.5 + 0.6) = 7.33$$

This is the same answer we would have got in the first place by adding up all the readings and dividing by 100. (Notice that we have shortened the averaging by multiplying the values by their frequencies). Now use equation (5) to compute the variance with $\bar{x} = 7.33$. Again choose $\mathbf{X} = 7$ for convenience.

$$S^2 = \frac{1}{9}(0.01 + 2\times0.04 + 3\times0.09 + 2\times0.16 + 0.25 + 0.36 - 10\times0.33^2)$$

$$= 0.022333$$

giving $\quad S = \sqrt{0.022333} = 0.14944$

(A somewhat better choice of origin in these two calculations would have been 7.3 since the numbers in the arithmetic would have been smaller).

Now return to your original data and compute the mean and standard deviation for the distribution of the mean of samples of size 4 and for the distribution of the mean of samples of size 10. The two mean values should equal the grand average initially computed. The two standard deviations, however, should not be the same. Which is smaller and why?

You may also, of course, consider your data simply as a sample of size 100 from the parent population and so estimate the mean and standard deviation of the parent population. The mean is, of course, the grand average. In our illustration suppose that 67 readings came out 7 cm and 33 came out 8 cm. Then the calculations would be (with $X = 7$)

$$\bar{x} = 7 + \frac{1}{100}(67 \times 0 + 33 \times 1) = 7.33$$

$$S^2 = \frac{1}{99}(67 \times 0^2 + 33 \times 1^2 - 100 \times 0.33^2) = 0.22333$$

$$S = 0.4725$$

The computation is like the previous ones but easier. How does this standard deviation compare with the other two? By a calculation similar to equation (1) we show that the true variance is actually $\theta(1-\theta)$. How does this compare with your value estimated from a sample size of 100? To what may the slight difference be ascribed?

We may form a general picture of the situation. In equation (2) the numbers $x_1, x_2, \ldots, x_N$ are called "random variables" because they will not be the same from one time to the next when the experiment is done. They follow a certain distribution which, in this case, happens to be the same for all of them. That is, x_1 is chosen at random from the distribution shown in Fig. AA-4, then x_2 is chosen from the same distribution, and so on. Because the x's are random, $\bar{x}$ is also a random variable since it varies from trial to trial. Thus $\bar{x}$ also follows a certain distribution, but not the same one. Its distribution for $N = 4$ and $N = 10$ was given. We can make some very general statements about the distribution of $\bar{x}$ for arbitrary N from equation (2). First of all, it is easily proved that the mean of a sum of random variables is equal to the sum of the individual means,

$$\overline{x_1 + x_2 + \cdots + x_N} = \bar{x}_1 + \bar{x}_2 + \cdots + \bar{x}_N \approx \mu + \mu + \cdots + \mu = N\mu \quad (6)$$

where each of the individual $\bar{x}$'s approach the same mean, called μ, equal to the true value $n + \theta$ in our case. By equation (2) the sample mean $\bar{x}$ is the sum of the x's divided by N so we have

$$\bar{\bar{x}} = \overline{\frac{1}{N}(x_1 + x_2 + \cdots + x_N)} = \mu, \text{ the same as any one of the x's.} \quad (7)$$

We are thus led to the apparently redundant theorem that the mean of the mean is equal to the mean. In other words, when we compute the mean of a sample, the answer will, on the average, be equal to the mean of the parent population.

It is a remarkable fact that the variance has a similar property: the variance of a sum of "independent" random variables is equal to the sum of the separate variances. The term independent means, as we would suppose, that one event does not affect the other. The measurements in your experiment are independent since the answer for one trial does not affect the answer for any other trial. Thus we can write

$$\text{var}(X_1 + X_2 + \cdots + X_N) = \text{var}\, X_1 + \text{var}\, X_2 + \cdots + \text{var}\, X_N$$
$$\approx \sigma^2 + \sigma^2 + \cdots + \sigma^2 = N\sigma^2 \tag{8}$$

where σ^2 is the variance of the parent population. But variance is a quadratic quantity as we saw and thus when you multiply (or divide) a random variable by a constant, you multiply (or divide) its variance by the square of that quantity. Thus

$$\text{var}\, \bar{X} = \text{var}\, \frac{1}{N}(X_1 + X_2 + \cdots + X_N) = \frac{1}{N^2}\text{var}(X_1 + X_2 + \cdots + X_N)$$
$$= \frac{1}{N^2} N\sigma^2 = \frac{\sigma^2}{N} \tag{9}$$

from which it follows that the standard deviation of $\bar{x}$ is

$$S_{\bar{X}} = \frac{\sigma}{\sqrt{N}} \tag{10}$$

For the sample size of 1 from an infinite sample we saw the true variance is $\theta(1-\theta)$ and the standard deviation is the square root of this quantity. By equation (10), the standard deviation of samples of size 4 should be $1/\sqrt{4}$, or half as big as the samples of size 1 for the deviation of the parent population. The standard deviation for samples of size 10 should be $1/\sqrt{10}$ as big, and for a sample of size 100, 1/10th as big. The histogram for samples of size 100 is shown in Fig. AA-7. Thus the single grand average computed at the beginning of your experiment was a sample of size 1 from this distribution. Because it is so narrow, it is rather difficult for the mean of a sample of size 100 to differ greatly from the true mean, and through Fig. AA-7, we can make probability statements about how likely we are to obtain answers within a given range. For example, the probability is about 66% that the mean of a sample of size 100 lies in the range 7.29 and 7.37.

Figure AA-7 illustrates another fact; this histogram is approaching a limiting form known as the "normal distribution." This is a curve shown in Fig. AA-8 which has the algebraic form

$$\text{frequency} = \frac{1}{\sqrt{2\pi}\,\sigma} e^{-(x-\mu)^2/2\sigma^2}$$

Clearly Fig. AA-8 bears a close resemblance to Fig. AA-7; the mean of a sample of size 100 is approximately normally distributed, and the mean and standard deviation are given by equations (7) and (10), with N 100. The mean of a sample of large size of any random variable (whatever the distribution of that variable may be), will be approximately normally distributed with mean and standard deviation given by equations (7) and (10); furthermore, the approximation will become better and better the larger N becomes. This is a special case of the Central Limit Theorem.

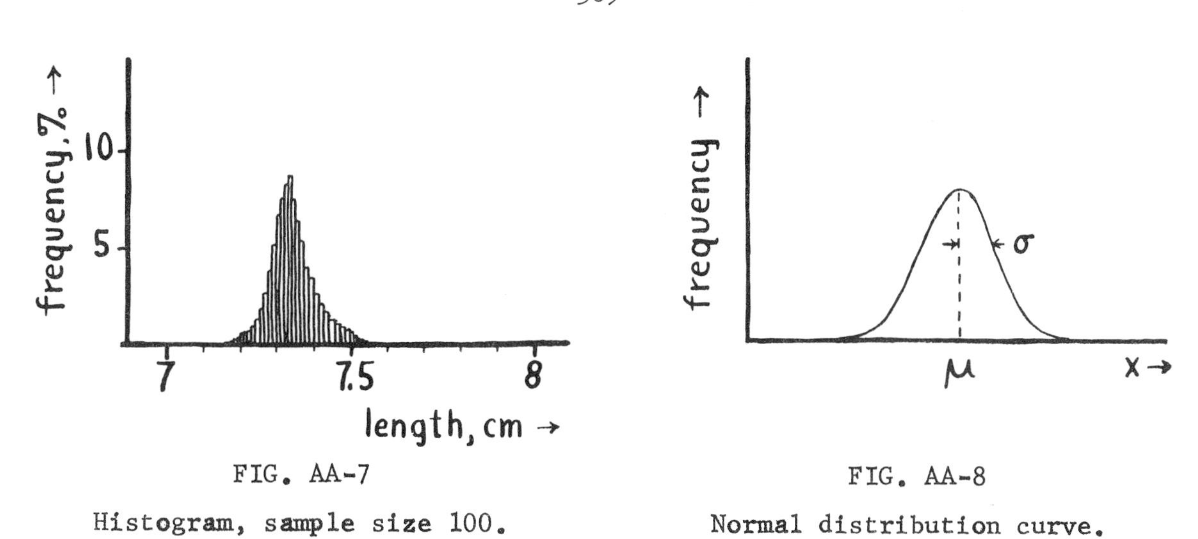

FIG. AA-7

Histogram, sample size 100.

FIG. AA-8

Normal distribution curve.

For the Central Limit Theorem to hold, all the x's need not even follow the same distribution, provided the various distributions are not so disparate that one of them completely masks the contributions of the others. The errors generally encountered in the use of experimental apparatus are often approximately normally distributed since they represent the overall effect of the superposition of a large number of random errors. Although the individual components of the error need not be distributed like a normal distribution, their sum will tend to be, by the Central Limit Theorem.

The normal distribution is tabulated in statistics texts and in mathematical tables of the Handbook of Physics and Chemistry (Chemical Rubber Publishing Co.). We mention a few properties easily read from such a table. For example, the probability that any given reading of a normally distributed random variable will lie within one standard deviation to the left or to the right of the mean is just over 68%, the probability that it will lie within two standard deviations of the mean is about 95 1/2%. Thus it is rare for a normally distributed random variable to fall more than two standard deviations away from the mean in either direction. The so-called probable error is defined as 0.6745 times the standard deviation because the probability that the reading falls within that distance of the mean is 1/2, the chances are even that it will fall within or outside this probable error.

To summarize, the basic definitions are:

(1) Mean value = sum of all readings in sample number of readings.

(2) Deviation from mean = reading minus mean value.

(3) Variance = sum of squares of deviations from sample mean ÷ number of readings less one.

(4) Standard deviation = square root of variance.

(5) Probable error = 0.6745 times standard deviation.

III. COMBINATIONS OF ERRORS

If two measured quantities are to be added to determine a third quantity, the question arises as to how to combine the separate probable errors. Clearly it would be unwise either to subtract the probable errors or to add them. The proper treatment on the basis of statistical theory is to take the square root of the sum of the squares of the probable errors if they are independent. Or, what is equivalent, the variance of the sum is the sum of the variances as we said before. If several quantities are to be added, the overall variance is still the sum of all the separate variances.

$$S^2 = S_1^2 + S_2^2$$

If we multiply or divide two measurements together to obtain the third quantity, we must combine the percentage errors rather than the absolute errors. The overall variance is still the sum of the separate variances expressed as percentages (provided these percentages are small, say less than 20%). In other words, the standard deviation is the square root of the sum of the squares of the individual probable percent errors. The errors, of course, can equally well be expressed in parts per thousand or parts per million, provided they are all expressed on a common basis.

Suppose we are required to take the square or square root of a quantity. Clearly, in forming the product x'x the errors are not independent if the reading is too large, the square must also be too large. The relative error in squaring is twice the error in the original quantity, and in taking the square root, one half. In terms of variance, the variance of the square is 4 times the variance of the measurement; the variance of the square root is 1/4 times the variance of the measurement.

IV. SYSTEMATIC AND OVERALL ERROR

In most experiments, there exists systematic error as well as statistical error. Systematic errors are not eliminated by taking many readings -- if the micrometer screw of the traveling microscope has an error in its thread, this error will affect each and every reading taken with that portion of the screw, and in just the same amount. With an adequate knowledge of the construction of the instrument and the method of use, it is possible to estimate the systematic errors. The problem is how to combine these estimated errors with the statistical errors. As far as possible, the estimated systematic errors should be put on the same basis as the statistical errors, and the estimated variance for systematic error added to the variance for statistical error by the rules described above.

In stating the overall error, it is important to indicate what the error means, i.e., on what basis it has been estimated. In conclusion, it should be pointed out that a quantitative experimental result is virtually worthless unless some attempt has been made to evaluate the errors.

APPENDIX B. SOME BASIC COMPONENTS FOR THE OPTICS LABORATORY.

One meter optical bench with several sliders and lens supports.
Adjustable slit to mount on the optical bench.
Ground glass observing screen for bench.
Student spectrometer.
Light sources
- 6-8 volt 21 cp. auto lamp and transformer.
- concentrated arc source.
- carbon arc source (ac or dc, 5 amp.).
- small flashlight.
- spectral lamps.
  - sodium vapor
  - mercury vapor (high pressure)
  - mercury vapor (low pressure)
  - cadmium vapor

  (George W. Gates and Co. supplies a convenient set of lamps).

Assorted lenses having various diameters, focal lengths, and including some achromatic doublets.
High power eyepiece for the spectrometer.
Traveling microscope (see Gaertner).

APPENDIX C. USEFUL LABORATORY MATERIALS.

Cements and waxes

Epoxy cement.
Beeswax and rosin (see pp. 113-114, for example).
Softseal Tackiwax (Central Scientific Co.).
Glyptal paint (General Electric Co.).

This paint is very useful for vacuum systems and as a cement for small delicate parts also; it comes as a clear, or black, or red paint.

Miscellaneous materials

Black poster paint.
White poster paint.
Flat black Krylon spray paint.
Cardboard.
White file cards.
Black paper.
Flock paper (black paper with black fuzz on one side to reduce reflectivity; it is used in lining telescope tubes for example).
Fluorescein powder or solution.
Polaroid sheet.
Brass shim stock.
Carborundum, coarse, medium, very fine. (Useful for making ground glass or beveling edges of glass plates).
Glass cutter.
Miscellaneous hand tools.
Dentist's mirror.

Infrared black paint, black even at long wavelengths, is made from a mixture of waterglass (sodium silicate) solution and lamp black or other extremely fine charcoal powder. The mixture is black and opaque from the visible to about 40 or 50 microns in the infrared, thereafter, it becomes translucent.

APPENDIX D. GLASS CUTTING AND GRINDING.

To cut glass sheets first scratch them with a glass cutter. It is often recommended (though it is not essential) that the cutter wheel be lubricated with kerosene or light oil, and that the glass be scratched only once. For good results, it is highly desirable to use a good straight edge to guide the cutter. Use moderate pressure to make a deep scratch from one edge of the glass to the other. It is easier to cut the glass so that the pieces are roughly divided in half; long narrow strips are difficult to cut. For glass 1/16 to 1/4 inch thick, lay the sheet of glass along a straight edge under the scratch so that pressure may be applied to produce tension in the glass along the scratch. In this way, it is generally possible to produce good clean breaks in the glass. If the cut to be made is more than about 3 or 4 inches long, use two blocks of wood (or books) to get more uniform distribution of pressure along the line of the scratch. If the glass is thick, the cut is best made by scratching with the cutter as before, and then placing the glass on a table, scratch down. Support the glass on a strip of cardboard or wood parallel to the scratch and several inches away from it on one side. Use a chisel or large screwdriver and a light hammer to crack the glass. Begin near one edge with the chisel edge over the scratch (on the underneath side of the glass), and tap gently with the hammer until a crack spreads out from the scratch. When the crack starts, lead it carefully across the plate until the glass is cut in two.

After cutting the glass, the edges may be beveled or the sides ground with carborundum. The same procedure is used to make frosted or ground glass. The glass is ground against another glass plate or tool of comparable size. Use coarse carborundum for edging or beveling or to make coarse ground glass. (Obviously very fine carborundum is used to make fine ground glass). Sprinkle a teaspoonful or less of the grit on the lower plate -- either the tool or the glass to be ground -- and about an equal amount of water. The glass may be ground using moderate to heavy pressure and random strokes whose length is perhaps 1/4 the dimensions of the glass. The glass will first become ground either at the center or at the edge, according to the method of grinding; continue to grind, perhaps changing the length of the stroke, until the glass is uniformly ground over the whole surface. After a short period, the noise of the grinding with coarse grit will decrease and the grit must be replaced. When fine carborundum is used the grit must, of course, be replaced periodically also, though in this case, there is little sound to use as a guide. After the glass is ground or is to be checked, the grit is washed off in running water and the glass dried for inspection. A magnifying glass is helpful for this examination.

The instructions given above apply only to the production of ground glass or for edging glass plates. If an accurate curve is to be ground, the procedure must be much more carefully controlled, and directions are given in the references listed on page 57 of this manual.

APPENDIX E. ISOLATION FILTERS FOR MERCURY ARC LINES.

Various filter combinations are possible and the best combination depends upon the source (high or low pressure), the detector, and the spectral purity requirements. For instance, if the purpose is to observe visually green interference fringes, it may be sufficient to use a light greenish filter to remove a substantial part of the yellow and blue. If fringes are to be photographed or studied with a photomultiplier, the requirements will be much more stringent. The following list of filter combinations will give satisfactory results in most cases for visual observations; the list should not be interpreted as being definitive or necessarily the best possible under all circumstances. Polished glass filters are much superior to unpolished filters -- they have better transmission and less scattering.

Color	Wavelength, Å	Filter Combination (Corning Glass Filters)	
		Color. Spec.	Code Number
Red	6907	2-64	2030
Yellow	5780	{ 3-66 4-76	3480 9780
Green	5461	{ 1-60 3-69 4-96	5120 (Didymium) 3486 9872 (Optional)
Blue	4358	{ 3-73 5-61 } 5-59 }	3389 5562 } (Alt.) 5850 }
Violet	4046	{ 7-51 3-75	5970 3060
Infrared	6900-27,500	7-56	2540

APPENDIX F. REFERENCE WAVELENGTHS.

Source	Wavelength Å		Wavelength Å		Wavelength Å
Mercury	5790.654 5769.59 5460.740 4358.35 4077.811 4046.561	Cadmium	6438.4696 5085.824 4799.918 4678.156	Helium	6678.149 5875.867 4471.477 4026.189 3888.646

APPENDIX G. SELECTED LIST OF SUPPLIERS OF OPTICAL APPARATUS.

American Optical Company
Southbridge, Mass.

Bausch and Lomb, Inc.
Rochester 2, New York.

Central Scientific Company (CENCO)
1700 Irving Park Road
Chicago 13, Illinois.

Corning Glass Works
Optical Sales Dept.
Corning, New York.

The Ealing Corp.
33 University Road
Cambridge 38, Mass.

Eastman Kodak Company
Rochester, New York.

Edmund Scientific Company
Barrington, New Jersey.

The Eppley Laboratory, Inc.
Newport, Rhode Island.

Gaertner
Chicago, Illinois.

George W. Gates and Co., Inc.
Hempstead Turnpike and Lucille Ave.
Franklin Square, L. I., New York.

A. Jaegers
691 Merrick Road
Lynbrook, L. I., New York.

J. Klinger (Leybold apparatus)
82-87 160th St.
Jamaica 32, New York.

Tube Light Engineering Co.
427 West 42nd St.
New York 36, New York.

APPENDIX H. SELECTED BIBLIOGRAPHY

M. Born and E. Wolf, Principles of Optics, (1959, Pergamon).

This book is a comprehensive treatise on the whole subject of optics from an advanced point of view. It begins with Maxwell's equations and treats both geometrical and physical optics on this basis. The new English edition includes a detailed discussion on coherence and partial coherence and rigorous diffraction theory.

R. W. Ditchburn, Light, (1953, Interscience).

The more usual electromagnetic theory treatment of light is interrelated to quantum mechanics by the author. The approach in the book is at an intermediate level.

L. C. Martin, Technical Optics, (Vol. I 1948, Vol. II 1950, new ed. 1960, Pitman).

An excellent description of the many fine points of a number of types of optical systems is given in addition to a good theoretical discussion.

R. A. Sawyer, Experimental Spectroscopy, (1946, Prentice-Hall).

Although not very new, this book is still a useful reference for visible and near ultraviolet spectroscopy.

Smith, Jones, and Chasmar, The Detection and Measurement of Infra-Red Radiation, (1957, Oxford University Press).

This is an excellent account of infrared spectroscopic techniques and related detector-amplifier-recorder methods. Numerous useful properties of gases, crystals, etc. are included.

J. Strong, Procedures in Experimental Physics, (1939, Prentice-Hall).

Although old, this classic still is invaluable in the field of experimental physics. Techniques such as glass working, mica splitting, silvering, etc. are described.

R. W. Wood, Physical Optics, 3rd ed. (1934, MacMillan).

As everyone knows this classic is a goldmine of information on ideas, techniques, and recipies in optics.

INDEX

- - - - -

References are to page number or to experiment number. To avoid undue repetition, author's works are listed only where first mentioned.

- - - - -